Student Workbook

Elementary and Intermediate Algebra

FOURTH EDITION

Alan S. Tussy
Citrus College

R. David Gustafson
Rock Valley College

Written and Prepared by

Maria H. Andersen
Muskegon Community College

Australia • Brazil • Japan • Korea • Mexico • Singapore • Spain • United Kingdom • United States

ISBN-13: 978-0-495-55478-3
ISBN-10: 0-495-55478-2

Brooks/Cole
10 Davis Drive
Belmont, CA 94002-3098
USA

Cengage Learning is a leading provider of customized learning solutions with office locations around the globe, including Singapore, the United Kingdom, Australia, Mexico, Brazil, and Japan. Locate your local office at: **international.cengage.com/region**

Cengage Learning products are represented in Canada by Nelson Education, Ltd.

For your course and learning solutions, visit **academic.cengage.com**

Purchase any of our products at your local college store or at our preferred online store **www.ichapters.com**

Printed in the United States of America
1 2 3 4 5 12 11 10 09 08

ED196

Table of Contents

Student Workbook: Chapter 1

Table of Contents: *An Introduction to Algebra*

Assessment 1A

Pretest and Diagnostic Tool: Introduction to the Language of Algebra

Directions: Complete this assessment without looking back at your notes or your book. You should **not** use a calculator on this assessment.

1. Add 8 and 3. 11

2. What is the answer to 7 minus 2?
5

3. Find the answer to 20 divided by 4.
5

4. What number is 2 more than 5? 7

5. What number is one-half of 12? 6

6. One factor-pair for 24 is $2 \cdot 12$. Write two more factor-pairs for 24.
4·6 3·8

7. One factor-pair for 30 is $5 \cdot 6$. Write two more factor-pairs for 30.
____ ____

8. What should be multiplied by 8 to get a result of 32? ____

9. Which is greater, 3 or 30? ____

10. On the number line below, which number is on the left, 3 or 5? 3

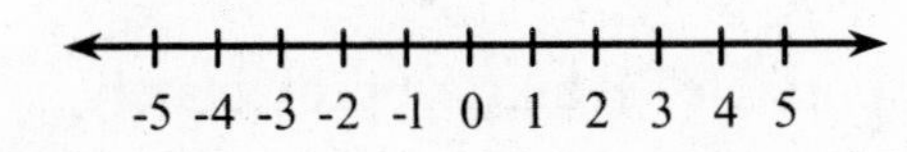

11. Add: $9+8$ ____.

12. Add: $0+7$ ____.

13. Add: $1.2+3.9$ ____.

14. Is the answer to $2+4$ the same as the answer to $4+2$?

Yes or no? ____

15. Subtract: $13-7-1$ ____.

16. Find the value: $13-7+1$ ____.

17. Subtract: $6-0$ ____.

18. Is the answer to $10-5$ the same as the answer to $5-10$?

Yes or no? ____

19. Multiply: $3 \cdot 2 \cdot 9$ ____.

20. Multiply: $12 \cdot 0$ ____.

21. Is the answer to $5 \cdot 3$ the same as the answer to $3 \cdot 5$?

Yes or no? ____

22. Divide: $48 \div 6$ ____.

23. Divide: $8 \div 8$ ____.

24. Is the answer to $20 \div 2$ the same as the answer to $2 \div 20$?

Yes or no? ____

25. Multiply: $7 \cdot 7$ ____.

26. Multiply: $2 \cdot 2 \cdot 2 \cdot 2$ ____.

27. Multiply 2 and 4; then add 1. ____

28. Add 4 and 1; then multiply by 2.
____

29. Add: 5 ft + 3 ft ________.

30. Subtract: 12 cm − 6 cm ________.

Guided Learning Activity

Section 1.1 Understanding Tables and Line Graphs

Example 1: Yuri makes $8 an hour at his part-time job. Use the table of values to construct a representative line graph.

Yuri's Pay

Hours worked	Pay (dollars)
0	0
5	40
10	80
15	120
20	160
25	200

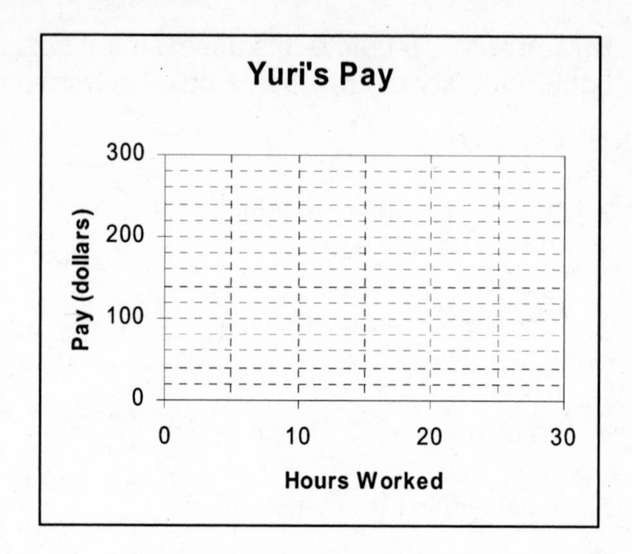

Example 2: Shelly, Yuri's neighbor, works at the same place as Yuri, but she has worked there longer and makes more per hour. She constructs the following line graph to represent her pay for a given number of hours worked. Fill in the table of values using the line graph.

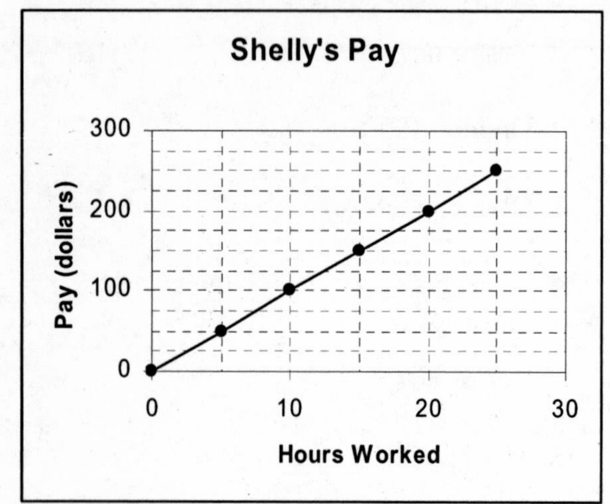

Shelly's Pay

Hours worked	Pay (dollars)
0	
5	
10	
15	
20	
25	

Example 3: Shelly and Yuri carpool to work whenever possible. It takes approximately 2 gallons of gas to make the drive round-trip if Shelly is driving her car, and it takes approximately 3 gallons of gas to make the round-trip drive if Yuri is driving his car (which has poorer gas mileage). Fill in the data table below to show how much the gas for a round-trip will cost, depending on the gas price and the car that is driven.

Cost of Gas (per gallon)	Cost to drive Shelly's Car	Cost to drive Yuri's Car
$2.00		
$2.50		
$3.00		
$3.50		
$4.00		

Student Activity A

Section 1.1 Learning the Notation of Mathematical Operations

Directions: For each line, fill in the missing boxes with the proper words or notation.

ADDITION: Written in words.	**Using the word *sum*.**	**Using the + sign.**
5 plus 9 is 14.		
		$3+8=11$
	The sum of 2 and x is 9.	

SUBTRACTION: Written in words.	**Using the word *difference*.**	**Using the – sign.**
12 minus 4 is 8.		
		$11-7=4$
	The difference of 9 and 3 is 6.	

MULTIPLICATION: Written in words.	**Using the word *product*.**	**Using a raised dot.**	**Using parentheses.**
3 multiplied by 5 is 15.			
		$8\cdot5=40$	
	The product of 2 and n is 14.		
			$(6)(7)=42$ or $6(7)=42$

DIVISION: Written in words.	**Using the division symbol.**	**Using long division notation.**	**Using the fraction bar.**
The quotient of 16 and 2 is 8.			
		$5\overline{)35}$ with quotient 7	
	$27\div3=9$		
			$\frac{32}{4}=8$

Student Activity B

Section 1.1 Translating Expressions and Equations

Directions: For each line, fill in the missing boxes with the proper words or notation. The first one has been done for you.

	Phrase or sentence	Expression or Equation?	Write the expression or equation.
a.	the number of feet, n, times 12	Expression	$12n$
b.			$17 - x = 3$
c.	The sum of the measures of angles x and y is 180°.		
d.	the quotient of m and 100		
e.			$z + 100$
f.	The product of y and 10 is 1000.		
g.			$20 \div 4$
h.			$u + v = 90°$
i.	The number of inches, n, divided by 12 is the number of feet, f.		
j.	100 less than x		
k.	the difference of 100 and x		

Student Activity A

Section 1.2: Factor Pairings

Directions: In each diagram, there is a number in the top box and exactly enough spaces beneath it to write all the possible factor-pairs involving whole numbers. The number 30 has been done for you. See if you can find all the missing factor-pairs.

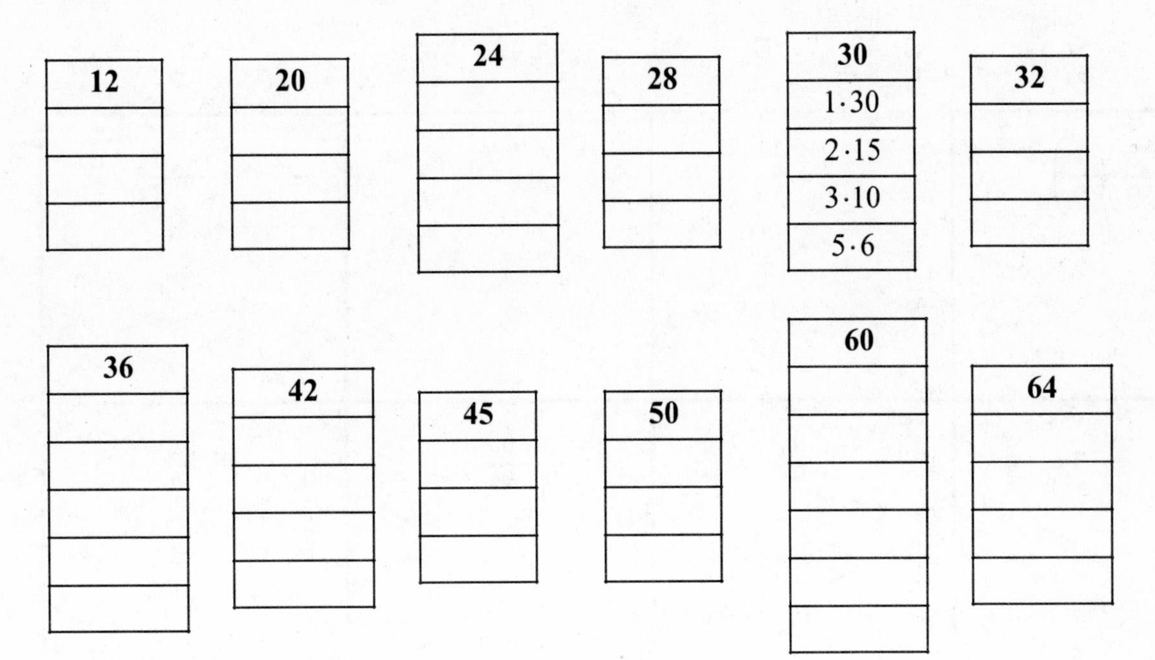

1. What is the largest number that is a factor of 20 and 30? _____

Simplify: $\frac{20}{30}$

2. What is the largest number that is a factor of 28 and 36? _____

Simplify: $\frac{28}{36}$

3. What is the largest number that is a factor of 36 and 60? _____

Simplify: $\frac{36}{60}$

4. What is the largest number that is a factor of 24 and 42? _____

Simplify: $\frac{24}{42}$

Student Activity B

Section 1.2: Match Up on Fractions

Match-up: Match each of the expressions in the squares of the table below with its simplified value at the top. If the solution is not found among the choices A through D, then choose E (none of these).

A 1 **B** $\frac{3}{4}$ **C** $\frac{7}{8}$ **D** 0 **E** None of these

$\left(\frac{3}{2}\right)\left(\frac{1}{2}\right)$	$\frac{7}{6} \div \frac{4}{3}$	$\frac{7}{8} \div 0$	$2\frac{1}{8} - \frac{5}{4}$
$\frac{15}{8} - 1$	$6\left(\frac{1}{6}\right)$	$\frac{63}{72}$	$\frac{2}{3} \cdot \frac{3}{2}$
$\frac{1}{2} + \frac{3}{8}$	$\frac{19}{12} - \frac{5}{6}$	$\frac{2}{3} \div \frac{8}{9}$	$0 \div \frac{3}{4}$
$\frac{1}{5} + \frac{8}{10}$	$\frac{3}{4}(0)$	$\frac{180}{240}$	$1\frac{1}{4} - \frac{10}{8}$
$\frac{1}{2} + \frac{2}{2}$	$\frac{1}{3} \div \frac{1}{3}$	$\frac{7}{2} \cdot \frac{3}{12}$	$\frac{9}{2} \div 6$

Student Activity C

Section 1.2: Fractions Using a Calculator

When you input fractions into a calculator, you must be careful to tell the calculator which parts are fractions. Each calculator has a set of algorithms that tell it what to do first (later on, we will learn the mathematical order of operations, which is similar). In order to ensure that fractions are treated as fractions, for now, you need to tell your calculator which parts ARE fractions.

1. For example, first show that $\frac{3}{4} \div \frac{2}{5}$ is $\frac{15}{8}$ by hand:

2. To get the decimal value of $\frac{15}{8}$ on the calculator, we type $15/8$ or $15 \div 8$ (depending on the calculator). Practice by finding the decimal values for:

$$\frac{15}{8} \qquad \frac{3}{20} \qquad \frac{1}{4} \qquad \frac{7}{8} \qquad \frac{2}{3}$$

3. Now try using your calculator to evaluate $\frac{3}{4} \div \frac{2}{5}$, but do it without using any parentheses. Do you get the decimal value equal to 15/8?

4. Find the button(s) on your calculator that allow you to input parentheses and write down how to use them on your calculator.

5. Try it on your calculator like this now: $\left(\frac{3}{4}\right) \div \left(\frac{2}{5}\right)$

On my calculators, I type $(3/4)/(2/5)$ or $(3 \div 4) \div (2 \div 5)$ to enter this expression. But each calculator is a little different. When you have done it correctly, you should get 1.875.

Write down how to do it on your calculator:

6. Now try these fraction problems *using parentheses* to tell your calculator which numbers represent fractions:

$$\frac{1}{2} \cdot \frac{4}{9} \qquad \frac{3}{4} \div \frac{2}{15} \qquad \frac{3}{8} + \frac{4}{5} \qquad \frac{7}{12} - \frac{1}{5}$$

7. The operation in mixed numbers is **addition**, so when you input $2\frac{3}{5}$ into your calculator, you must treat it like $2 + (3/5)$. What is $2\frac{3}{5}$ as a decimal? ____

Guided Learning Activity

Section 1.3: Charting the Real Numbers

The set of **natural numbers** is $\{1, 2, 3, 4, 5, ...\}$.

The set of **whole numbers** is $\{0, 1, 2, 3, 4, 5, ...\}$.

The set of **integers** is $\{..., -4, -3, -2, -1, 0, 1, 2, 3, 4, ...\}$.

The set of **rational numbers** consists of all numbers that can be expressed as a fraction (or *ratio*) of *integers* (except when zero is in the denominator). Note that all rational numbers can also be written as decimals that either terminate or repeat.

The set of **irrational numbers** consists of all *real* numbers that are *not* rational numbers.

The set of **positive numbers** consists of all the numbers *greater* than zero.

The set of **negative numbers** consists of all the numbers *less* than zero.

Part I: Using the definitions above, we will categorize each number below. For each of the numbers in the first column, place an "X" in any set to which that number belongs.

		Natural	Whole	Integer	Rational	Irrational	Positive	Negative
a.	5	X	X	X	X		X	
b.	1							
c.	0							
d.	−2							
e.	$\sqrt{3}$							
f.	$\frac{2}{3}$							
g.	$-1.\overline{4}$							
h.	0.75							
i.	$\frac{\pi}{3}$							

Part II: Now we'll do it backwards. Given the checked properties, find a number (try to use one that is different from one of the numbers in the previous table) that fits the properties. If it is not possible to find a number with all these properties, write "impossible" instead.

	Number	Natural	Whole	Integer	Rational	Irrational	Positive	Negative
a.				X	X			X
b.					X		X	
c.			X	X	X			
d.						X		X
e.					X			X
f.				X		X	X	
g.		X	X	X	X		X	
h.						X	X	

Student Activity A

Section 1.3: Venn Diagram of the Real Numbers

Directions: Place each number below in the *smallest* set in which it belongs. For example, -1 is a real number, a rational number, and an integer, so we place it in the "Integers" box, but not inside the whole numbers or natural numbers.

$8 \quad \frac{7}{3} \quad -1.3 \quad \pi \quad 2.175 \quad 0 \quad -7 \quad \sqrt{2} \quad 1000 \quad 0.00005$

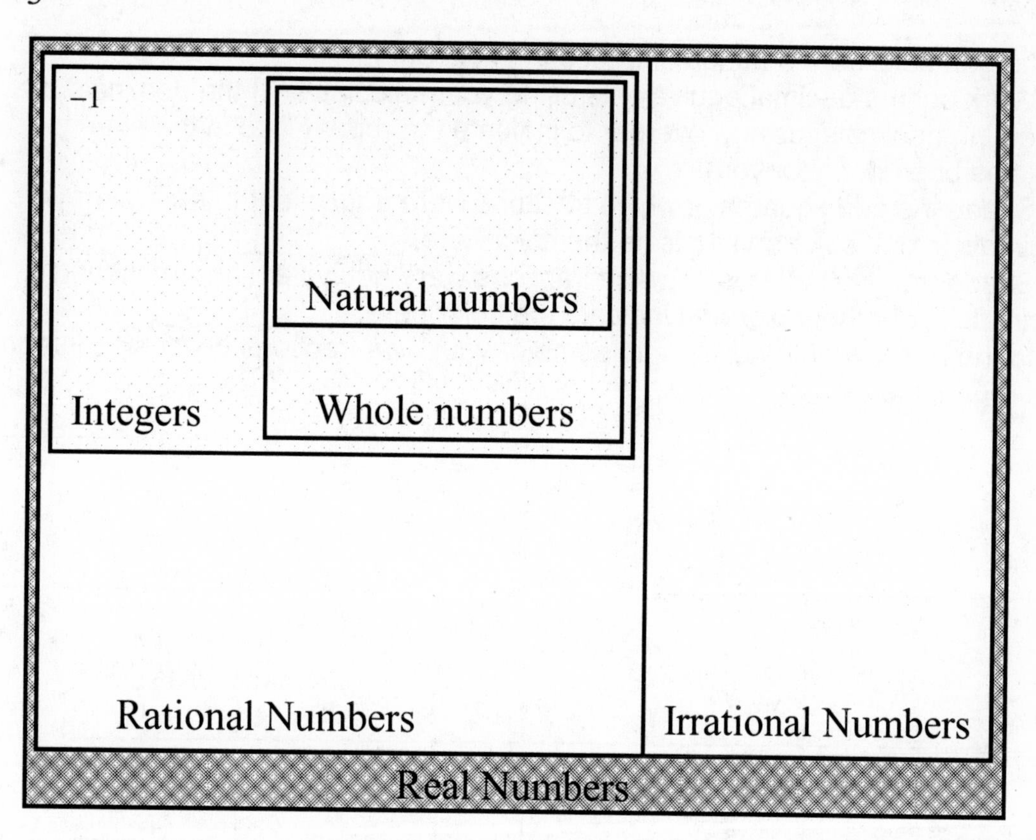

1. Given all possible real numbers, name at least one number that is a whole number, but not a natural number: __________

2. Can a number be both rational and irrational? _____ If yes, name one: ______

3. Can a number be both rational and an integer? _____ If yes, name one: ______

4. Given all possible real numbers, name at least one number that is an integer, but not a whole number: _______

Side note: *Just for the record, this diagram in no way conveys the actual size of the sets. In mathematics, the number of elements that belong to a set is called the* ***cardinality*** *of the set. Technically (and with a lot more mathematics classes behind you) it can be proven that the cardinality of the irrational numbers (uncountable infinity) is actually larger than the cardinality of the rational numbers (countable infinity). Another interesting fact is that the cardinality (size) of the rational numbers, integers, whole numbers, and natural numbers are all equal. This type of mathematics is studied in a course called Real Analysis (that comes after the Calculus sequence).*

Student Activity B

Section 1.3: Linking Rational Numbers with Decimals

Let's investigate why we say that decimals that terminate and repeat are really rational numbers. You will need a calculator and some colored pencils for this activity.

Rational numbers consist of all numbers that can be expressed as a fraction (or *ratio*) of *integers* (except when zero is in the denominator).

In the grid below are a bunch of fractions of integers.

1. Work out the decimal equivalents using your calculator. If the decimals are repeating decimals, use an overbar to indicate the repeat (like in the example that has been done for you).

2. Shade the grid squares in which fractions were equivalent to repeating decimals in one color and indicate the color here: ___________.

3. Shade the grid squares in which fractions were equivalent to terminating decimals in another color and indicate the color here: ___________.

4. In the last row of the grid, write some of your own fractions built using integers and repeat the steps above.

$\frac{2}{3} = 0.\overline{6}$	$\frac{1}{4} =$	$\frac{7}{8} =$	$\frac{5}{9} =$
$\frac{1}{2} =$	$\frac{4}{9} =$	$\frac{8}{5} =$	$\frac{17}{25} =$
$\frac{1}{1000} =$	$\frac{5}{12} =$	$\frac{3}{4} =$	$\frac{7}{27} =$
$\frac{1}{5} =$	$\frac{12}{5} =$	$\frac{4}{3} =$	$\frac{1}{6} =$

5. Are there any fractions in the grid that were not shaded as either terminating or repeating?

6. If you write one of these fractions as its decimal equivalent, what kind of decimal do you get?

Guided Learning Activity

Section 1.4: Using Addition Models

Part I: The first model for addition of real numbers that we look at is called the "colored counters" method. Traditionally, this is done with black and red counters, but we make a slight modification here to print in black and white.

 Solid counters (black) represent positive integers, $+1$ for each counter.

 Dashed counters (red), represent negative integers, -1 for each counter.

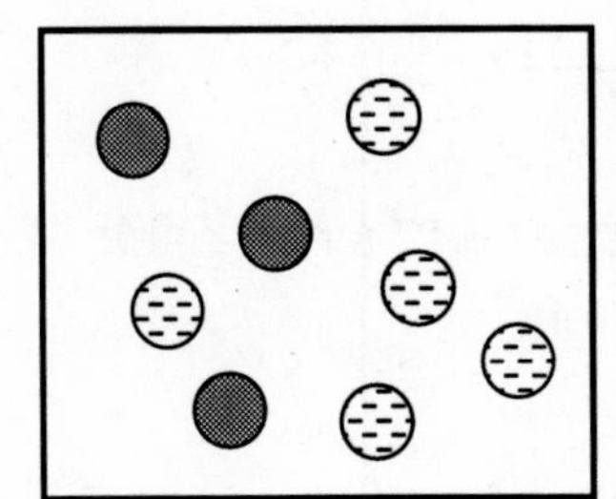

When we look at a collection of counters (inside each rectangle) we can write an addition problem to represent what we see. We do this by counting the number of solid counters (in this case 3) and counting the number of dashed counters (in this case 5). So the addition problem becomes $3+(-5)=$____.

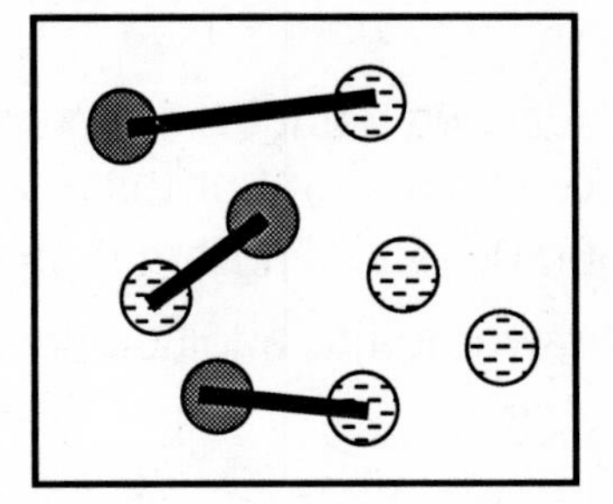

To perform the addition, we use the Additive Inverse Property, specifically, that $1+(-1)=0$. By matching up pairs of positive and negative counters until we run out of matched pairs, we can see the value of the remaining result. In this example, we are left with two dashed counters, representing the number -2. So the collection of counters represents the problem $3+(-5)=-2$.

Now try to write the problems that represent the collections below.

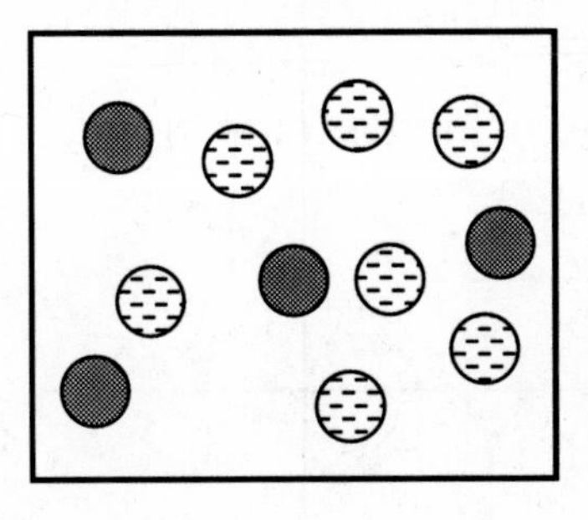

a. ____ + ____ = ____

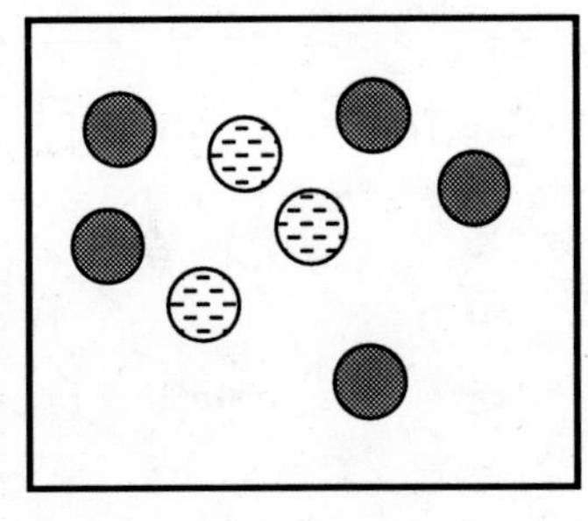

b. ____ + ____ = ____

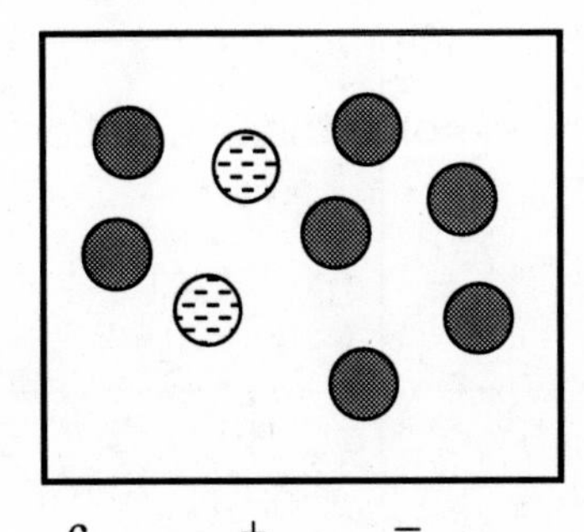

c. ____ + ____ = ____

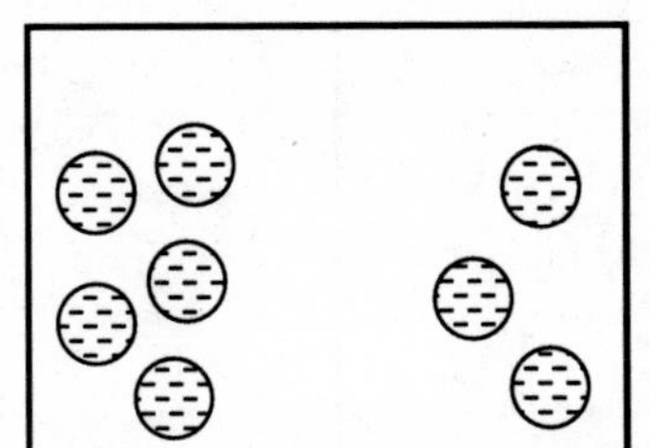

d. ____ + ____ = ____

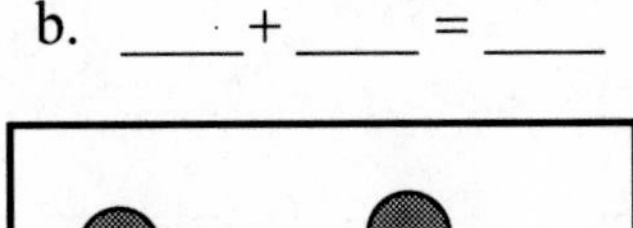

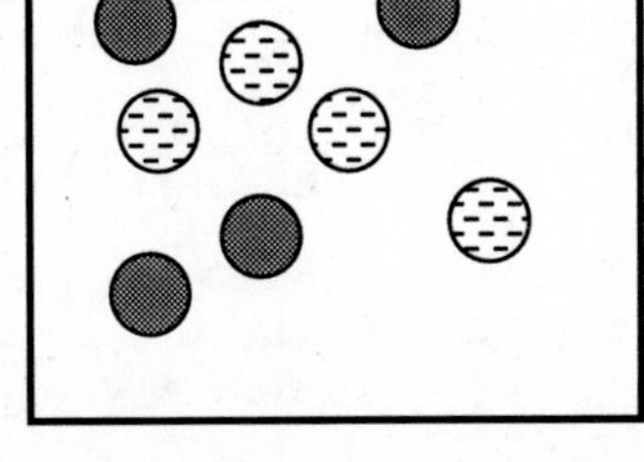

e. ____ + ____ = ____

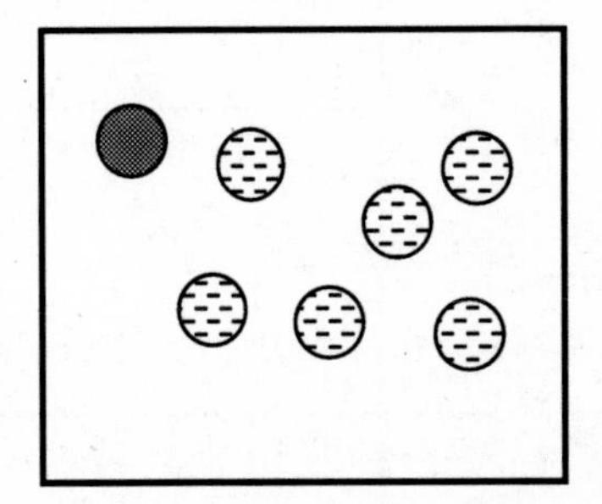

f. ____ + ____ = ____

Part II: The second model for addition of real numbers that we look at is called the "number line" method. We use directional arcs to represent numbers that are positive and negative. The length that the arc represents corresponds to the magnitude of the number.

When a directional arc indicates a positive direction (to the right), it represents a positive number. In the diagram below, each arc represents the number 2, because each arc represents a length of two and each arrow points to the right.

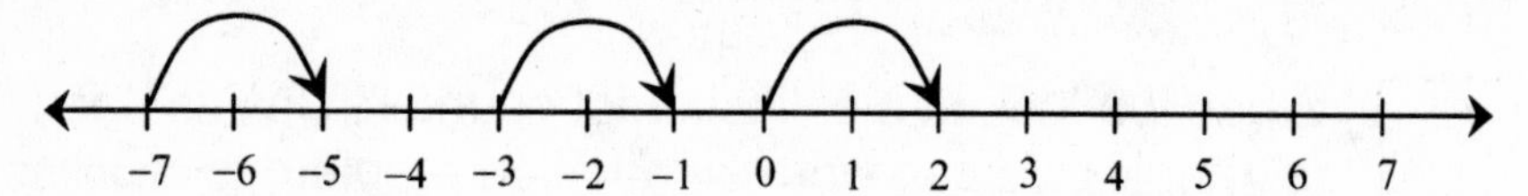

In the next diagram, each directional arc represents the number -5, since each arc represents a length of five and each arrow points to the left.

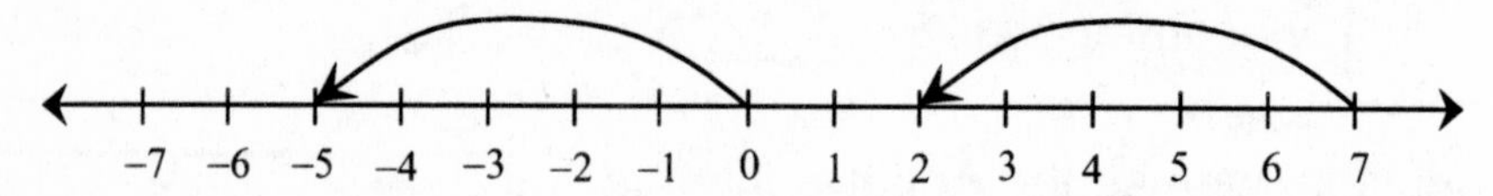

When we want to represent an addition problem, **we start at zero**, and travel from each number to the next using a new directional arc. Thus, the following number line diagram represents $2+(-5)=-3$. The final landing point is the answer to the addition problem.

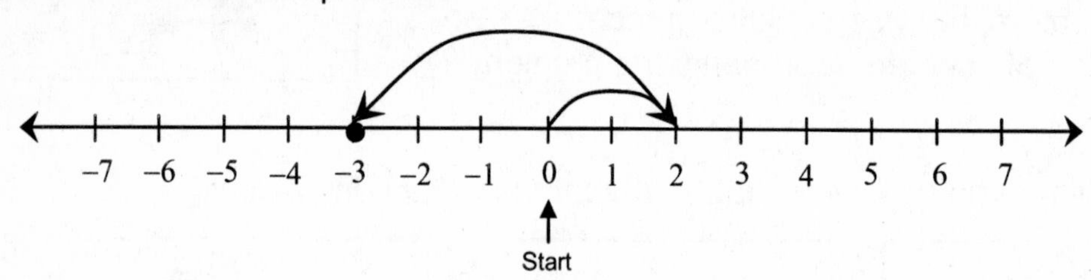

Now try to solve these addition problems on a number line using directional arcs.

a. $-3+7=$ _____

b. $-2+(-3)=$ _____

c. $5+(-5)=$ _____

d. $-6+2=$ _____

e. $-2+(-2)+(-1)=$ _____

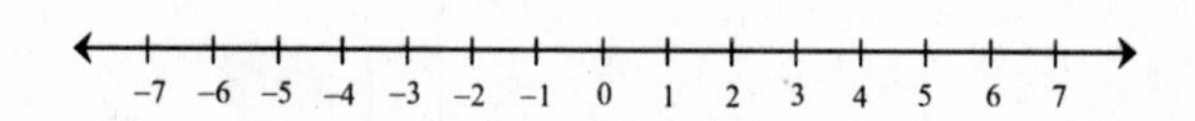

f. $4+(-8)+3=$ _____

Student Activity A

Section 1.4: Match Up on Addition of Real Numbers

Match-up: Match each of the expressions in the squares of the table below with its simplified value at the top. If the solution is not found among the choices A through D, then choose E (none of these).

A 3 **B** -6 **C** -1 **D** 1 **E** None of these

$-8+2$	$-9+8$	$-7+10$	$-3+(-3)$
$-2+(-1)+6$	$3+(-3)$	$-10+(-4)+8$	$-\frac{1}{2}+\left(-\frac{1}{2}\right)$
$6+(-7)$	$5+(-6)$	$-1+(-1)+(-1)+4$	$-11+12$
$-4+\left(-\frac{1}{2}\right)+\frac{1}{2}$	$8+(-5)+(-9)$	$10+(-6)+(-1)$	$-6+12$
$0+(-6)$	$-1+7$	$30+(-27)$	$1+\frac{1}{4}+\frac{1}{2}+\left(-\frac{3}{4}\right)$

Student Activity B

Section 1.4: Signed Numbers Magic Puzzles

Directions: In these "magic" puzzles, each row and column adds to be the same "magic" number. Fill in the missing squares in each puzzle so that the rows and columns each add up to be the given magic number.

Magic Puzzle #1

–2	8	
	–9	4

Magic Number = 5

Magic Puzzle #2

		4
–5		
8		–6

Magic Number = 0

Magic Puzzle #3

		$2\frac{1}{4}$
	$2\frac{1}{2}$	$-2\frac{1}{2}$
$\frac{1}{4}$		

Magic Number = $\frac{1}{2}$

Magic Puzzle #4

–2		9	–7
	–9		4
8	–4		–5
		–6	

Magic Number = 1

Magic Puzzle #5

–8	–3		7
	–11	2	6
8	–2		–7

Magic Number = – 2

Guided Learning Activity

Section 1.5: Language of Subtraction

How do you interpret the – sign? It is a minus sign if it is **between** two numbers as a mathematical operation. Otherwise, it is a negative.

Other ways to signify minus: difference, less than, subtract ... from ...
Other ways to signify negative: opposite

How do you tell if *less* means < or –? Look for the distinction between "**is** less than" and "less than." See the two examples in the table below.

	Expression? Equation? Or Inequality?	**Equivalent statement or phrase in words**
$-(-4)$	Expression	the opposite of the opposite of 4 the opposite of negative 4
$8-3$	Expression	the difference of 8 and 3 subtract 3 from 8 8 minus 3 3 less than 8
$8-3=5$	Equation	The difference of 8 and 3 is 5. Subtract 3 from 8 to get 5. 8 minus 3 is 5. 3 less than 8 is 5.
$-2-4$	Expression	the difference of negative 2 and 4 subtract 4 from negative 2 negative 2 minus 4 4 less than negative 2
$-5<-2$	Inequality	Negative 5 is less than negative 2.
$9-(-2)$	Expression	9 minus negative 2 subtract negative 2 from 9 the difference of 9 and negative 2

Note that expressions are represented in words by phrases (no period) and equations and inequalities are represented by sentences (with a period).

Now try these! For any problem with subtraction, find at least two ways to write it in words.

		Expression? Equation? Or Inequality?	**Equivalent statement or phrase in words**
a.			Zero is less than 8.
b.	$10-2=8$		
c.			the difference of 2 and negative 5
d.	$-(-10)$		
e.	$5-(-10)=15$		
f.			Negative 6 is less than negative 3.
g.			The opposite of negative 5 is 5.
h.	$3<-(-6)$		
i.	$\frac{1}{2}-\frac{3}{4}$		
j.	$-19<-18$		

Student Activity A

Section 1.5: Match Up on Subtracting Real Numbers

Match-up: Match each of the expressions in the squares of the table below with its simplified value at the top. If the solution is not found among the choices A through D, then choose E (none of these).

A 3 **B** −6 **C** −1 **D** 1 **E** None of these

$-1-(-4)$	$0-(-1)$	$\frac{1}{2}-\frac{6}{4}$	$\frac{1}{2}-\left(-\frac{1}{2}\right)$
$2-(-1)$	$\frac{5}{2}+4-2\frac{1}{2}-3$	$3-(-3)$	$-9-(-3)$
$-2-(-1)$	$0-(-1)$	$0-1$	$3-6$
$-7+1$	$12-(-6)$	$0-(-3)$	$-3+(-3)$
$5\frac{2}{3}-\left(-\frac{1}{3}\right)+(-3)$	$-6-(-3)$	$-30-(-25)+(-1)$	$-\frac{9}{4}-\left(-\frac{1}{2}\right)+\frac{3}{4}$

Student Activity B

Section 1.5: Scrambled Addition and Subtraction Tables

Here are simple addition and subtraction tables with natural number inputs.

Addition:

+	1	2	3	4
1	2	3	4	5
2	3	4	5	6
3	4	5	6	7
4	5	6	7	8

Subtraction:

−	1	2	3	4
1	0	−1	−2	−3
2	1	0	−1	−2
3	2	1	0	−1
4	3	2	1	0

Notice that the subtraction order is: column number – row number

Directions: The following tables are addition or subtraction tables involving integer inputs. Fill in the missing squares with the appropriate numbers.

+	−3	−2	−1	0	1	2	3
−3							
−2							
−1							
0							
1							
2							
3							

−	−3	−2	−1	0	1	2	3
−3							
−2							
−1							
0							
1							
2							
3							

Directions: The following tables are ***scrambled*** addition or subtraction tables with missing information (this means that the numbers do not increase nicely like 1, 2, 3, 4). Fill in the missing squares with the appropriate numbers.

+	2			3	
−5					−5
−2			−3		
				7	
	8				
		−4		3	

−		−1	1		4
−2				−2	
			2		
		1			
6	1				
					−8

Student Activity C

Section 1.5: Signed Numbers Using a Calculator

When you input expressions with signed numbers into a calculator, you must be careful to tell the calculator which "–" signs represent a minus, and which represent a negative.

1. The minus button on your calculator looks like this: [–]. It is found with the addition, multiplication, and division functions. The button on your calculator that is used to denote a negative may look like [+/–] or [(–)], or it may be above one of the keys, accessed with a 2nd function [2nd]. Locate where your calculator input for a negative is, and draw it here:

2. On some calculators, the negative is typed before the number, and on some it is typed after the number. We need to figure out which type you have. We'll calculate $-2+5$ (which should be ___). Try it both ways. Write down exactly how to do $-2+5$ on your calculator here:

3. Let's try something more complicated now. How would we write $-8-3$ in words using the word minus? ______________________________

What should the answer be? ___ Now write down the keystrokes for inputting this expression into your calculator here:

4. Work out each of these expressions by hand, then write down how to express them in words, and finally, write down how to input the keystrokes properly into your calculator.

Expression	Answer	In words (using negative and/or minus)	Keystrokes
$-6+3$			
$5-(-4)$			
$-19-6$			
$-1.25-0.25$			
$10-(-8)$			
$-3-(-3)$			

Assessment 1B

Chapter 1: Mid-chapter Assessment for Understanding

For each of the following, describe the type of problem and the strategies and key steps to remember while doing the problem. You do **not** have to complete the problems.

	Type of Problem	Strategies and Key Steps
1. Add: $-3+(-7)$.		
2. Use < or > to make this statement true: $-3 \square -4$		
3. Divide: $\frac{3}{4} \div \frac{1}{6}$.		
4. Write $10-8=2$ in words.		
5. Subtract: $-2-(-4)$.		
6. Is 1.5 rational or irrational?		
7. Add: $\frac{1}{3}+\frac{4}{5}$.		
8. Fill in the missing value to make the statement true: $\frac{2}{5}=\frac{\square}{20}$		
9. Is -3 a natural number?		
10. Write $x \div 3$ using different symbols.		

Student Activity A

Section 1.6: What's Wrong with Division by Zero?

Let's spend some time investigating why division by zero is undefined. You will need a calculator for this activity.

1. First, let's see what your calculator thinks. Try the following division problems on your calculator and write down the results:

$0 \div 5$ $\qquad$ $5 \div 0$ $\qquad$ $\frac{12}{0}$ $\qquad$ $\frac{0}{12}$

2. Even your calculator will reject the idea of division by zero, so let's try dividing by numbers *close* to zero. Looking at the number line below, name and label some numbers that are really close to zero (on both sides of zero).

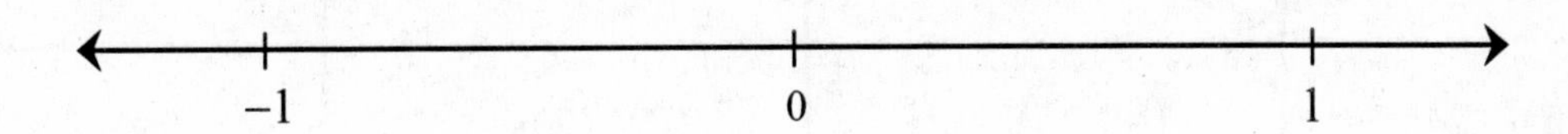

3. Find the following quotients using your calculator.

$5 \div 0.1$ $\qquad$ $5 \div 0.01$ $\qquad$ $5 \div 0.001$ $\qquad$ $5 \div 0.0001$

4. What is happening to the quotients in #3 as the divisor gets closer to zero?

5. Find the following quotients using your calculator.

$5 \div (-0.1)$ $\qquad$ $5 \div (-0.01)$ $\qquad$ $5 \div (-0.001)$ $\qquad$ $5 \div (-0.0001)$

6. What is happening to the quotients in problem **5** as the divisor gets closer to zero?

7. Using the results from problems **4** and **6**, why do you think that division by zero is undefined?

Student Activity B

Section 1.6: Match Up on Multiplying and Dividing Real Numbers

Match-up: Match each of the expressions in the squares of the table below with its simplified value at the top. If the solution is not found among the choices A through D, then choose E (none of these). Note that some of the expressions involve other operations besides multiplication and division, so be careful!

A 2 **B** 0 **C** –8 **D** 12 **E** None of these

$\frac{-6}{-3}$	$-\frac{4}{3}\left(-\frac{3}{2}\right)$	$(-4)(-2)(-1)$	$-2 \div 0$
$(-0.5)(-24)$	$(-0.25)(-8)$	$(0)(-2)(-1)$	$32 \div (-4)$
$36 \div (-3)$	$-3\left(\frac{4}{6}\right)$	$2 \div \left(-\frac{1}{2}\right)$	$-36\left(-\frac{2}{6}\right)$
$4-(-3)+5$	$5 \div \left(-\frac{5}{8}\right)$	$\left(-\frac{4}{3}\right)(6)$	$0(-8)-(-2)$
$-2(-2)(-3)(-1)$	$-\frac{1}{3}(-6)$	$-\frac{3}{2} \div \left(-\frac{1}{8}\right)$	$\frac{0}{-5}$

Student Activity A

Section 1.7: Exponent Trios

Directions: In each of the "trios" below, place three equivalent expressions of the following format:

Expanded Expression using multiplication	
Compact exponential expression	Simplified exponential expression

The first one has been done for you. Sometimes there are two possible trios for a simplified exponential expression, so you will see some of these listed twice.

$(-3)(-3)$	
$(-3)^2$	9

-3^2	

$3 \cdot 3$	

$-(2)(2)(2)(2)$	

$(-2)^4$	

$2 \cdot 2 \cdot 2 \cdot 2$	

	8

$(-2)(-2)(-2)$	

-2^3	

$-(4)(4)$	

$(-4)^2$	

-10^3	

$2 \cdot 2 \cdot 2 \cdot 2 \cdot 2 \cdot 2 \cdot 2 \cdot 2$	

$(-3)^4$	

$(-9)^2$	

$\left(-\frac{1}{2}\right)^2$	

$-\left(\frac{1}{2}\right)\left(\frac{1}{2}\right)$	

	$\frac{1}{4}$

Student Activity B

1.7 Order of Operations: What's First?

Directions: With a colored pencil, shade the numbers and operation that comes first in the order of operations in each problem. For example, for $7-3\cdot 2$, you would lightly shade over $3\cdot 2$. If more than one operation could be done first (at the same time), shade both. Do not simplify the expressions until you are sure that you have chosen the correct first step (your instructor will either go over those, or ask you to collaborate with your classmates). Once you are certain that the first steps are correct, then simplify each expression.

Order of Operations

1. Perform calculations inside **parentheses** (fraction bars, and absolute values) working from innermost to outermost pairs.
2. Evaluate **exponential expressions**.
3. Perform **multiplications** and **divisions** as they occur from left to right.
4. Perform **additions** and **subtractions** as they occur from left to right.

1. $5+3\cdot 5-2$

2. $12\div 2\cdot 6+1$

3. $4-3^2+6$

4. $6\cdot 2-8\div 2+4$

5. $-12-(4+3)$

6. $5-2(4\cdot 3)-5$

7. $9\cdot 2\div 6-5(2)^2$

8. $\dfrac{2-4}{(-3)^2+1}$

9. $\left[4+2(2-5)^2\right]-3$

10. $\dfrac{3}{4}-\left(\dfrac{2}{3}\right)\left(\dfrac{1}{2}\right)^2$

Guided Learning Activity

Section 1.8: Language of Algebraic Expressions

The **terms** of an algebraic expression are separated by addition or subtraction. We rewrite subtraction as addition, like this: $3x-5$ becomes $3x+(-5)$ and we call each part that is separated by addition a **term**. Thus, $3x$ is a term and -5 is a term.

If a term involves both a variable and a numerical factor, then the number part is called the **coefficient.** Thus, for the term $3x$, 3 is the coefficient of the term. A term that consists of a single number is called the **constant term.**

Expression	**How many terms?**	**Coefficient of... (or constant term)**			
		1st term	**2nd term**	**3rd term**	**4th term**
$5x^2-2x+7$	3	5	−2	7	None
$x-2.5$					
$a^2+\frac{1}{2}a-1$					
$b^3+\frac{b^2}{3}-\frac{b}{2}+12$					
$x^2-3xy-10y^2$					
$2\ell+2w$					

If an algebraic expression is held together by multiplication, then we call the multiplied parts **factors**.

Expression	**How many factors?**	**1st factor**	**2nd factor**	**3rd factor**
$7x$	2	7	x	None
$(x+3)(x-3)$				
$-3ab$				
$h(b_1+b_2)$				
$x(x+2)(x-3)$				

Words that indicate a mathematical operation

Addition	Subtraction	Multiplication	Division
add sum plus more than greater than increased by exceeds	subtract minus difference less than decreased by reduced by less	multiply product times twice, thrice, … double, triple fraction words like three-fourths	divide quotient ratio split into equal parts

Now try these! If indicated, use the desired variable. The first one has been done for you.

	Phrase	Indicated Operation	Algebraic expression
a.	the ratio of a to h	division	$\frac{a}{h}$
b.	10 added to x		
c.	x less than y		
d.	one-half of t		
e.	j increased by 25		
f.	triple the length, l		
g.	d split into 6 equal parts		
h.	the product of x and 27		
i.	T reduced by 50		
j.	f divided by 5280		
k.	the sum of u and v		
l.	the quotient of c and 100		
m.	twice g		
n.	exceeds R by t		
o.	x greater than 32		
p.	five-sevenths of h		

Student Activity A

1.8 Evaluating Algebraic Expressions

Because expressions like $(-3)^2$ and -3^2 yield different results (9 and -9), it is important to use parentheses in your notation whenever it is appropriate. But how do you know when to insert parentheses in your evaluation of an expression? If you use the method below, you will always have parentheses in the right places.

Skeleton Method: To evaluate an expression, it is helpful to "see" the expression without the variables. To make the "skeleton" of the expression, replace each variable with an empty parentheses "skeleton". Here are two examples:

Evaluate x^2+4x+3 at $x=-2$.	Evaluate x^2-2xy at $x=5$ and $y=-3$.
$(\quad)^2+4(\quad)+3$ (skeleton) $(-2)^2+4(-2)+3$ $4-8+3$ -1	$(\quad)^2-2(\quad)(\quad)$ (skeleton) $(5)^2-2(5)(-3)$ $25+30$ 55

Now try these! For each problem, first write a "skeleton," then evaluate.

Evaluate $6x-3x^2$ at $x=-1$.	Evaluate b^2-4ac at $a=2$, $b=-5$, and $c=-8$.

Evaluate $\lvert x-4\rvert+3x$ for $x=1$.	Evaluate $\dfrac{3xy+4}{x+y^2}$ for $x=3$ and $y=-2$.

Evaluate $\lvert a^2+b^2\rvert$ for $a=-3$ and $b=4$.	Evaluate $-y-y^3$ for $y=2$.

Student Activity B

Section 1.8: Answering Questions with Algebra

Directions: Fill in the tables, completing each table with an algebraic expression.

Number of days	Number of hours
1	
2	
3	
D	

Number of dimes	Value in cents
1	
2	
3	
d	

Number of centimeters	Number of meters
200	
100	
1	
x	

Number of \$6 tickets	Total value
5	
10	
20	
n	

1. A movie ticket that normally costs \$6 is on sale for \$2 off. How much will it cost for a family of x people to go to a movie?

2. A roll of dimes from the bank holds 50 dimes. What is the value, in cents, of x rolls of dimes?

3. Steve works twice as many hours as Adam. Justine works three hours less than Adam. Let x represent the number of hours that Adam works.

a. Write expressions to represent the number of hours that the other two work:

Steve ___________ Justine ___________

b. Suppose Adam works 25 hours in a week. How many hours do Steve and Justine work that week?

Steve ___________ Justine ___________

4. A Honda Civic gets 3 more than twice as many miles per gallon as a Ford F-150.

a. Let x represent the fuel efficiency (in miles per gallon) of one of the vehicles. Write an expression for the fuel efficiency (in miles per gallon) of the other vehicle.

b. A Ford F-150 gets 15 miles per gallon of gasoline. How many miles per gallon does a Honda Civic get?

Student Activity A

Section 1.9: Match Up on Like Terms

Match-up: Match each of the expressions in the squares of the lower table with an equivalent expression from the top. If the solution is not found among the choices A through D, then choose E (none of these).

A $4x+16-12y$ **B** $5+4x+6y$ **C** $x+y-z$ **D** $3x+3y+3z$ **E** None of these

$3(x+y)+z$	$5+2x+6y+2x$	$3y+3(x+z)$	$2(8+2x-6y)$
$(x-z)+y$	$4(x-3y+4)$	$4(x+3y)+16$	$6y+5+4x$
$4(x+4)-12y$	$2(2x+8-6y)$	$(5+6y)+4x$	$\frac{1}{3}(-3z+3y+3x)$
$5+2(3y+2x)$	$2y-(z+y)+x$	$4x+12y-16$	$(4x+16)-12y$
$2(2x+8)+6y$	$3x+3(z+y)$	$6x+6y-3(x+y-z)$	$x-(z-y)$
$5x+3y+3z-2x$	$6y+5+4x$	$(z+y)-z$	$4x-(12y-16)$

Student Activity B

Section 1.9: Working with Algebraic Language and Parentheses

For each problem below either write out the mathematical expression in words, or write the missing expression. Then simplify the expressions. Remember to use words like "the quantity of" or "the sum of" or "the difference of" when you describe more than one term in parentheses.

	Expression	Write the expression in words
1.		8 times the difference of $5y$ and 3
2.	$8-5(y-3)$	
3.		8 minus the quantity of $5y$ minus 3
4.	$8-5y-3$	
5.		the difference of 8 and $5y$ times negative 3
6.	$-6-2(x+7)$	
7.		negative 6 times the sum of negative $2x$ plus 7
8.	$-6-(2x+7)$	
9.		negative 6 minus $2x$ plus 7
10.	$(-6-2x)\cdot 7$	

Assessment 1C

Chapter 1: End-of-chapter Assessment for Understanding

For each of the following, describe the type of problem and the strategies and key steps to remember while doing the problem. You do **not** have to complete the problems.

	Type of Problem	Strategies and Key Steps
1. Write an expression to represent: y increased by 10.		
2. Multiply: $-\frac{3}{4}\left(\frac{2}{3}\right)$.		
3. Evaluate: $(-5)^2$.		
4. Divide: $\frac{-36}{-9}$.		
5. Name the property that is shown here: $2+(5+x)=(2+5)+x$.		
6. Name the property that is shown here: $(3+x)(2)=2(3+x)$.		
7. Evaluate: $5-2(-3\cdot4)$.		
8. Multiply: $-2(5x-4)$.		
9. Evaluate: $\frac{-2+\lvert 3-7\rvert}{9-2^2}$.		
10. Simplify: $6x+8-x-3$.		

Assessment 1D

Chapter 1: Metacognitive Skills Assessment

Metacognitive skills refer to the ability to judge how well you have learned something and to effectively direct your own learning and studying. This is a self evaluation tool designed to help you focus your studying and to improve your metacognitive skills with regards to this math class.

Fill the 1st column out before you begin studying.
Fill the 2nd column out after you study and before you take the test.
Go back to this page after your test and circle any of the ratings that you would now change – this identifies the "disconnects" between what you think you know well and what you actually know well.

Use the scale below to assign a number to each topic.

5 *I am confident I can do any problems in this category correctly.*
4 *I am confident I can do most of the problems in this category correctly.*
3 *I understand how to do the problems in this category, but I still make a lot of mistakes.*
2 *I feel unsure about how to do these problems.*
1 *I know I don't understand how to do these problems.*

Topic or Skill	Before Studying	After Studying
Answering a question about the information in a table or graph.	4	
Distinguish between equations and expressions.	1	
Using an equation to construct a table of data.	4	
Multiplying or dividing (signed) fractions and simplifying the result.	5	
Adding and subtracting (signed) fractions and simplifying the result.	2	
Working with mixed numbers in mathematical expressions.	4	
Finding the prime-factored form of a number; finding a factor-pair for a number.	5	
Knowing the difference between an "opposite" and a "reciprocal" number.	5	
Categorizing numbers in different number sets (Real, rational, natural, etc.)	1	
Evaluating expressions involving absolute value.	4	
Graphing numbers on a number line.	5	
Adding or subtracting real numbers (signed numbers).	5	
Identifying which addition or multiplication property has been used in a statement.	2	
Solving application problems that involve subtraction of signed numbers.	5	
Multiplying or dividing real numbers (signed numbers).	5	
Understanding the division properties of zero.	5	
Evaluating or rewriting exponential expressions.	5	
Distinguishing between exponential expressions like $(-3)^2$ and -3^2.	3	
Knowing the order of operations.	4	
Applying the order of operations to an expression.	4	
Applying the order of operations when it involves absolute value or a fraction bar.	4	
Evaluating an algebraic expression for a given number or numbers.	3	
Finding the average of a set of data.	2	
Identifying the number of terms in an expression, its coefficients, or its factors.	1	
Identifying the mathematical operation that is involved from problem words.	1	
Translating a phrase or sentence into a mathematical expression or equation.	1	
Answering an application question where the answer involves algebra.	2	
Simplifying expressions involving like and unlike terms.	2	
Applying the distributive property.	4	
Applying a negative that is in front of a set of parentheses.	3	

Student Workbook: Chapter 2

Table of Contents: *Equations, Inequalities, and Problem Solving*

Assessment 2A

Pretest and Diagnostic Tool: Equations, Inequalities, and Problem Solving

Directions: Complete this assessment without looking back at your notes or your book. **Do not use a calculator on this assessment.**

1. Evaluate $4x+7$ for $x=-2$. _____

2. Evaluate $5-x$ for $x=-3$. _____

Fill in the blank for problems 3-14:

3. $-2+(-2)=$ ____

4. $3+(-3)=$ ____

5. $-\frac{1}{5}+\frac{1}{5}=$ ____

6. $5+$ ____ $=0$

7. $-\frac{1}{3}+3=$ ____

8. $(-4)\div 4=$ ____

9. $-3\cdot\left(-\frac{1}{3}\right)=$ ____

10. $-5\div$ ____ $=1$

11. $\frac{2}{3}\cdot$ ____ $=1$

12. $-\frac{4}{5}\cdot\frac{5}{4}=$ ____

13. $\frac{2}{3}\cdot 24=$ ____

14. $\frac{5}{3}\cdot 3=$ ____

15. Evaluate: $12\left(\frac{3}{4}\right)-12\left(\frac{2}{3}\right)$.

16. Simplify: $4x-2y+7y$.

17. Simplify: $x+(x+1)+(x+2)$.

18. Simplify: $20(x-4)-5x$.

19. What is 50% of 80? ____

20. Would 125% of 50 be bigger or smaller than 50? __________

21. A price of $120 is reduced by $15. What is the new price? _______

22. Multiply: $0.075(82)=$______.

23. Find and simplify an algebraic expression for the sum of 5, *x*, and 2*x*. ___________

24. Write an algebraic expression to represent the product of *x* and 3, decreased by 4. ___________

25. When we describe the amount of water in a pool, is this a length, area, or volume? _________

26. Write three consecutive integers starting with 40: ___, ___, ___.

27. If the length of a board is 20 ft., and it is cut into two pieces, one measuring x feet, what is the length of the other piece? _________

Use $<$ or $>$ to make each statement true:

28. 0 ____ -2

29. -3 ____ -5

30. -0.75 ____ -0.5

Student Activity A

Section 2.1: Is it a Solution?

Tic-tac-toe Directions: If the number in the square **IS** a solution of the equation, then put an **O** on the square. If it **IS NOT** a solution, then put an **X** on the square.

$x+5=9$ 4	$\lvert y-3\rvert=5$ 2	$4.3+x=7.7$ 3.4
$10-\frac{x}{2}=4$ 6	$3z+7=-1$ −2	$x^2-3x-4=0$ 4
$0.2x=3$ 0.6	$\lvert 6-a\rvert=9$ 15	$x^2-3x-4=0$ −1

Student Activity B

Section 2.1: Checking Solutions with a Calculator

When you check the solution to an equation, we hope to see an equality (a true statement) when we are finished. However, when the solution is a rounded decimal value, you may only see an approximate equality. To denote this, use the symbol $\approx$ instead of $=$.

For example, we check two solutions in the equation $7x-2=4$.

Check $x=\frac{6}{7}$:	Check $x=0.86$:
$7x-2=4$ $7\left(\frac{6}{7}\right)-2\overset{?}{=}4$ $6-2\overset{?}{=}4$ $4=4$ True.	$7x-2=4$ $7(0.86)-2\overset{?}{=}4$ $6.02-2\overset{?}{=}4$ $4.02\approx 4$ True.

Note that $\frac{6}{7}=0.8571428...\approx 0.86$, so both answers should work, but in the second check, we only get only an approximate equality. These numbers are close enough for us to believe that $6/7$ is a solution of the equation.

Directions: In each box, determine if the given number creates an equality, an approximate equality, or is not a solution. Use the symbols $=$, $\approx$, and $\neq$ where appropriate.

$8x=4-x$ 0.44	$2y+1=4$ 1.5	$3u^2+5u-2=0$ 0.33	$\lvert 2s+3\rvert=3.5$ -0.25
$w^2=\frac{9}{16}$ 0.75	$x=\frac{x+2}{4}$ 0.67	$3z+3=2-z$ -0.12	$a^3=7$ 1.91

Student Activity C

Section 2.1: Solving One-step Equations

Match-up: Match each of the equations in the squares in the lower table with its solution from the top. If the solution is not found among the choices A through D, then choose E (none of these).

A -2 **B** 8 **C** -1 **D** 0 **E** None of these

$2 = x + 4$	$-5a = -40$	$\frac{m}{4} = 0$	$-3 + t = 5$
$\frac{5r}{2} = 5$	$\ell - 5 = -7$	$-\frac{2}{3}u = \frac{2}{3}$	$1 = h + 3$
$8 = -8 + g$	$6.2 + k = 4.2$	$14b = 28$	$32 = 4d$
$-7 = n - 6$	$-5 = c - 7$	$-2 = \frac{w}{-4}$	$v = -3 + 7 + (-4)$
$\frac{3}{4}f = -\frac{3}{2}$	$-6 = 6y$	$-x = -8$	$z - \frac{1}{6} = -\frac{7}{6}$

Student Activity A

Section 2.2: Match Up on Solving Equations

Match-up: Match each of the equations in the squares of the table below with its solution from the top. If the solution is not found among the choices A through E, then choose F (none of these).

A $x=-2$ **B** $x=3$ **C** $x=2$ **D** $x=0$ **E** $\varnothing$ **F** None of these

$-2(1-4x)=3x+8$	$5x-4=3x+2$	$5x-3x=0$	$5x+4x=x-4$
$5-4x=-1-x$	$\frac{1}{3}-x=2-x$	$5=5-3x$	$2x+6=-x$
$6x+(-3)=2x-3$	$3x=4-3x$	$\frac{x}{2}+2=\frac{7}{2}$	$\frac{1}{3}x+2=\frac{4}{3}$
$2(x-1)+4=x+5$	$\frac{1}{2}=\frac{x}{4}+1$	$2x-4=-8$	$3x-1=2x+x$
$4x-3=2(4+2x)$	$7=6-x$	$\frac{3x}{2}-4=-1$	$-3x+5=-4$

Guided Learning Activity

Section 2.2: Clearing the Fractions from Equations

Description: We multiply both sides of the equation by a number that "clears" all the fractions by reducing them.

Strategy:

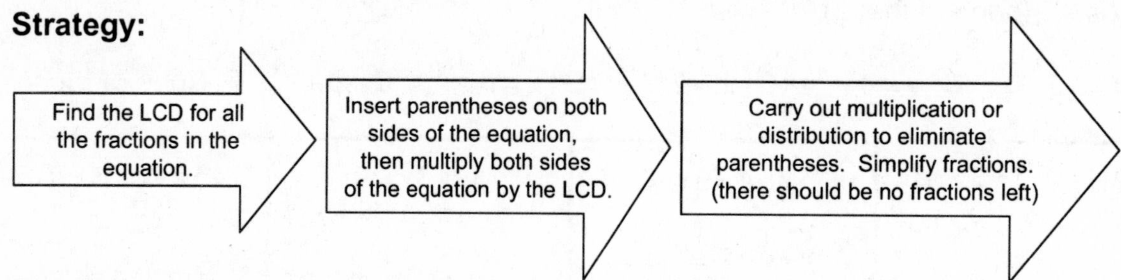

Examples: Use a colored pencil to write in the parentheses and the multiplication by the LCD.

1. Solve: $\frac{x}{3}+\frac{2x}{5}=-4$

$$15\cdot\left(\frac{x}{3}+\frac{2x}{5}\right)=15\cdot(-4)$$

$$15\cdot\frac{x}{3}+\frac{2x}{5}\cdot 15=15\cdot(-4)$$

3. Solve: $\frac{2+t}{3}-3t=4-\frac{t+5}{2}$

$$\frac{2+t}{3}-3t = 4-\frac{t+5}{2}$$

$$\frac{2+t}{3}-3t = 4-\frac{t+5}{2}$$

2. Solve: $\frac{1}{4}x+2 = x+\frac{1}{2}$

$$\frac{1}{4}x+2 = x+\frac{1}{2}$$

$$\frac{1}{4}x+2 = x+\frac{1}{2}$$

4. Solve: $\frac{9-2t}{7}=t$

$$\frac{9-2t}{7} = t$$

QUESTION: You cannot clear fractions or decimals in an expression, this method only works in equations. *Why?*

Student Activity B

Section 2.2: Equations vs. Expressions

Tic-Tac-Toe: For the problems below, decide whether it is an expression or an equation – if it's an expression, simplify the expression and put an **X** in the box. If it's an equation, then solve the equations and circle the solution (thus making an **O**).

$4x-(6x+2)$	$5x+3=7$	$3-x=9$
$3(x-2)-2x$	$5x+3-7$	$\frac{2}{5}x-\frac{1}{3}+\frac{x}{5}=\frac{2}{3}$
$3(x-2)=2x$	$4x-(2x-4)=8$	$\frac{2}{5}x-\frac{1}{3}+\frac{x}{5}-\frac{2}{3}$

QUESTION: Do the answers to equations look any different than the answers to expressions? Why or why not?

Student Activity A

Section 2.3: Writing and Solving Percent-sentences

Directions: Rewrite each sentence as a percent-sentence, then translate to an equation, then solve. Use x or p to represent the unknown quantity in the equations. Remember percents must be expressed as decimals in equations.

1. What is 25% of 50?

____ is ____ % of _____

____ = _____ · _____

Solution: _________

2. 20 is what percent of 60?

____ is ____ % of _____

____ = _____ · _____

Solution: _________

3. What percent of 80 is 5?

____ is ____ % of _____

____ = _____ · _____

Solution: _________

4. What is 60% of 200?

____ is ____ % of _____

____ = _____ · _____

Solution: _________

5. 12 is 6% of what number?

____ is ____ % of _____

____ = _____ · _____

Solution: _________

6. 0.5 is what percent of 60?

____ is ____ % of _____

____ = _____ · _____

Solution: _________

7. What is 2% of 50?

____ is ____ % of _____

____ = _____ · _____

Solution: _________

8. 8 is 40% of what number?

____ is ____ % of _____

____ = _____ · _____

Solution: _________

9. What is 120% of 200?

____ is ____ % of _____

____ = _____ · _____

Solution: _________

10. What percent of 80 is 100?

____ is ____ % of _____

____ = _____ · _____

Solution: _________

Student Activity B

Section 2.3: Working with Percent Increase or Decrease

Directions: Calculate the amount of increase or decrease for each problem, then find what is being asked.

(Percent of increase)·(Original amount) = Amount of increase

(Percent of decrease)·(Original amount) = Amount of decrease

$$\text{Percent of increase (or decrease)} = \frac{\text{Amount of increase (or decrease)}}{\text{Original amount}}$$

---

1. The population of a small town is 890. If the population decreases 10% after a factory closing, what is the new population of the town?

   By how many people does the population decrease? _______

   What is the new population? _______

2. The assessed value of a \$120,000 home increases by 5% in one year. What is the new assessed value of the home?

   By how much does the home value increase? ________

   What is the new assessed value of the home? ________

3. A stock was valued at \$34.50 per share. If the stock value decreases by 20% after a poor sales report, what is the new stock price?

   By how much does the stock decrease? ________

   What is the new stock price? ________

4. An elderly woman receives \$582.00 in social security each month. After a recent increase, she began receiving \$616.92 per month. What was the percent increase in her social security check?

   By how much did the check amount increase? ________

   Calculate the percent increase:

5. A laptop computer was advertised for \$1699. When you call to inquire about purchasing the laptop, the phone salesperson says they will give you the laptop for \$1449 instead. What was the percent decrease in the cost of the computer?

   By how much did the computer cost decrease? ________

   Calculate the percent decrease:

Student Activity A
Section 2.4: Circle Scavenger Hunt

Materials: Your instructor will provide you with a long piece of string and a ruler.

Directions: You have 10 minutes to find five circular objects to measure on your campus. Use the string and ruler to measure the circumference and diameter of each circle. Measure the first two objects in inches and the rest in centimeters. Once you have found five circles, return to your classroom to do the rest of the calculations.

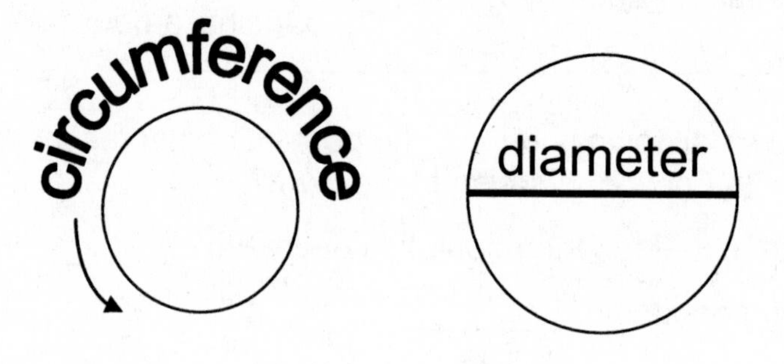

Object	Circumference	Diameter	Calculate: $\frac{\text{circumference}}{\text{diameter}}$
1. (measure in inches)			
2. (measure in inches)			
3. (measure in centimeters)			
4. (measure in centimeters)			
5. (measure in centimeters)			
			Average:

Conclusion:

Student Activity B

Section 2.4: Is it Length, Area, or Volume?

Measuring Length

$P = 2w + 2\ell$ (rectangle)

$P = 4s$ (square)

$C = 2\pi r = \pi d$ (circle)

$P = a + b + c$ (triangle)

English units: in, ft, yd, mi
Metric units: mm, cm, m, km

Measuring Area

$A = \ell w$ (rectangle)

$A = s^2$ (square)

$A = \pi r^2$ (circle)

$A = \frac{1}{2} b h$ (triangle)

English units: in^2, ft^2, yd^2, mi^2
Metric units: mm^2, cm^2, m^2, km^2

Measuring Volume

$V = \ell w h$ (rectangular solid)

$V = s^3$ (cube)

$V = \frac{4}{3} \pi r^3$ (sphere)

$V = \pi r^2 h$ (cylinder)

English units: in^3, ft^3, yd^3, mi^3
Metric units: mm^3, cm^3, m^3, km^3

Directions: Read the problems and decide whether you will have to find a length, area, or volume. Then identify the appropriate formula to use and decide what the resulting units will be. You do not have to solve the problems.

Problem	Length, Area, or Volume?	Formula	Units of answer
1. Mark is going to paint a large rectangular wall in a museum that measures 20 feet by 9 feet. How many square feet will Mark be painting?			
2. Alice wants to build a fence around her circular garden. If the diameter of the garden is 6 meters, how long does the fence need to be?			
3. The cylindrical mixing barrel on a cement truck measures 4 yards long and has a radius of 0.8 yards. How much cement can the truck hold in its mixing barrel?			
4. A rectangular fish tank measures 24 inches by 10 inches by 10 inches. How much water can the tank hold?			
5. A real-estate agent is pricing a New York studio apartment for someone. If the apartment is square, with 15 feet on each side, and apartments cost approximately \$6.00 per square foot per month, how much will this one rent for?			

Assessment 2B

Chapter 2: Mid-chapter Assessment for Understanding

For each of the following, describe the type of problem and the strategies and key steps to remember while doing the problem. You do **not** have to complete the problems.

	Type of Problem	Strategies and Key Steps
1. Change 0.25% to a decimal.		
2. Solve: $3x+5=11$.		
3. 25 is what percent of 60?		
4. Solve: $-4x=12$.		
5. If the circumference of a circle is 14π cm, find the radius.		
6. Solve: $\frac{3x}{4}-2=\frac{x}{6}$.		
7. Find 2.5% annual simple interest on \$500.		
8. Find the volume of a sphere if the radius is 4 mm.		
9. Is -3 a solution to $4x+6=-2$?		
10. Solve: $3(-2n+4)=-30$.		

Guided Learning Activity

Section 2.5: Using the Problem Solving Strategy

Problem 1: In a *sprint* triathlon, the participants swim 0.75 km, bike 20 km, and then complete the triathlon with a run. If the total distance of the triathlon is 25.75 km, then what is the running distance?

Analyze the problem: What is given? What are you being asked to find? Can you draw a picture or diagram? Can you construct a table?

Form an equation: Define the variable and write the equation.

Solve the equation:

State the conclusion:

Check the result:

Problem 2: The terminal bus at an airport drives a 4.2 mile loop between the three terminals (A, B, and C). If Terminal A is 1.6 miles from Terminal B, and Terminal B is 0.9 miles from Terminal C, then how far is the drive from Terminal C back to Terminal A?

Analyze the problem: What is given? What are you being asked to find? Can you draw a picture or diagram? Can you construct a table?

Form an equation: Define the variable and write the equation.

Solve the equation:

State the conclusion:

Check the result:

Problem 3: Three consecutive even integers sum to 78. What are the three integers?

Analyze the problem: What is given? What are you being asked to find? Can you draw a picture or diagram? Can you construct a table?

Form an equation: Define the variable and write the equation.

Solve the equation:

State the conclusion:

Check the result:

Problem 4: In April, the water level in a mountain reservoir increased 0.6 m as the snow began to melt. In May, the water level increased another 0.3 m. In June the water level remained constant, but in July the water level decreased by 0.2 m because of a dry spell. At the end of July, the water level was 14.2 m; what was the water level at the beginning of April?

Analyze the problem: What is given? What are you being asked to find? Can you draw a picture or diagram? Can you construct a table?

Form an equation: Define the variable and write the equation.

Solve the equation:

State the conclusion:

Check the result:

Problem 5: Two water tanks hold a total of 600 liters of water. One tank holds sixty liters less than twice the amount in the other. How many liters does each tank hold?

Analyze the problem: What is given? What are you being asked to find? Can you draw a picture or diagram? Can you construct a table?

Form an equation: Define the variable and write the equation.

Solve the equation:

State the conclusion:

Check the result:

Problem 6: The campsite fee at a state park is $27 per day plus $8 to make your campsite reservation online. Based on a budget of $170 to make the reservations and pay for camping, how many days can a family camp at the state park?

Analyze the problem: What is given? What are you being asked to find? Can you draw a picture or diagram? Can you construct a table?

Form an equation: Define the variable and write the equation.

Solve the equation:

State the conclusion:

Check the result:

Student Activity A

Section 2.5: Working with Consecutive Integers

Fill in the table below:

Let $x = ...$	$x+1$	$x+2$	$x+3$
5			
10			
22			
a			
$2n$			

a. How would you represent the two integers **after** x: ______ and ______

b. How would you represent the two integers **before** x: ______ and ______

Fill in the table below:

Let $x = ...$	$2x$	$2x+1$	$2x+2$	$2x+3$	$2x+4$
1					
2					
3					
18					
25					
a					
Odd or even?					

c. How do we represent an unknown number that must be an even integer? ________

d. How do we represent an unknown number that must be an odd integer? ________

e. How would you represent the two integers that follow $2x$? ______ and ______

f. How would you represent the two **even** integers that follow $2x$? _____ and ______

g. How would you represent the two **odd** integers that follow $2x+1$? _____ and ______

Student Activity B

Section 2.5: What was the Problem?

For each of the solutions below, **write a problem** that could go with the solution. Luckily for you, each student has used at least part of the problem-solving strategy.

Problem 1:

Form an equation:

Let $x =$ the measure of one of the isosceles angles. Then $2x+20$ is the measure of the large angle.

Equation: $x+x+(2x+20)=180$

Solve the equation:

$$4x+20=180$$
$$4x=160$$
$$x=40$$

State the conclusion: The angles measure $40°$, $40°$, and $100°$.

Check the result:

$$40+40+(2\cdot 40+20)\stackrel{?}{=}180$$
$$180=180 \quad \text{True.}$$

Problem 2:

Form an equation:

Let $C =$ commission and $P =$ selling price of house., then $C=0.07P$.

The owner would receive $P-C$ or $P-0.07P$.

Equation: $P-0.07P=250{,}000$

Solve the equation:

$$P-0.07P=250{,}000$$
$$0.93P=250{,}000$$
$$P\approx \$268{,}817$$

State the conclusion: The selling price of the house should be $268,817.

Check the result:

$$268{,}817-0.07(268{,}817)\stackrel{?}{=}250{,}000$$
$$250{,}000=250{,}000 \quad \text{True.}$$

Problem 3:

Form an equation:

Susan has a budget of \$65. For 30 pages, the scrapbook price is \$30.

Let $p =$ the number of extra pages in the scrapbook.

Then $1.75p$ is the cost of the extra pages.

Equation: $30 + 1.75p = 65$

Solve the equation:

$$\begin{aligned} 30 + 1.75p &= 65 \\ 1.75p &= 35 \\ p &= 20 \end{aligned}$$

State the conclusion: Susan can get 20 extra pages, for a total of 50 pages in the scrapbook.

Check the result:

$$\begin{aligned} 30 + 1.75(20) &\stackrel{?}{=} 65 \\ 65 &= 65 \quad \text{True.} \end{aligned}$$

Student Activity A

Section 2.6: Understanding the Smaller Pieces

Directions: Write out how you calculated the answer to each problem. Include the proper units in your answers. The scrambled numerical answers (without units) are found at the bottom of the page.

1. The operator of an after-school tutoring program needs to purchase some tables and chairs. For each table, he needs four chairs. Tables cost $250 and chairs cost $60. If the program requires 24 tables, how much will he spend total on tables and chairs?

2. The grandparents of a new baby invest $15,000 in a fund that pays simple annual interest with a rate of 5.2%. How much interest will the fund earn during the first year?

3. A ship heads west at a speed of 25 mph. If the ship travels at this speed for 20 minutes, how far will it travel?

4. A runner goes a distance of 7.2 miles in 1 hour and 10 minutes. What was the runner's average speed in mph?

5. A chemist has 80 liters of 20% acid solution. How many liters of the solution is acid?

6. A bulk-foods department purchases 50 pounds of almonds for $238.99. What is the price per pound for the almonds?

7. The monthly cost to store big-screen TVs at a warehouse is $7.50 per TV per month. How much would it cost to store 200 big-screen TVs for 3 months?

8. A concept-mapping software package sells for $179. If a salesman has a monthly goal of $15,000 in sales, how many software packages does he have to sell to exceed his goal?

9. If a search-and-rescue team heads north traveling at 2.5 mph, how many hours will it take them, traveling at the same speed, to travel 12 miles?

10. Two gallons of paint that costs $21 per gallon is mixed with one gallon of primer that costs $12 per gallon. What is the new cost per gallon of the mixed paint?

Scrambled answers: 16 4.8 4.78 11760 18 8.33 4500 780 84 6.17

Guided Learning Activity

Section 2.6: Organizing Information into Tables

For each problem:

a) Fill in each table with numbers in the appropriate spaces (there will be a few unfilled spaces).
b) One of your unfilled spaces should be filled with the unknown variable or expressions involving the unknown variable (like x, $2x$, $x+5$, etc.).
c) The last column of each table is calculated using the previous columns. Go ahead and fill this in.

1. A man decided to invest the $15,000 inheritance he received so that he could use the annual interest earned to pay the annual taxes on his home, which are $1,200. The highest bank rate that he could find was 6% annual simple interest, but this is not high enough to make the $1,200. So, instead of investing all the money at the bank, he invested some of the money in a riskier, but more profitable, investment offering a 10% return. How much should the man invest in each account to make exactly $1,200 in one year?

Investment	**P**	**r**	**t**	**Interest**
Bank				
Riskier investment				

2. A chemistry experiment calls for a 40% hydrochloric acid solution. If the lab supply room has only 60% and 10% hydrochloric acid solutions, how much of each should be mixed to obtain 12 liters of the 40% acid solution?

Solution	**Solution Amount**	**Strength**	**Acid Amount**
Strong Acid			
Weak Acid			
Mixed Acid			

3. A bulk-foods department purchaser wants to mix up 20 pounds of trail mix consisting of peanuts, chocolate candies, and raisins. The mixture should contain twice as many peanuts as raisins (by weight) and the amount of raisins and chocolate candies should be equal. If peanuts cost $2.40 per pound, chocolate candies cost $1.80 per pound, and raisins cost $3.20 per pound, how many pounds of each ingredient should be used in order to create the trail mix and how much will it cost per pound?

Ingredient	**Amount**	**Price per pound**	**Total Value**
Peanuts			
Raisins			
Chocolate			
Trail Mix			

4. A feed store sells sunflower seeds for \$2.50 per pound and feed corn for \$0.60 per pound. The manager wants to create 100 pounds of small animal feed that costs \$1.50 per pound by mixing these two ingredients. How much of each should be used?

Ingredient	**Amount**	**Price per pound**	**Value**
Sunflower seed			
Feed corn			
Small Animal Feed			

5. Two runners leave traveling in opposite directions on a 22-mile loop. The first runner sets a pace of 5 mph and the second runner sets a pace of 6 mph. How long will it take them to meet back up on the loop?

	Rate	**Time**	**Distance**
First Runner			
Second Runner			

6. Two tractor trailer trucks are 540 miles apart and their speeds differ by 5 mph. Find the speed of each truck if they are traveling toward each other and will meet in 4 hours.

	Rate	**Time**	**Distance**
First truck			
Second truck			

7. At an electronics store, the number of 37-inch HDTVs that is expected to be sold in a month is double that of 58-inch HDTVs and 50-inch HDTVs combined. Sales of 58-inch and 50-inch HDTVs are expected to be equal. 58-inch HD-TVs sell for \$4,800, 50-inch HD-TVs sell for \$2,000, and 37-inch HDTVs sell for \$1800. If total sales of \$168,800 are expected this month, how many of each HDTV should be stocked?

	Number sold	**Price**	**Sales \$**
37-inch HDTV			
50-inch HDTV			
58-inch HDTV			

Student Activity A

Section 2.7: Tic-Tac-Toe with Inequalities

Tic-tac-toe #1: If the inequality in the square is true, then put an **O** on the square. If it is false, then put an **X** on the square.

$-5 < -1$	$8 > 8$	$-6 > -2$
$2 - 3 < -1$	$\frac{1}{2} < \frac{1}{3}$	$7 \leq 7$
$a \geq a$	$0 < -4$	$\lvert -3 \rvert > 1$

Tic-tac-toe #2: If the number in the square is a solution to the inequality, then put an **O** on the square. If it is false, then put an **X** on the square.

$x - 2 < 7$ 9	$x + 4 \geq -2$ -7	$x - 3 \leq -1$ 1
$-2x < 4$ -1	$\frac{1}{2}x \geq 2$ 6	$-1 \leq 2 - x$ 3
$8 > 4 - x$ -5	$-x < 4$ -3	$2x + 3 < 3x$ 2

Guided Learning Activity

Section 2.7: Graphing Inequalities and Using Interval Notation

Think of interval notation as a way to tell someone how to draw the graph, from left to right, giving them only a "begin" value and an "end" value for each interval.

- Use $-\infty$ and ∞ to denote the "ends" of the number line (as shown in the number line below) Remember that it would read $(-\infty, \ldots$ or $\ldots, \infty)$ at the ends of an interval.

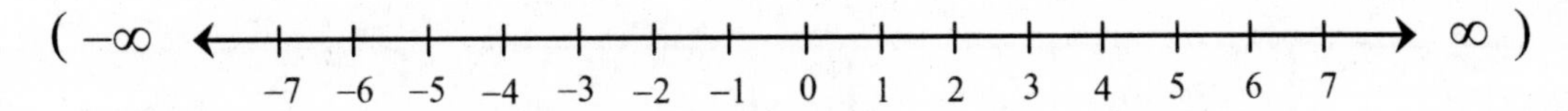

- Always give intervals from LEFT to RIGHT on the number line.
- Use (or) to denote an endpoint that is approached, but not included.
- Use [or] to denote an endpoint that may be included.
- The parenthesis or bracket needs to open in the direction of the true part of the inequality.

	Set notation	Graph	Interval Notation
1.	$x > 2$	-5 -4 -3 -2 -1 0 1 2 3 4 5	
2.	$x \leq -3$	-5 -4 -3 -2 -1 0 1 2 3 4 5	
3.	$x \geq -1$	-5 -4 -3 -2 -1 0 1 2 3 4 5	
4.		-5 -4 -3 -2 -1 0 1 2 3 4 5	
5.		-5 -4 -3 -2 -1 0 1 2 3 4 5	
6.	$-2 < x \leq 3$	-5 -4 -3 -2 -1 0 1 2 3 4 5	
7.		-5 -4 -3 -2 -1 0 1 2 3 4 5	
8.		-5 -4 -3 -2 -1 0 1 2 3 4 5	$(-\infty, 4)$
9.		-5 -4 -3 -2 -1 0 1 2 3 4 5	$(-2, 0]$
10.		-5 -4 -3 -2 -1 0 1 2 3 4 5	$(-\infty, \infty)$

Student Activity B

Section 2.7: Match Up on Solving Inequalities

Match-up: Match each of the inequalities in the squares of the table below with an equivalent inequality from the top. If the solution is not found among the choices A through D, then choose E (none of these).

A $x > 2$ **B** $x < 3$ **C** $x > -1$ **D** $x < -3$ **E** None of these

$2x < -6$	$-3x < -6$	$-5x < 5$	$x + 2 < 5$
$1 > x - 1$	$-1 < x$	$8 < 4x$	$15 < -5x$
$\frac{3}{2}x > 3$	$6 > -6x$	$2x + 3 < -3$	$\frac{x}{2} + 1 < \frac{5}{2}$
$5 - 2x < 11$	$2 - 3x < 5$	$\frac{x}{3} < 1$	$x - 7 > -8$
$2x < -2$	$\frac{x+1}{2} > \frac{1}{2}$	$1 - x > 4$	$-9 > x - 6$

Assessment 2C

Chapter 2: End-of-Chapter Assessment for Understanding

For each of the following, describe the type of problem and the strategies and key steps to remember while doing the problem. You do **not** have to complete the problems.

	Type of Problem	Strategies and Key Steps
1. If the sales tax on an item is 6%, and the price is $129.95, find the total paid by the customer.		
2. Write $-2 < x \leq 4$ in interval notation.		
3. Solve: $\frac{b}{2} - \frac{1}{3} = \frac{3}{4}$.		
4. Solve: $-4x \leq 24$.		
5. How many liters of a 2% acid solution must be added to 30 liters of a 10% acid solution to dilute it to an 8% acid solution?		
6. A car averaged 60 mph for part of a trip and 70 mph for the remainder. If the 7 ½ -hour drive covered 500 miles, for how long did the car average 60 mph?		
7. Solve $A = bh + Bh$ for B.		
8. If the vertex angle of an isosceles triangle is $24°$, find the measure of each base angle.		
9. The cost of a stereo to an electronics store is $268. If the markup is 20%, what is the sales price of the stereo?		
10. Is $x = -2$ a solution of $2x + 1 < -3$?		

Assessment 2D

Chapter 2: Metacognitive Skills Assessment

Metacognitive skills refer to the ability to judge how well you have learned something and to effectively direct your own learning and studying. This is a self evaluation tool designed to help you focus your studying and to improve your metacognitive skills with regards to this math class.

Fill the 1st column out before you begin studying.
Fill the 2nd column out after you study and before you take the test.
Go back to this page after your test and circle any of the ratings that you would now change – this identifies the "disconnects" between what you think you know well and what you actually know well.

Use the scale below to assign a number to each topic.
5 I am confident I can do any problems in this category correctly.
4 I am confident I can do most of the problems in this category correctly.
3 I understand how to do the problems in this category, but I still make a lot of mistakes.
2 I feel unsure about how to do these problems.
1 I know I don't understand how to do these problems.

Topic or Skill	Before Studying	After Studying
Checking the solution to an equation.		
Applying the addition or subtraction property of equality to solve an equation.		
Applying the multiplication or division property of equality to solve an equation.		
Setting up an equation to represent an application problem.		
Setting up a percent-sentence and then translating it into an equation.		
Setting up and solving application problems that involve percents, including those with a percent increase or decrease.		
Solving linear equations like $ax + b = c$.		
Solving linear equations that involve parentheses and like terms.		
Clearing the fractions from equations involving fractions.		
Understanding the difference between solving an equation and simplifying an expression.		
Solving a formula for a selected variable.		
Choosing a formula and solving a problem involving business or science formulas.		
Choosing the formula and solving a problem involving length, area, or volume.		
Placing the correct units on the results to application problems.		
Setting up a table to organize the information in an application problem.		
Solving problems involving consecutive integers, even integers, or odd integers.		

Continued on next page.

Use the scale below to assign a number to each topic.

5 *I am confident I can do any problems in this category correctly.*
4 *I am confident I can do most of the problems in this category correctly.*
3 *I understand how to do the problems in this category, but I still make a lot of mistakes.*
2 *I feel unsure about how to do these problems.*
1 *I know I don't understand how to do these problems.*

Topic or Skill	**Before Studying**	**After Studying**
Solving application problems involving right angles, straight angles, or triangles.		
Solving application problems involving distance, rate, and time.		
Solving application problems involving interest.		
Solving application problems involving percent mixtures or value mixtures.		
Deciding whether an inequality is true or false.		
Solving an inequality.		
Graphing an inequality.		
Writing the answer to an inequality in interval notation.		
Solving a compound inequality (like $-3 < x+1 < 5$).		
Solving an application involving inequalities.		

Student Workbook: Chapter 3

Table of Contents:

Graphing Linear Equations and Inequalities in Two Variables; Functions

Assessment 3A

Pretest and Diagnostic Tool: Graphing Linear Equations and Inequalities

Directions: Complete this assessment without looking back at your notes or your book. **Do not use a calculator on this assessment.**

1. Plot $-2\frac{1}{2}$ on the number line.

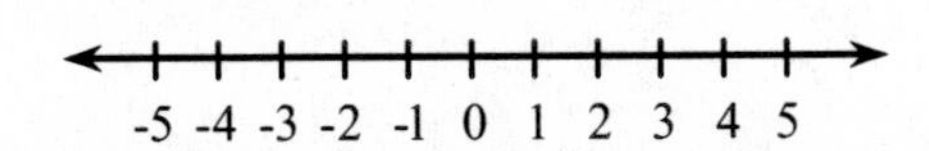

2. When you move your pencil up and down on the line to the right, does that signify a horizontal direction or a vertical direction? ____________

3. Evaluate $2x-3$ for $x=1$. _____

4. Evaluate $-\frac{1}{2}x+3$ for $x=-2$. _____

5. Evaluate $3x+2y$ for $x=-2$ and $y=5$. _____

6. $-3=4(-2)+5$ True or false? ________

7. Simplify: $(-3)^2-(-3)+5$.

8. Solve: $3x+(0)=-6$.

9. Solve: $0-10y=-5$.

10. $-\dfrac{2}{3}=\dfrac{-2}{-3}$ True or false? ________

11. $3=\dfrac{3}{1}$ True or false? ________

12. $-\dfrac{1}{3}=\dfrac{-1}{3}$ True or false? ________

13. Solve $2x+y=3$ for y.

14. Solve $x-2y=4$ for y.

15. Evaluate: $\dfrac{1-(-2)}{-1-2}$.

16. Evaluate: $\dfrac{2-(-2)}{4-3}$.

17. Simplify: $-2(x-(-4))+6$.

18. Simplify: $\dfrac{1}{3}(x-6)+2$.

19. $-4<-2$ True or false? ________

20. $0<-4$ True or false? ________

Student Activity A

Section 3.1: Where is the Point?

Directions: Capital letters are often used to name specific points. For each point that is graphed below, write down the letter and coordinates of the point and then place it in the appropriate category in the table below (individual quadrant or axis). Point A has been done for you.

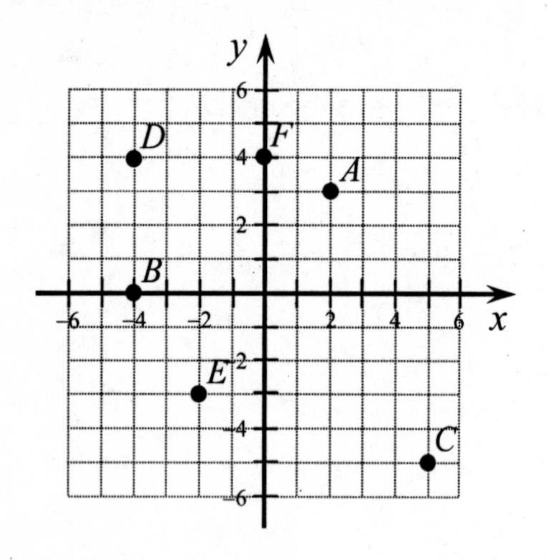

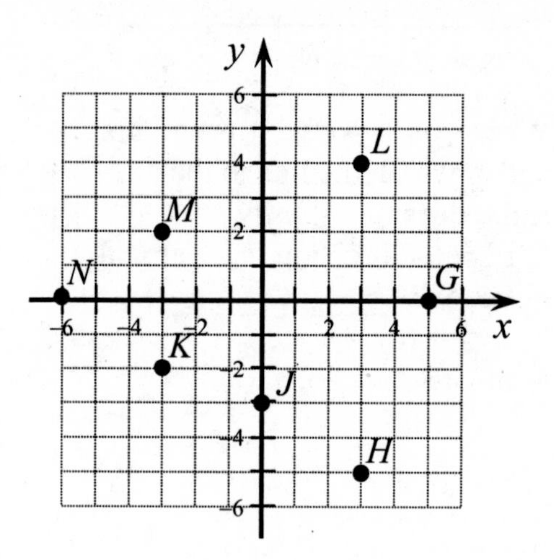

Quadrant I	Quadrant II	Quadrant III	Quadrant IV	x-axis	y-axis
$A(2,3)$					

Directions: Now plot these points on the axis provided. Indicate each point with it's letter on the graph.

$P(2,3)$ $G(-3,4)$

$F(-3,-4)$ $H(4,5)$

$A(0,0)$ $R(-2,2)$

$U(0,-3)$ $N(3,-4)$

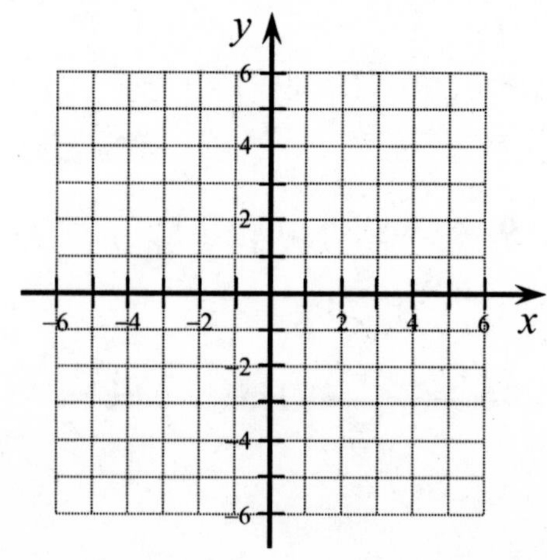

Guided Learning Activity

Section 3.1: Graphing Paired Data

Example 1: Gas mileage
In 2007, the most fuel-efficient mid-sized car for city driving was the Toyota Prius, which was rated at 60 miles per gallon. The least efficient mid-sized car for city driving was the Ferrari 612 Scaglietti, rated at 10 miles per gallon. (www.fueleconomy.gov)

In the table, each row of data represents an ordered pair. Plot the data using two different colors (one for each type of car) and then draw a straight line through each set of plotted points.

Toyota Prius

gallons	miles
1	60
2	120
3	180
4	240

Ferrari 612

gallons	miles
2	20
4	40
6	60
8	80

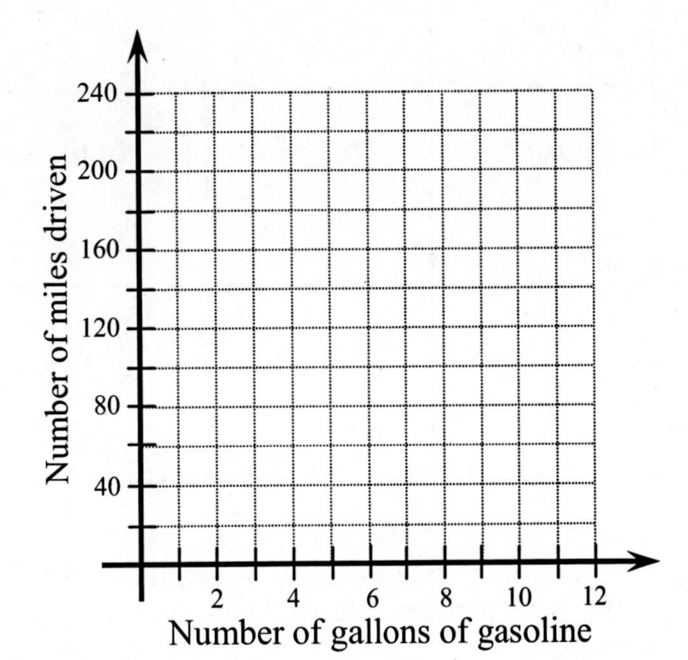

Example 2: Temperature Scales
Although we still measure temperature in the U.S. using the Fahrenheit scale for non-scientific purposes, the rest of the countries in the world use the Celsius Scale. In the data table below, we see paired data that show both Fahrenheit and Celsius temperatures. To read some interesting historical theories about why the Fahrenheit scale is set up the way it is, look at the Wikipedia entry for Fahrenheit under "History." (www.en.wikipedia.org)

°F	°C
20	–6.7
40	4.4
60	15.6
80	26.7
100	37.8
120	48.9

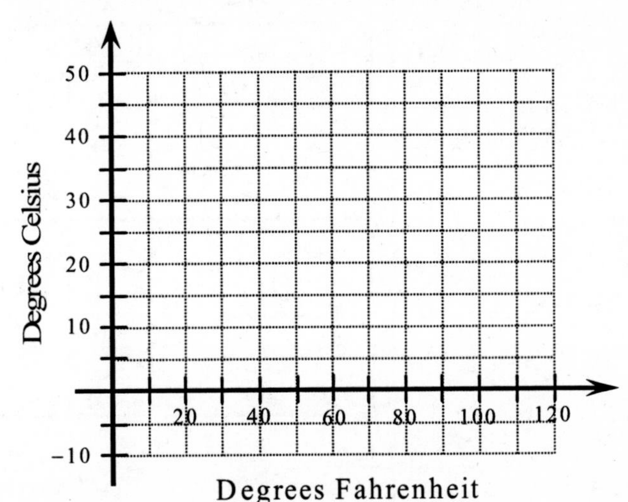

Student Activity A

Section 3.2: Tic-Tac-Toe with Ordered Pairs

Tic-tac-toe #1: If the ordered pair in the square **is** a solution to the equation $y = 2x - 5$, then circle it (thus placing an **O** on the square). If the ordered pair is **not** a solution, then put an **X** over it.

$(-10,-30)$	$(-10,-25)$	$\left(\frac{2}{3},-\frac{11}{3}\right)$
$(26,47)$	$(0,-5)$	$(0,0)$
$(5,0)$	$\left(-\frac{1}{2},-4\right)$	$(1,3)$

Tic-tac-toe #2: If the ordered pair in the square **is** a solution to the equation $4x - 6y = 12$, then circle it (thus placing an **O** on the square). If the ordered pair is **not** a solution, then put an **X** over it.

$(-9,-8)$	$\left(1,-\frac{4}{3}\right)$	$(3,0)$
$(-2,0)$	$(0,0)$	$\left(-\frac{3}{2},-3\right)$
$(1,1)$	$\left(\frac{2}{3},-\frac{14}{9}\right)$	$(9,-4)$

Guided Learning Activity

Section 3.2: Graphing Linear Equations

Example 1: Fill in the missing values in the table of solutions for the linear equation, $y = -x + 4$, plot the points, and then draw the line through the points.

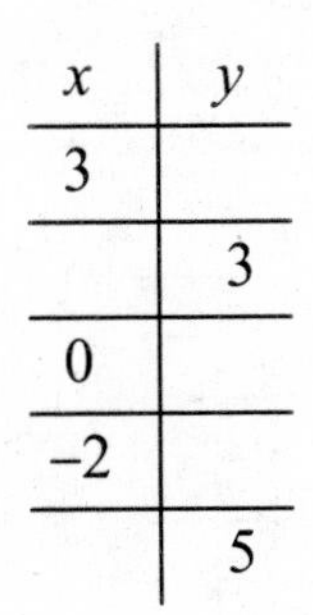

x	y
3	
	3
0	
–2	
	5

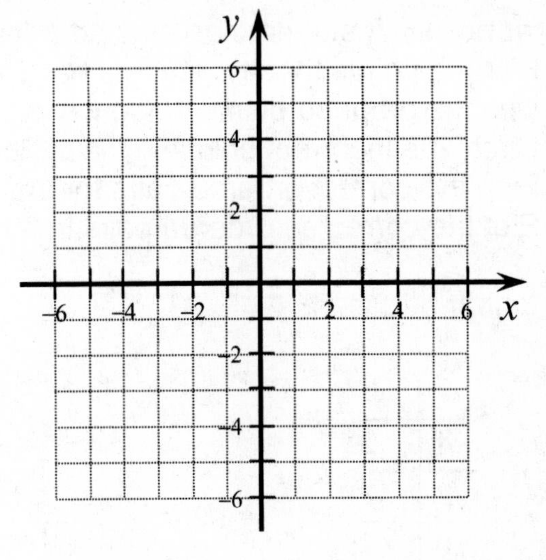

If $x = 3$, find y.	If $y = 3$, find x.	If $x = 0$, find y.	If $x = -2$, find y.	If $y = 5$, find x.

Example 2: Fill in the missing values in the table of solutions for the linear equation, $3x - y = -2$, plot the points, and then draw the line through the points.

x	y
	5
0	
	–1
–2	
	3

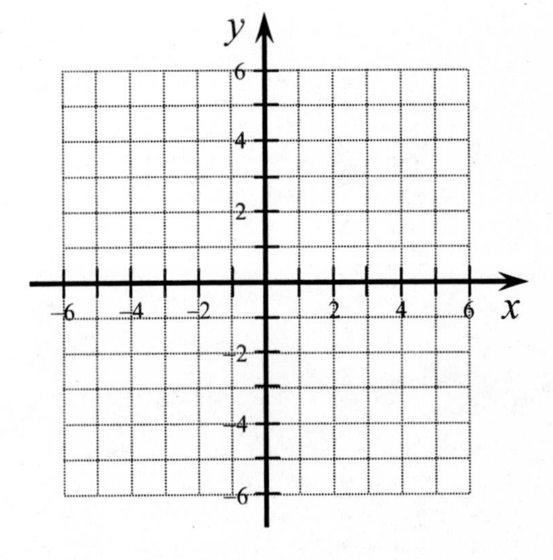

If $y = 5$, find x.	If $x = 0$, find y.	If $y = -1$, find x.	If $x = -2$, find y.	If $y = 3$, find x.

Student Activity B

Section 3.2: Out-of-line Suspects

Directions: A student has created a table of solutions for each **linear** equation below.
1) Plot the points that the student has found.
2) Use the graphed points to identify which points (if any) are most likely incorrect.
3) Circle the incorrect points in the table of values.
4) Find the correct ordered pairs for any that are circled.
5) Plot the corrected ordered pairs in a different color.

1. Linear equation: $y = 3x - 1$

x	y
-1	2
-2	3
0	-1
$\frac{1}{3}$	0
2	5

2. Linear equation: $y = \frac{1}{2}x + 2$

x	y
-2	0
-1	$\frac{3}{2}$
2	5
0	2
4	4

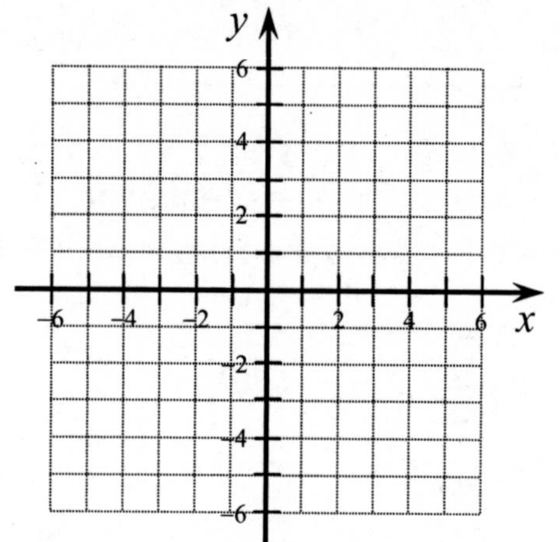

3. Linear equation: $y = 4 - x$

x	y
-2	6
-6	1
1	3
2	-4
6	-2

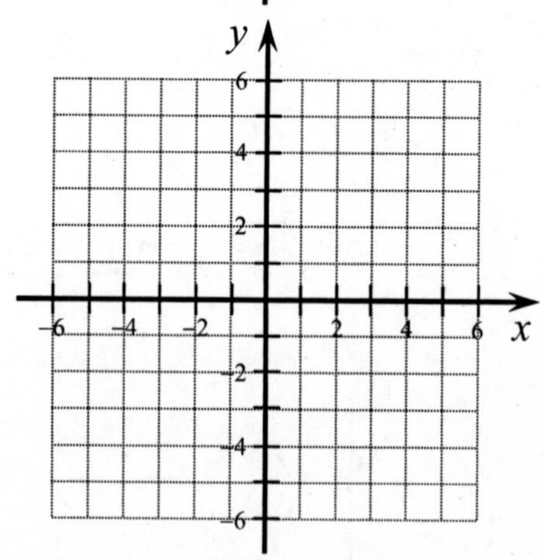

Student Activity C

Section 3.2: The Cost of College – A Linear Function

When you pay for your college education, you have typically have three different types of charges:

1) ***Variable cost*** – Tuition, which is paid per credit hour (or unity of study). There may also be fees that vary depending on the number of credit hours (for example, an additional $6 per credit hour).
2) ***Fixed cost*** – Student fees, which are paid regardless of how many or what types of classes you take, (for example, a $25 Registration Fee). These are fixed costs.
3) Course fees or lab fees, which are paid for certain courses only (we will ignore these in this model). Your instructor may tell you to ignore some other fee or tuition oddities to allow a linear model to be used.

Our linear model of college costs will only consider the first two types of charges. Your instructor will help you locate the information necessary to build the linear model – perhaps they have asked you to bring in a copy of your tuition bill!

How much do you pay in **fixed** costs for your enrollment at your college? __________

Per credit hour, how much do you pay in tuition (and **variable** fees)? ______

Fill in the table below for a student at your college:

Credits (or units) taken by student, *n*	Fixed student fees	Tuition and other variable costs for this number of credits (or units)	Total cost for this number of credits, *C*
4			
8			
12			
16			

Create a graph containing the data points above.

Write a linear equation to model the cost C, for taking a certain number of credits n.

Show how you could use the linear equation you just wrote to estimate the cost of taking 9 credits (or units).

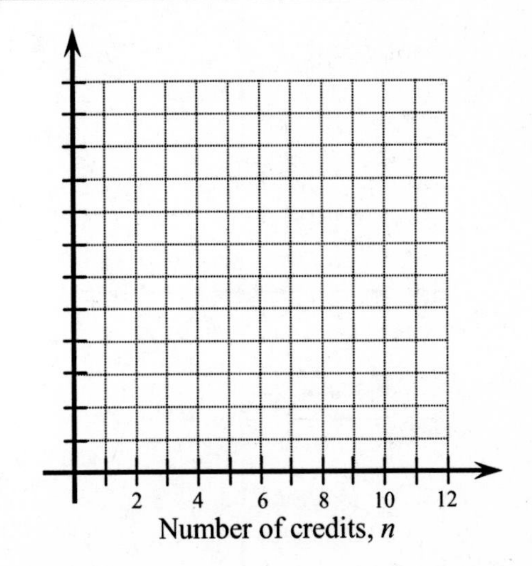

Student Activity A

Section 3.3: ID the Intercepts

Directions: For each graph, identify the x- and y-intercepts (if they exist) using **both** coordinates of the point.

1. x-intercept: (,)

y-intercept: (,)

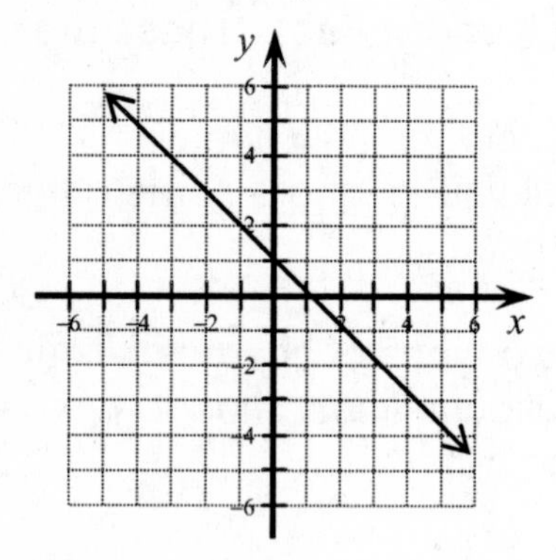

2. x-intercept: (,)

y-intercept: (,)

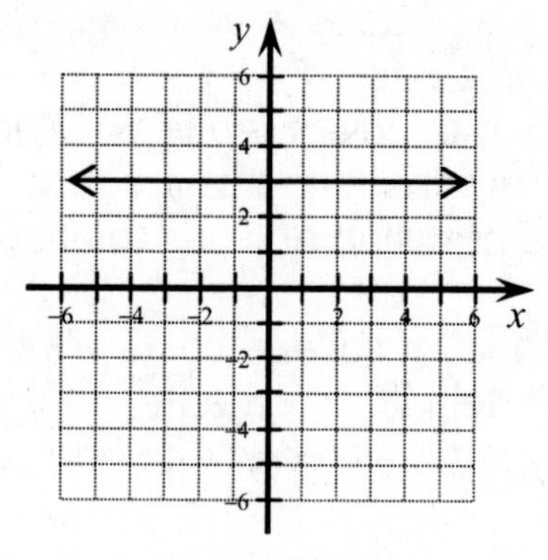

3. x-intercept: (,)

y-intercept: (,)

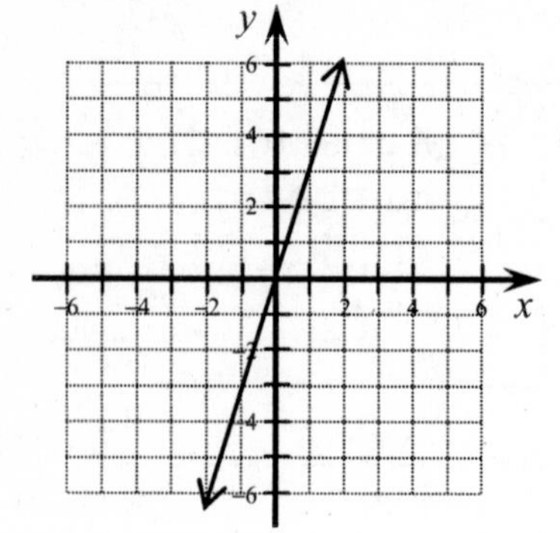

4. x-intercept: (,)

y-intercept: (,)

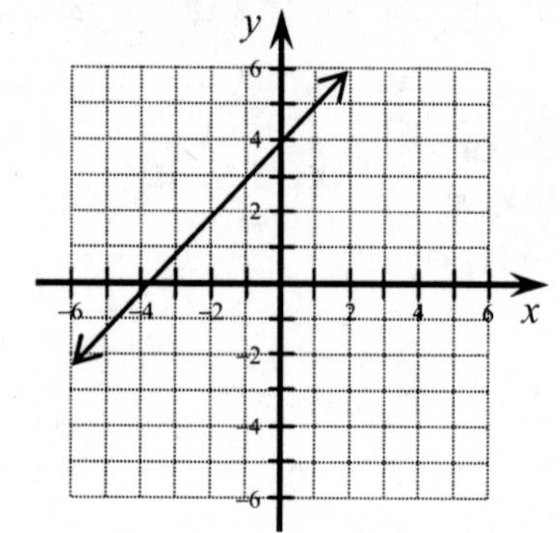

5. x-intercept: (,)

y-intercept: (,)

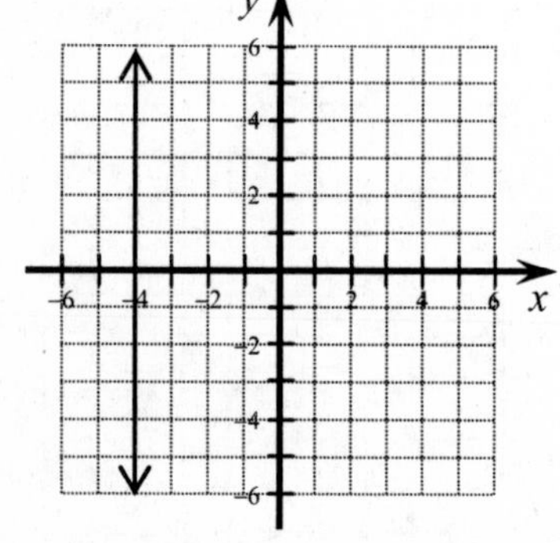

6. x-intercept: (,)

y-intercept: (,)

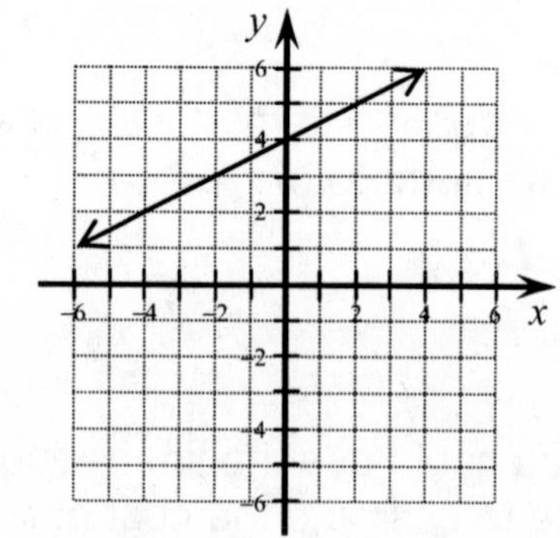

Student Activity B

Section 3.3: A Web of Lines

Directions: First, use the method of intercepts to find the x- and y-intercepts for each equation below. **Then** draw each of the lines on the graph provided using a straight-edge. (Use different colors for each line if you have colored pencils.)

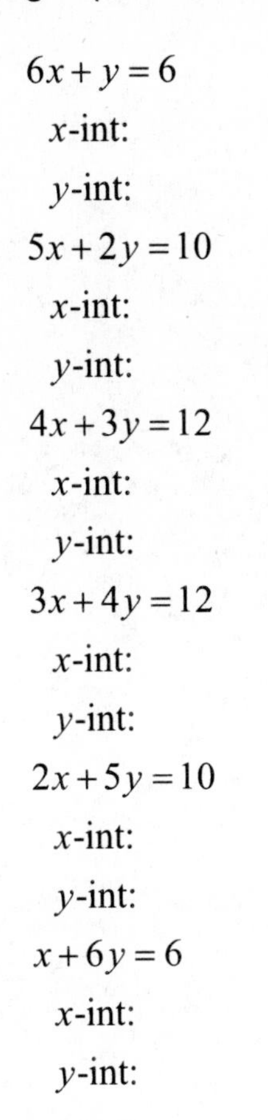

$6x + y = 6$

x-int:

y-int:

$5x + 2y = 10$

x-int:

y-int:

$4x + 3y = 12$

x-int:

y-int:

$3x + 4y = 12$

x-int:

y-int:

$2x + 5y = 10$

x-int:

y-int:

$x + 6y = 6$

x-int:

y-int:

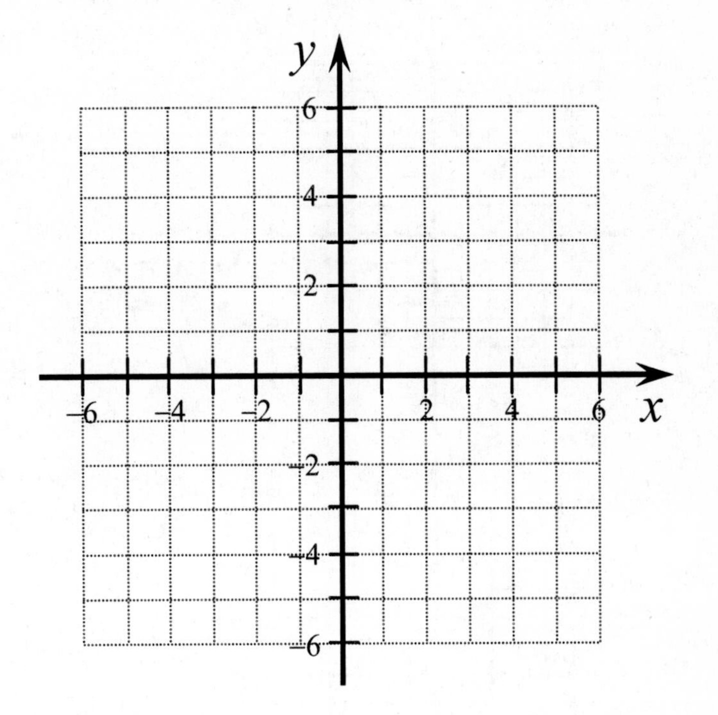

Now, if you're really a good math "detective" you can construct and graph a set of your own linear equations to make a similar "web" in **Quadrant III**. Look for patterns in the equations above to help you.

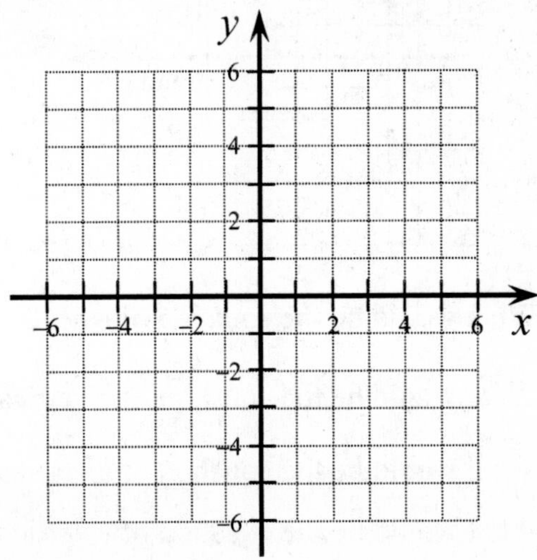

Guided Learning Activity

Section 3.3: Clues to the Equation

Directions: For each graphed line, construct a possible table of values that could accompany the line. Then, together with your class, you will write a linear equation to describe the line mathematically.

1.

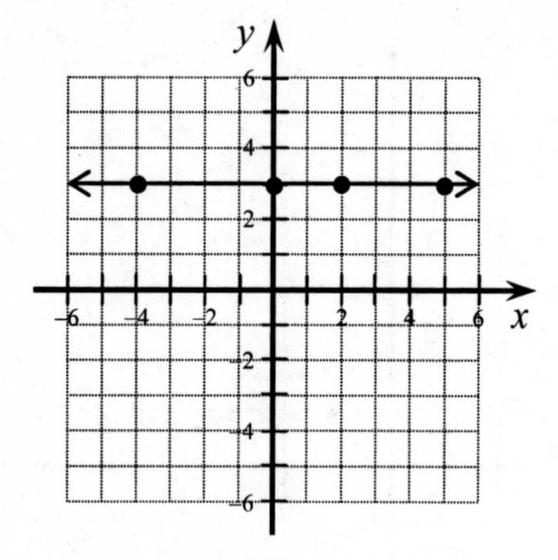

Clue: Table of Solutions

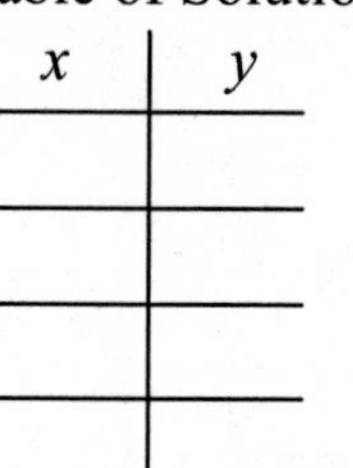

Equation: ___________

2.

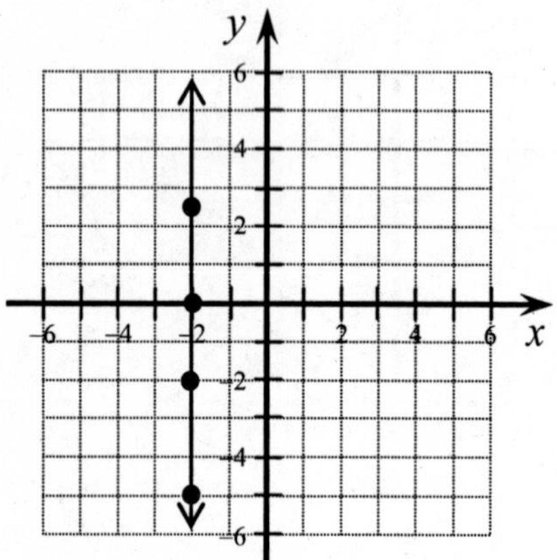

Clue: Table of Solutions

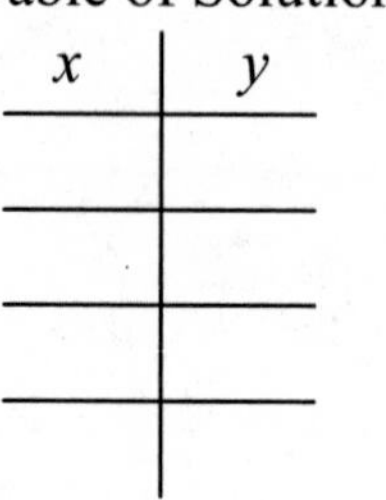

Equation: ___________

3.

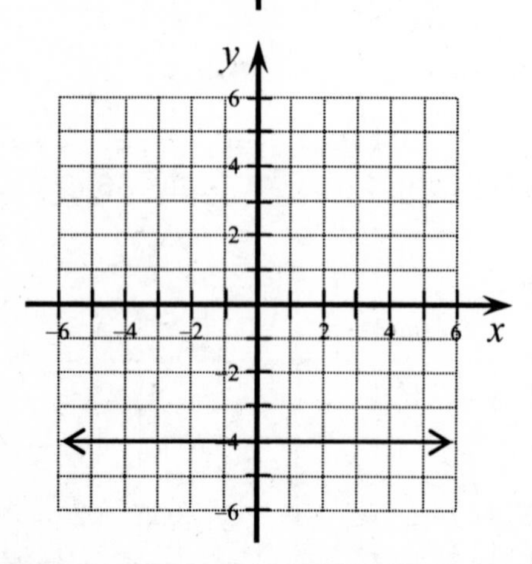

Clue: Table of Solutions

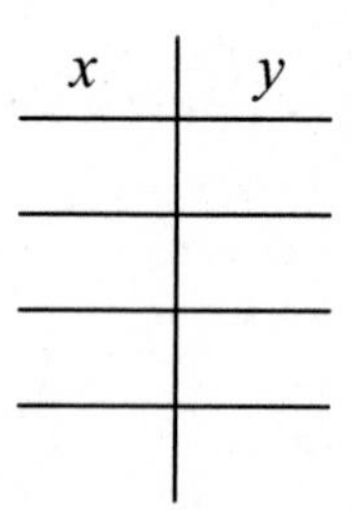

Equation: ___________

4. What would be the equation of a horizontal line that passes through $(2,3)$? _______

5. What would be the equation of a vertical line that passes through $(-2,4)$? _______

6. What would be the equation of a vertical line with an x-intercept of 5? _______

7. What would be the equation of a vertical line with an x-intercept of 0? _______

Student Activity C

Section 3.3: Graphing Linear Equations with a Calculator

You can use your graphing calculator to graph linear equations. Each model of calculator will have a different set of keystrokes that will allow you to graph these equations. Before you start, it is a good idea to guess what your graph should look like (in case you enter something into your calculator incorrectly). The standard viewing window on most graphing calculators goes from -10 to 10 on the x-axis and from -10 to 10 on the y-axis.

1. Sketch a graph of the equation $2x+3y=3$ on the axes below.

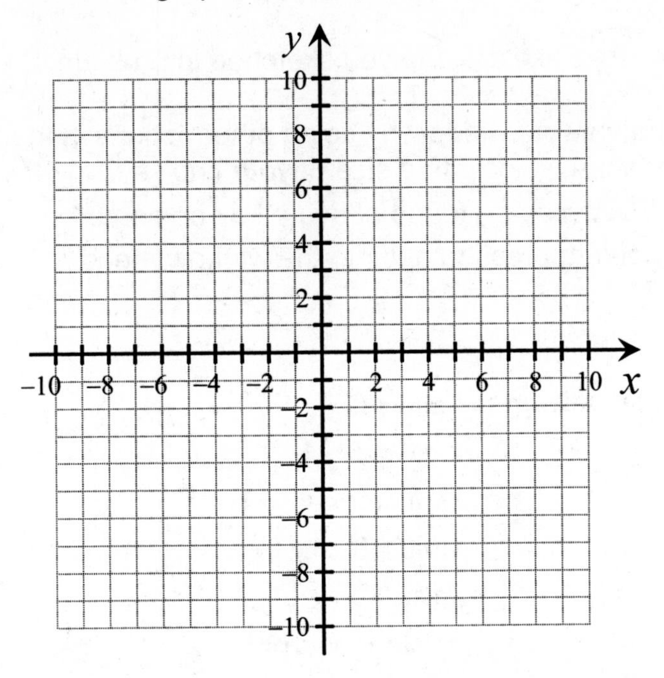

Graphing calculators can only accept equations beginning with $y=$.

2. Solve the equation $2x+3y=3$ for y.

3. Now find the key on your calculator that takes you to the $y=$ screen. It will usually look like [Y =]. It may be a function above another key, so you may have to press the *function* or [2nd] key first. Below, draw the keys you need to press.

Once you reach the $y=$ screen, you should see your cursor next to $Y_1=$. **Enter the equation you found in problem 2.** Note that you do not need to type $y=$ since it is already there.

4. Next, find your graph key. It may say [GRAPH] or it may be a function above another key. Select graph. Draw the keys you need to press.

5. Often times, it is convenient to view a graph in a **standard** viewing window. Find your calculator's zoom menu. It may be a key that says [ZOOM] or it may be a function above another key. If you have it as an option, select **ZoomStd** or **Zoom Standard**. Draw the keys you need to press to do this.

Hopefully your calculator's graph looks like the one you sketched in problem 1.

6. You can change the size of the viewing window by using other options in the zoom menu. You can also change the window to a size of your choice using the [WINDOW] key. Note that WINDOW may be found *above* a key on some calculators. Draw the keys you need to press to get into the window menu.

Now change the window to these settings: $\text{xmin} = 0$

$\text{xmax} = 5$

$\text{xscale (or xscl)} = 1$

$\text{ymin} = 0$

$\text{ymax} = 5$

$\text{yscale (or yscl)} = 1$

Using the [GRAPH] key or function, look at the graph again.

7. What has happened to the view of the graph on the calculator screen?

8. What window settings would show you only the second Quadrant?

$\text{xmin} =$

$\text{xmax} =$

$\text{xscale (or xscl)} =$

$\text{ymin} =$

$\text{ymax} =$

$\text{yscale (or yscl)} =$

Guided Learning Activity

Section 3.4: Slope Sleuth

Directions: For each line that is graphed below, determine the slope of the line using either a slope triangle or the slope formula.

1.

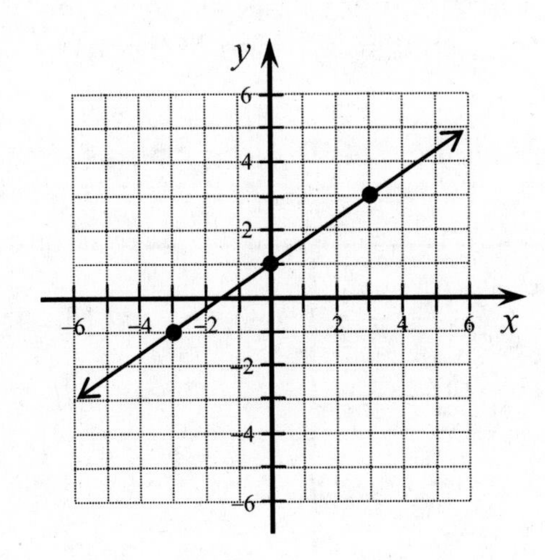

2.

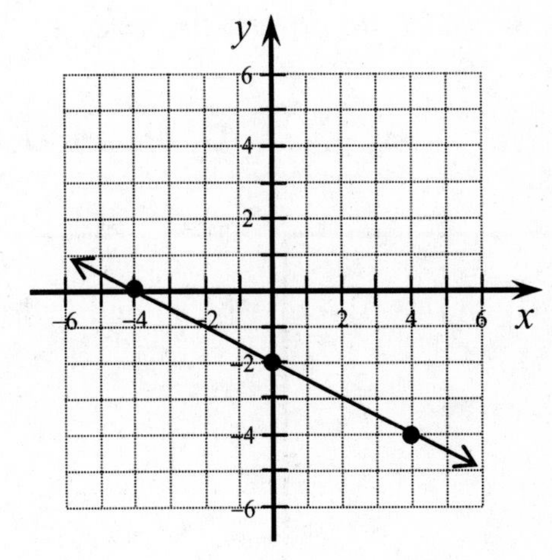

3.

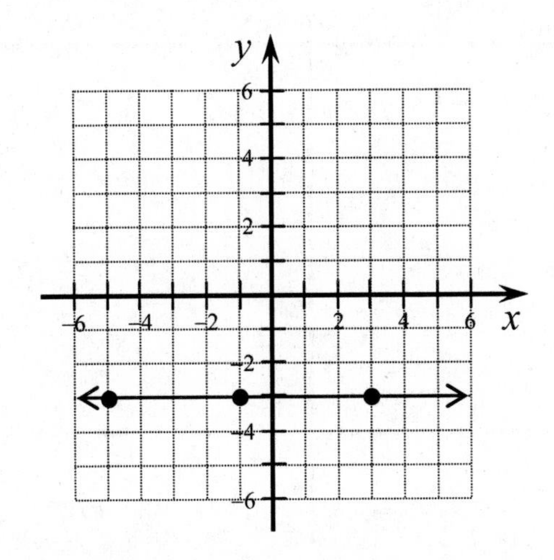

4.

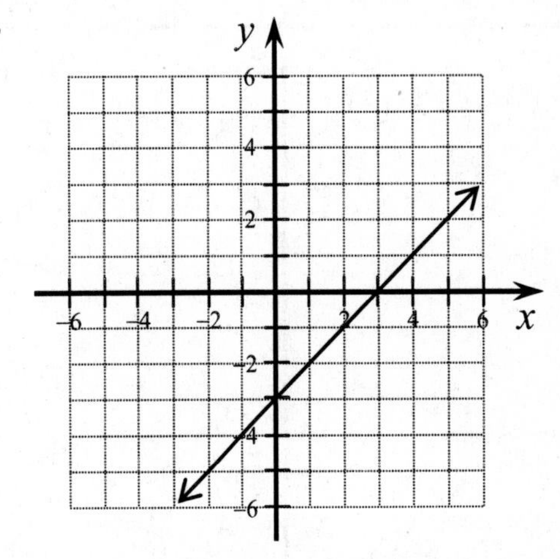

5.

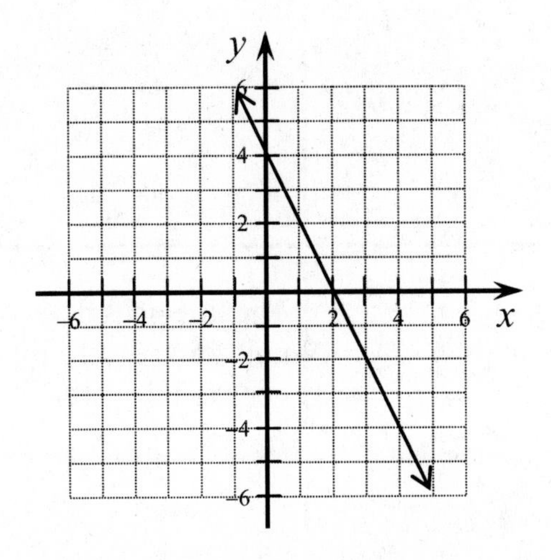

6.

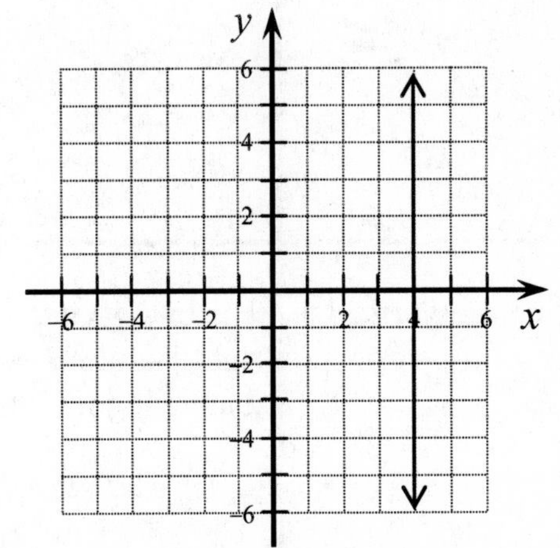

Student Activity A

Section 3.4: Match Up on Slopes

Match-up: In each box of the grid below, you will find either a pair of points or the description of a line or slope. If it is a pair of points, find the slope of the line that passes through those points. If it is a description, match it with the appropriate slope. If the slope is not found among the choices A through E, then choose F (none of these).

A 2 **B** $\frac{1}{2}$ **C** -1 **D** 0 **E** undefined **F** None of these

$(0,0)$ and $(3,0)$	$(4,4)$ and $(-4,0)$	$\left(\frac{3}{2},\frac{1}{2}\right)$ and $\left(\frac{3}{2},\frac{10}{2}\right)$	$(-3,-6)$ and $(3,-4)$
$(2,5)$ and $(3,4)$	$\left(\frac{7}{6},\frac{1}{9}\right)$ and $\left(\frac{1}{6},\frac{10}{9}\right)$	$(3.14, 2.72)$ and $(1.14, 1.72)$	$(-6,-1)$ and $(-8,0)$
$(-100,300)$ and $(-100,50)$	$(9,7)$ and $(7,9)$	$(1,1)$ and $(-5,1)$	$(8,8)$ and $(7,6)$
The line is a vertical line.	The line goes to the right 1 unit for every 2 units it goes up.	The line is a horizontal line.	The line goes down 1 unit for every 1 unit it goes to the right.

Assessment 3B

Chapter 3: Mid-chapter Assessment for Understanding

For each of the following, describe the type of problem and the strategies and key steps to remember while doing the problem. You do **not** have to complete the problems.

	Type of Problem	Strategies and Key Steps
1. Find the intercepts of the graph of $2x-8y=8$.		
2. Find the slope of a line perpendicular to the one that passes through $(0,0)$ and $(-3,4)$.		
3. What is the slope of the line $x=2$.		
4. Graph the points $A\left(\frac{1}{2},-1\right)$ and $B(3,0)$.		
5. Find the slope between $(-2,3)$ and $(1,-5)$.		
6. Graph $2x+4y=8$ by creating a table of values.		
7. Graph $3x-y=6$ using the intercepts.		
8. Given the graph of a line, find the slope of the line.		
9. Is $(-2,1)$ a solution of the linear equation $y-2x=5$?		
10. Find the intercept(s) of the horizontal line $y=2$.		

Student Activity A

Section 3.5: Match Up on Slope-Intercept Form

Match-up: In each box of the grid below, you will find either the equation of a line, a pair of points, or the description of a line. For each, determine the slope and the y-intercept and match it with the appropriate letters. If the slope is not found among the choices A through D or cannot be determined from the information given, then choose E (none of these). If the y-intercept is not found among the choices J through M or cannot be determined from the information given, then choose N (none of these).

Slope: **A** 2 **B** $\frac{1}{2}$ **C** -1 **D** 0 **E** None of these or cannot be determined

y**-Intercept:** **J** 3 **K** -2 **L** 0 **M** 1 **N** None of these or cannot be determined

$y=\frac{1}{2}x-3$	$y=-x+3$	$x+y=0$	The line is a vertical line passing through $(-2,3)$.
$6x-3y=6$	$y=1-3x$	The line passes through $(0,3)$ and $(2,1)$.	The line passes through $(0,-2)$ and $(2,-1)$.
The line is a vertical line passing through $(2,1)$.	The line is a horizontal line passing through $(-2,3)$.	The line has intercepts $(0,1)$ and $(-2,0)$.	The line is parallel to a horizontal line and passes through $(0,1)$.
The line is perpendicular to a line with a slope of -2 and passes through $(0,3)$.	$10x-5y=10$	$4y-2x=0$	The line is parallel to a line with a slope of 2 and passes through the origin.

Guided Learning Activity

Section 3.5: Graphing with Slope-Intercept Form

Slope-Intercept Form: $y = mx + b$ where m is the slope and $(0, b)$ is the y-intercept.

To graph using slope-intercept form:

1. Graph a point (the y-intercept).
2. Use $m = \dfrac{\text{rise}}{\text{run}}$ to move from that point to locate another point on the line.

1. Graph: $y = \frac{1}{2}x + 3$

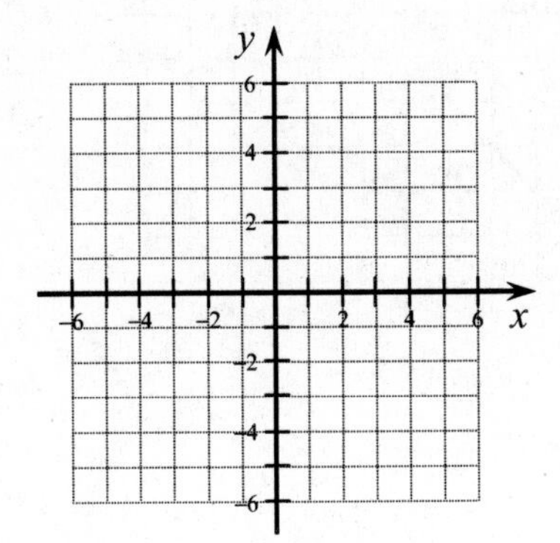

2. Graph: $y = 3x - 2$

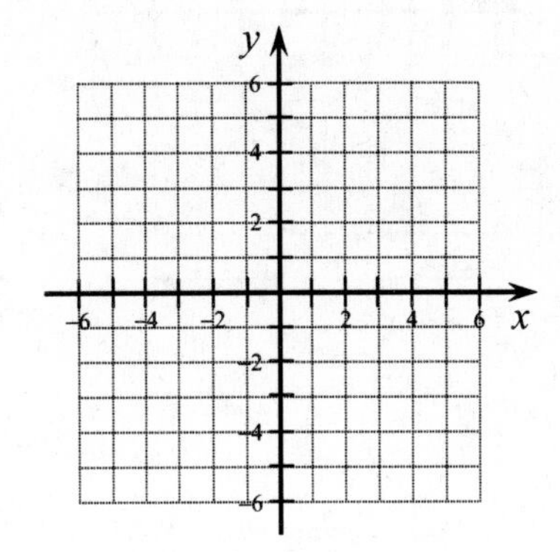

3. Graph: $y = -\frac{3}{2}x + 3$

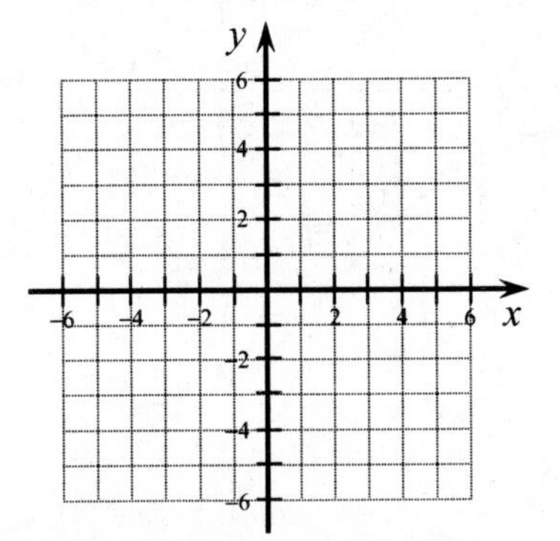

4. Graph: $y = -2x$

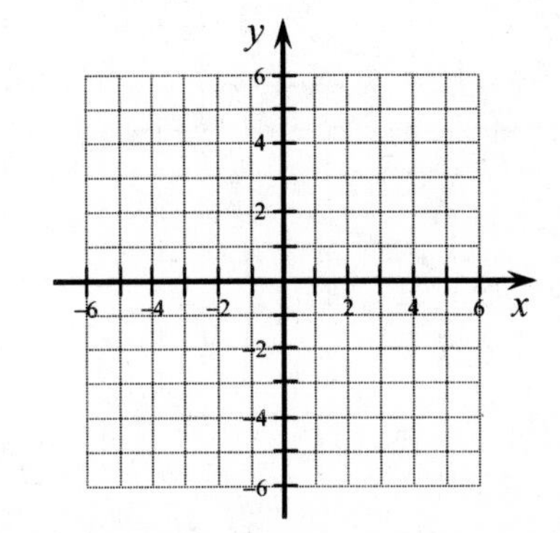

5. Graph: $3x - 4y = -4$

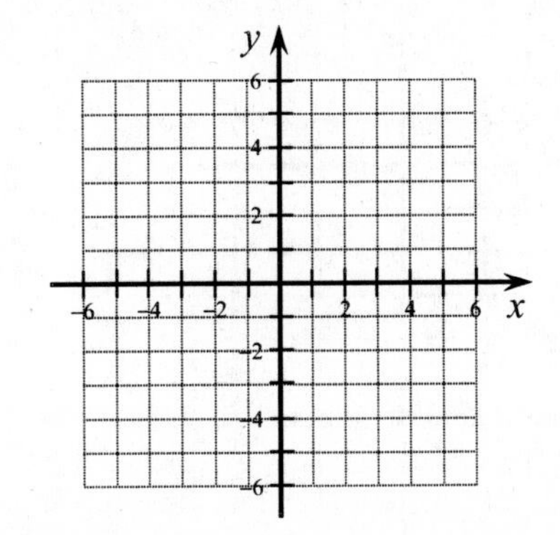

6. Graph: $y - x = 2$

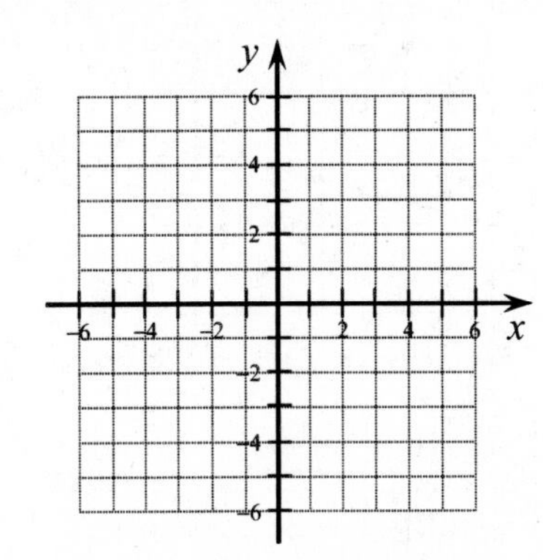

Student Activity B

Section 3.5: Evidence from the Graph

Directions: For each line that is graphed below, determine the equation of the line and write it in slope-intercept form.

1.

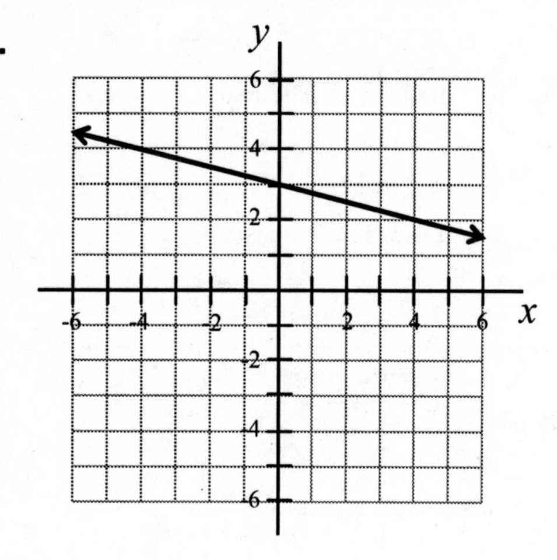

2.

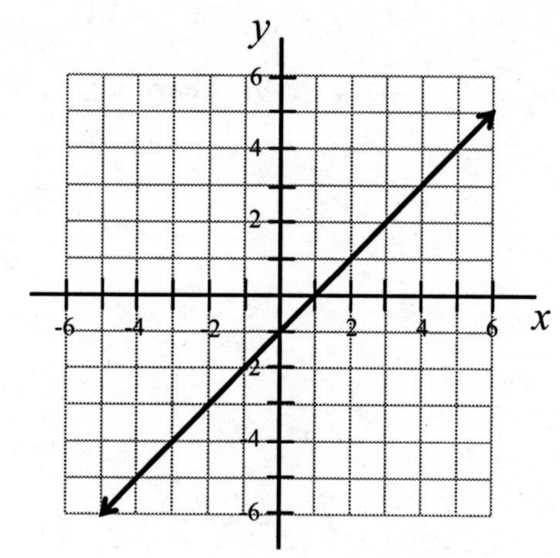

3.

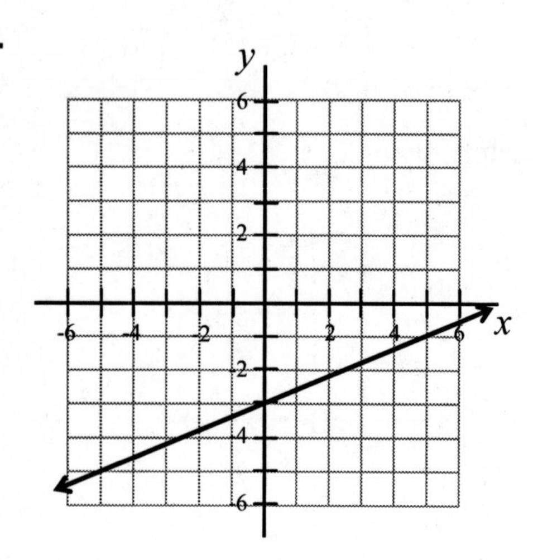

4.

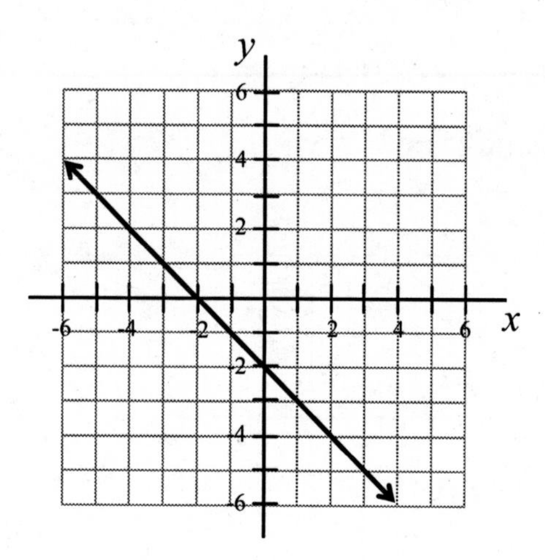

5.

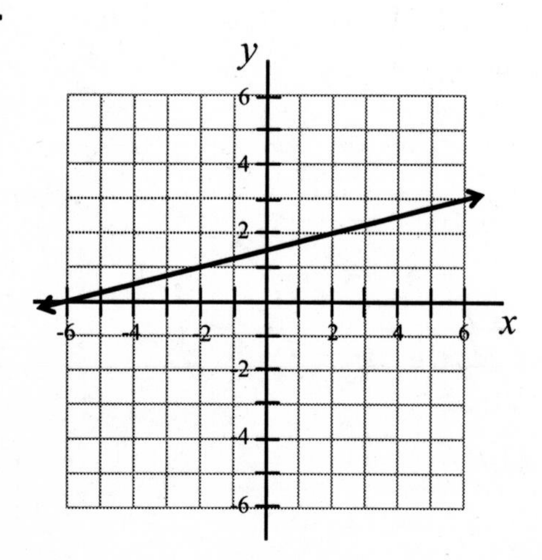

6.

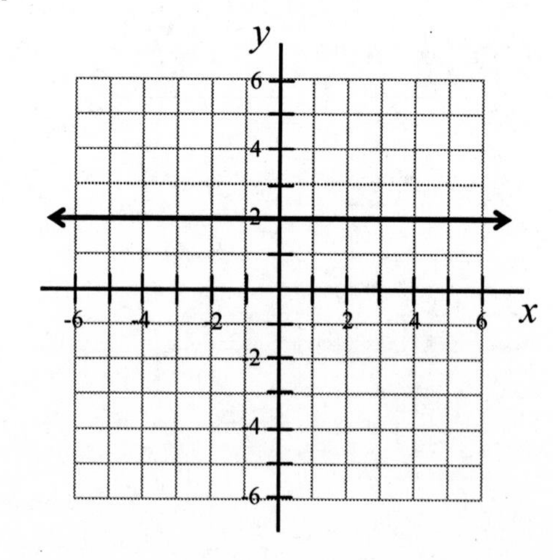

Student Activity C

3.5 What's the Verdict, Parallel or Perpendicular?

Directions: Each of the "accused" pairs of lines is going to trial. Your task is to assign each pair a verdict: parallel or perpendicular based on your graphs of the lines. If the pair of lines is neither parallel nor perpendicular, you may declare a mistrial.

1. $y=\frac{1}{2}x+3;\ y=\frac{1}{2}x+1$

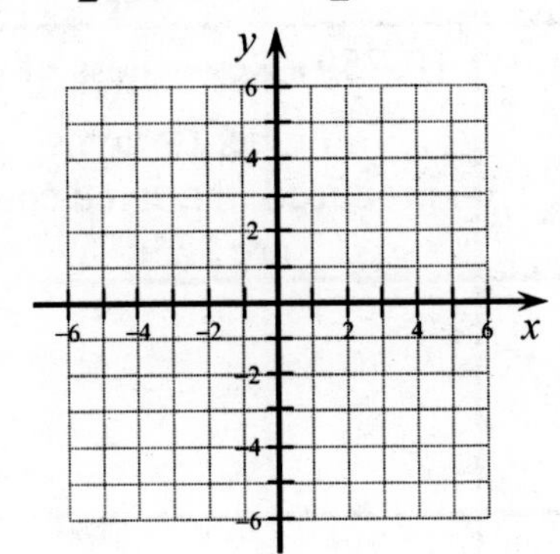

2. $y=\frac{1}{3}x-1;\ y=3x$

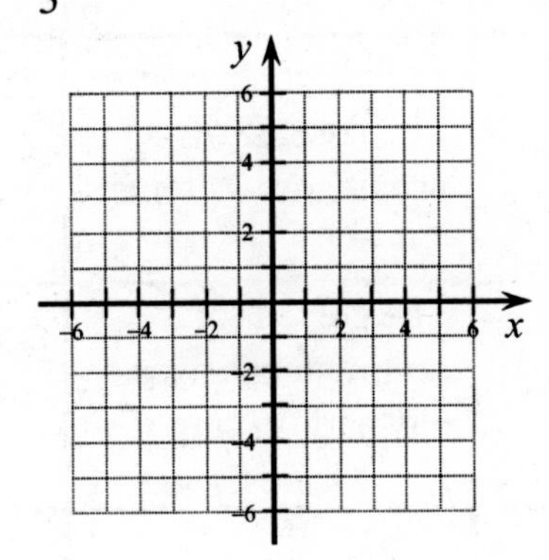

3. $y=\frac{1}{2}x+1;\ y=-\frac{1}{2}x+2$

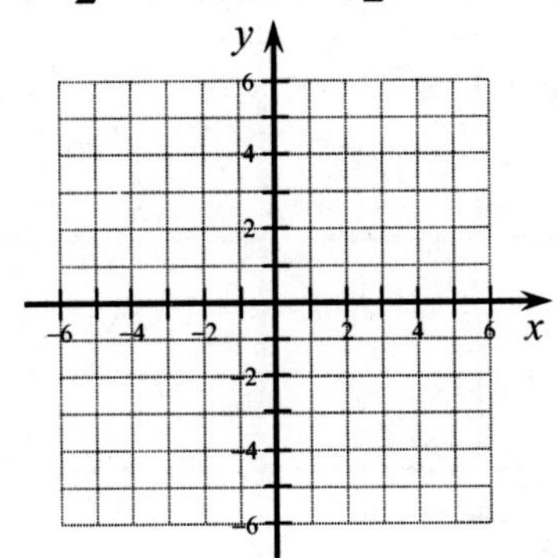

4. $y=2x-3;\ y=-\frac{1}{2}x+4$

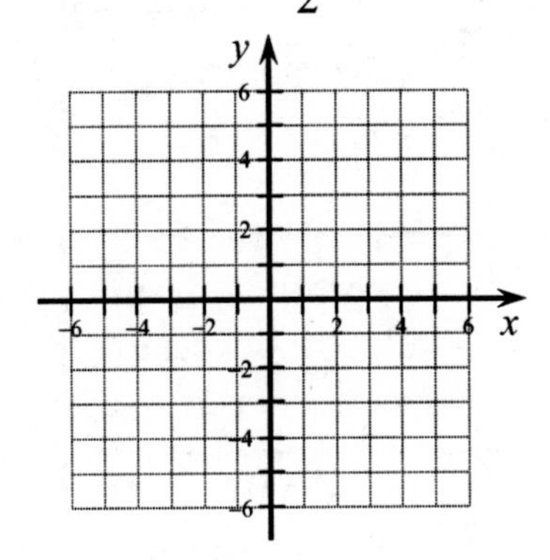

5. $y=3;\ y=-1$

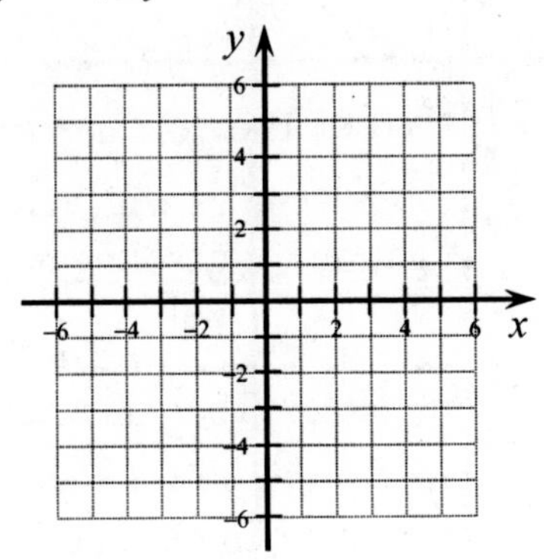

6. $y=-x+4;\ y=x-2$

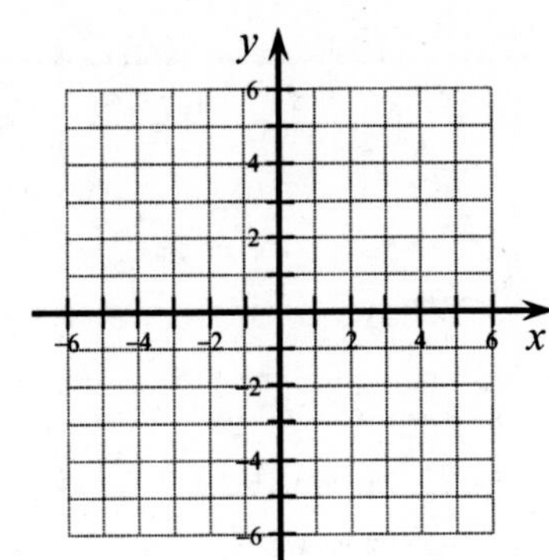

Student Activity D

Section 3.5: Modeling Data with Linear Equations

Directions: Fill in the missing information in the table below. The first one has been done for you.

Mathematical model	Description	Slope and meaning of the slope	y-intercept and meaning of the y-intercept
$C = 4.5x + 250$	The cost C of producing x items.	$m = 4.5$ It costs \$4.50 to produce each item.	$(0, 250)$ Regardless of the number of items produced, the fixed costs are \$250.
$C = 0.10n + 25$	The cost C of renting a car for a day and driving it for n miles.		
$R = 8.5t$	The revenue R from selling t movie tickets.		
$P = 10h$	The weekly pay P from working h hours.		
$P = 25t - 500$	The profit P from selling t tickets to a benefit concert.		
$v = 32t$	The velocity v of a falling object (in feet per second) t seconds after it is dropped.		
	The distance d a Toyota Prius hybrid car can travel on g gallons of gasoline.	$m = 55$ A Toyota Prius gets 55 miles per gallon of gasoline. (Source: www.toyota.com/prius/specs.html)	$(0, 0)$ The car can travel zero miles with zero gallons of gasoline in the tank.
	The cost C of taking n credit hours at a community college.	$m = 66$ The cost is \$66 per credit hour.	$(0, 25)$ There is a \$25 registration fee regardless of the number of credit hours taken.

Student Activity A

Section 3.6: Double Trouble: Two Ways to Find an Equation of a Line

1. Write an equation for the line that passes through $(2,5)$ and $(-1,-1)$ by following the directions below:

a) Find the slope between these two points.

b) Use the slope (from part **a**), the point $(2,5)$, and point-slope form, to find an equation of the line. Solve the equation for y.

c) Use the slope (from part **a**), the point $(-1,-1)$, and point-slope form to find the equation of the line. Solve the equation for y.

d) What do you notice about parts **b** and **c**? Why does it work?

2. Write an equation for the line that passes through $(-3,-1)$ and $(1,2)$ by following the directions below.

a) Find the slope between these two points.

b) Use the x- and y-coordinate values from the point $(1,2)$ and the slope (from part **a**) to solve the equation $y = mx + b$ for b.

c) Now write the equation of the line using the slope (found in a), the y-intercept (found in **b**), and the slope-intercept form of a line.

d) Use the slope (from part **a**), the point $(1,2)$, and the point-slope form to find the equation of the line. Solve the equation for y.

e) What do you notice about the results from part **c** and **d**? Why does it work?

3. Write an equation for the line that passes through $(0,4)$ and $(2,1)$ by following the directions below.

a) Find the slope between these two points.

b) Use the point-slope form to find the equation of the line and solve the equation for y.

c) Use the slope-intercept form to find the equation of the line.

d) In this case, which was easier? Why?

Student Activity B

Section 3.6 Aliases to a Line Problem

1. Graph the line $2x-4y=12$ by completing the table of solutions and plotting points..

x	y
–2	
–1	
0	
1	
2	

2. Find the x- and y-intercepts of the line $2x-4y=12$ and use the intercepts to graph.

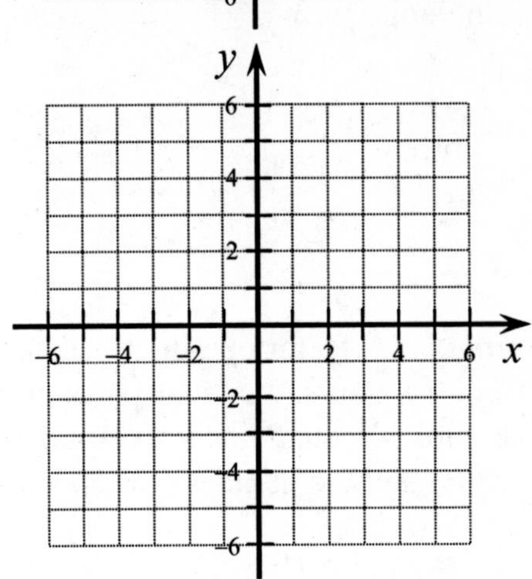

3. Write the equation $2x-4y=12$ in slope-intercept form:

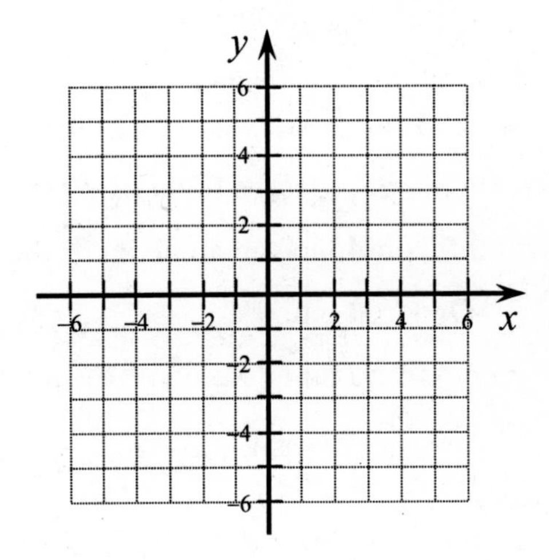

What is the slope of the line? ____

What is the y-intercept of the line? ____

Use this information to graph the line.

1. If a line were parallel to $2x-4y=12$, what would be its slope? ____

2. If a line were perpendicular to $2x-4y=12$, what would be its slope? ____

Guided Learning Activity

Section 3.6: Writing Equations that Model Data

Example 1: In year 1 of a population study, the population of flamingoes at a zoo was 25. In year 5 of the study, the population of flamingoes was 45. Find a linear equation to model the population of flamingoes, *P*, in the year t.

Declare the variables: t = year of study, P = number of flamingoes

In terms of these variables, what form do ordered pairs have? (t, P)

Rewrite the point-slope form using the new variables: $P - P_1 = m(t - t_1)$

What information are you given in the problem?

Ordered pairs(s): $(1, 25)$ and $(5, 45)$ **Slope:** *(none)*

Strategy to find an equation of the line: Find the slope and then use a point and the slope in the point-slope form.

Finish:

Example 2: A factory produces ethanol at a rate of 4000 liters every 5 hours. After the factory had run for 8 hours, the factory manager noted that there were 10,000 liters of ethanol in the storage tank. Write a linear equation relating the time *t* in hours since the factory began production and the number of liters *L* of ethanol in the storage tank.

Declare the variables: t = ______________________

L = ______________________

In terms of these variables, what form do ordered pairs have? (,)

Rewrite the point-slope form using the new variables: ______________

What information are you given in the problem?

Ordered pair(s): ______________ **Slope:** __________

Strategy to find an equation of the line: ______________________

____

Finish:

Student Activity A

Section 3.7: Is it a Solution of the Inequality?

Directions for Tic-tac-toe #1: If the ordered pair in the square **IS** a solution of the inequality, then circle the ordered pair (thus putting an **O** on the square). If it **IS NOT** a solution, then put an **X** over the ordered pair on the square.

$x+y<5$ $(2,3)$	$y<x+6$ $(9,3)$	$y>0$ $(2,-1)$
$2x-3y\le 0$ $\left(\frac{1}{2},\frac{1}{3}\right)$	$x>y+4$ $(4,0)$	$x\le 5$ $(4,9)$
$3x+3\ge 4y+4$ $(2,1)$	$3-x<-4$ $(8,1)$	$y-2x\le 3x-y$ $(1,5)$

Directions for Tic-tac-toe #2: If the ordered pair in the square **IS** a solution of the graphed inequality, then circle the ordered pair (thus putting an **O** on the square). If it **IS NOT** a solution, then put an **X** over the ordered pair on the square.

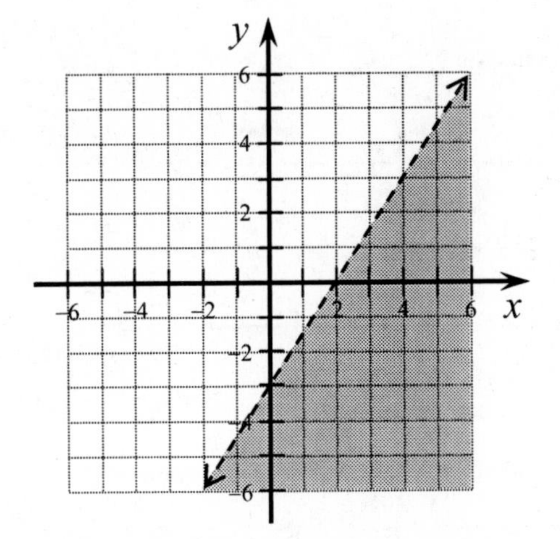

$(4,1)$	$(4,3)$	$(-2,-4)$
$(1,-3)$	$(-0.5,-6)$	$(0,0)$
$(-1,-4)$	$(-3,-6)$	$(7,-7)$

Guided Learning Activity

Section 3.7: Shady Inequalities

Graphing an Inequality in two variables:

1. Draw the boundary line (remember to draw it either dashed or solid).
2. Using a test point, determine which side of the boundary line to shade.
3. Shade the half-plane that shows the solution of the inequality.

---

1. Graph: $y \le \frac{1}{2}x - 1$

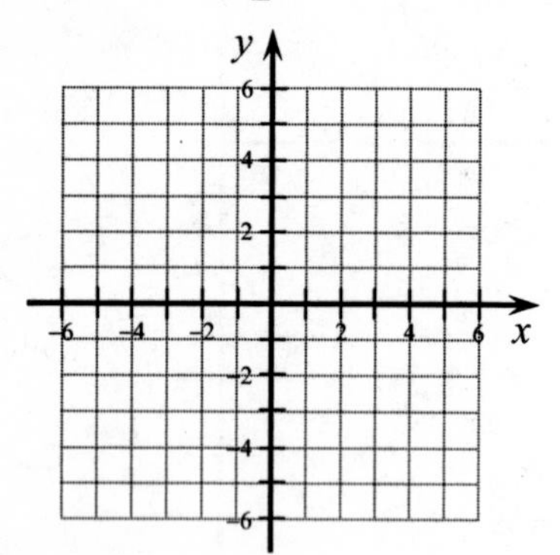

2. Graph: $x + y > 3$

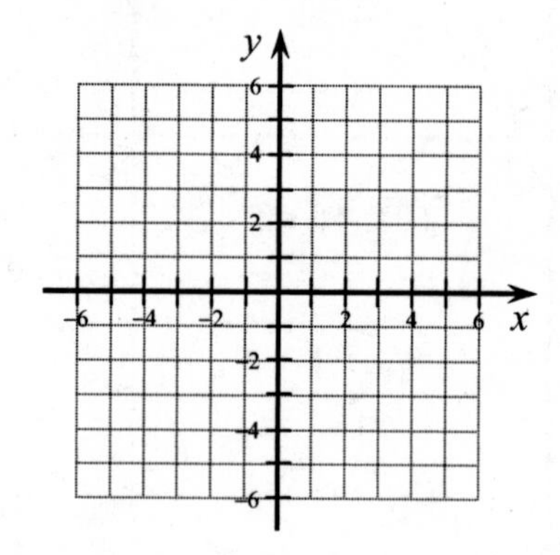

3. Graph: $2y + 3x < 6$

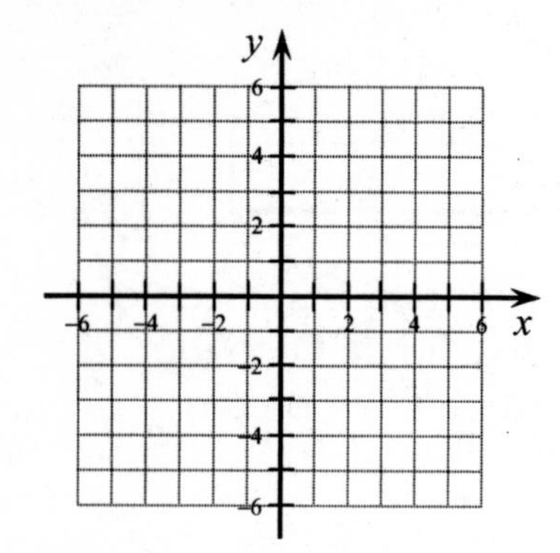

4. Graph: $y \ge 4$

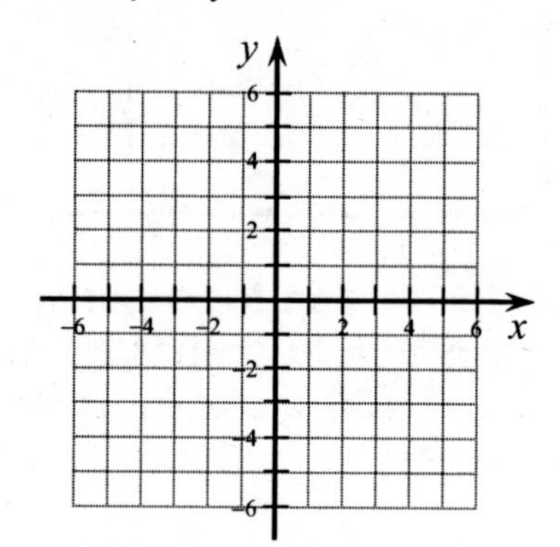

5. Graph: $x < -1$

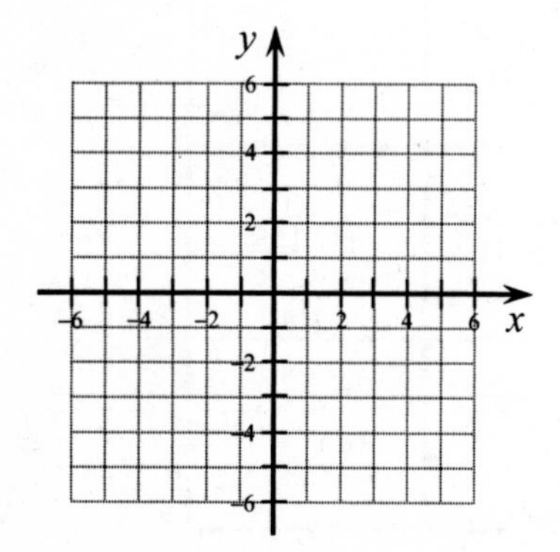

6. Graph: $x - 2y > 4$

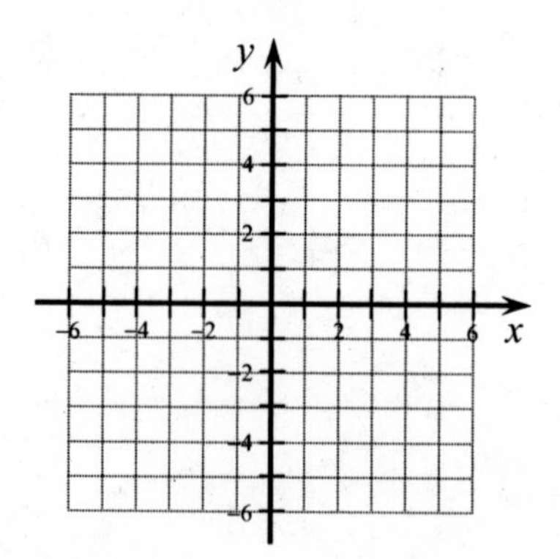

Student Activity B

Section 3.7: Following the Clues Back to the Inequality

Directions: In each "crime-scene" below, you are shown the graph of an inequality. Use your mathematical powers of reasoning (and detective skills) to determine what the inequality must have been to result in this graph.

1.

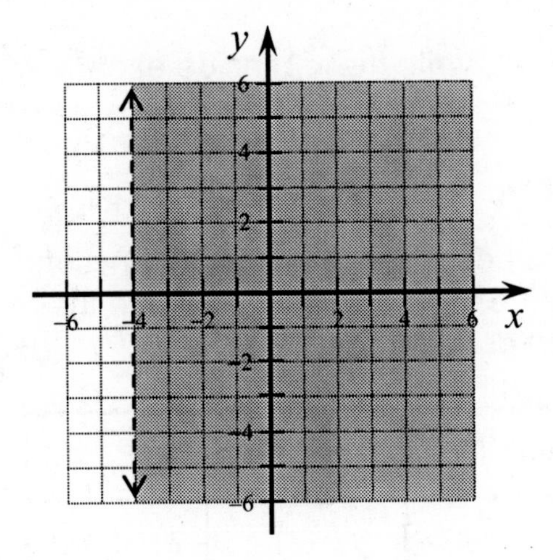

2.

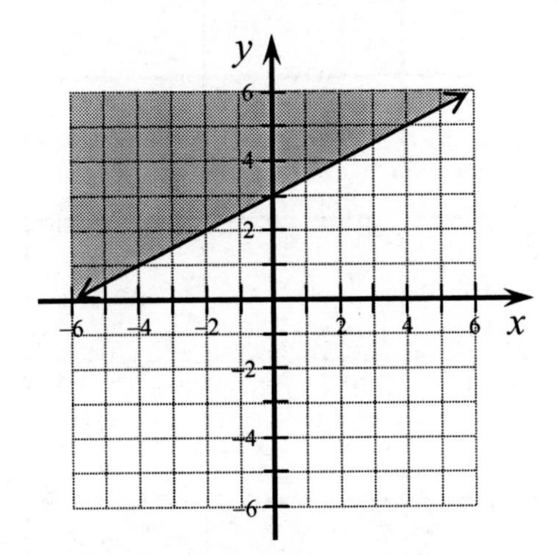

3.

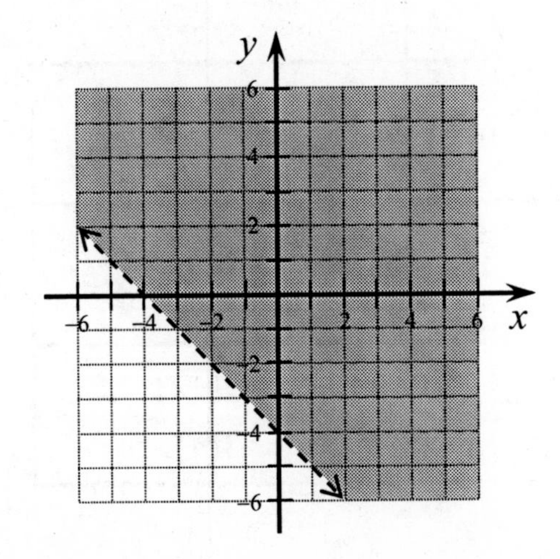

Student Activity A

Section 3.8: Finding the Skeletons

Skeleton Method: To evaluate a function that uses function notation, it is helpful to "see" the function without the variables. To make the "skeleton" of the problem, replace the variables with an empty parentheses "skeleton".

For example, $f(x)=x^2+4x+3$ becomes $f(\ )=(\ )^2+4(\ )+3$.

Now if you have to evaluate the expression for $x=-2$, write the -2 inside each parentheses skeleton and evaluate.

So $f(\ )=(\ )^2+4(\ )+3$ becomes $f(-2)=(-2)^2+4(-2)+3$.

Directions: For each row of the table, a function is described in words, in function notation, or as a parentheses skeleton. Complete the table and then evaluate the function for the given value. The first one has been done for you.

Word Description	$f(x)$	Skeleton	$f(-2)$
Multiply by 3 and then subtract 5	$f(x)=3x-5$	$f(\ )=3(\ )-5$	$f(-2)=3(-2)-5$ $=-6-5=-11$
Add 2 then multiply by 5			
		$f(\ )=(\ )^2+3$	
	$f(x)=(x-1)^2$		
Subtract 4 then take the absolute value			
		$f(\ )=(\ )^2+6(\ )$	
	$f(x)=4-x$		
			$f(-2)=-3(-2)^2+5$ $=-3(4)+5=-7$

Guided Learning Activity

Section 3.8: Investigating Functions

Definition of a function: a relation in which each **first component** corresponds to exactly one **second component**.

Vertical Line Test: If a vertical line intersects a graph in more than one point, the graph is **not** the graph of a function.

1. Graph: $f(x) = |x|$

x	$f(x)$

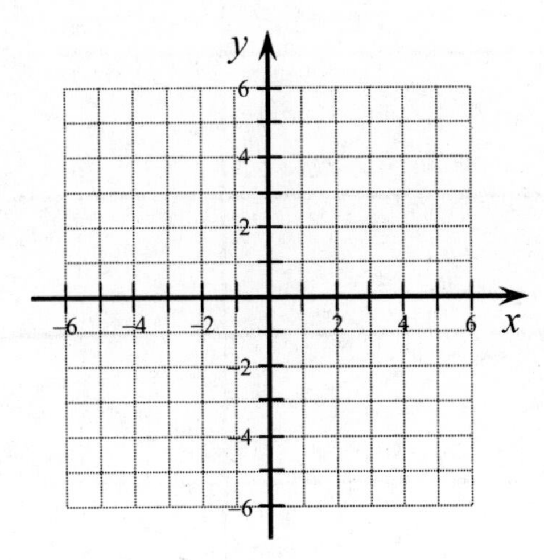

Function?

2. Graph: $f(x) = x - 3$

x	$f(x)$

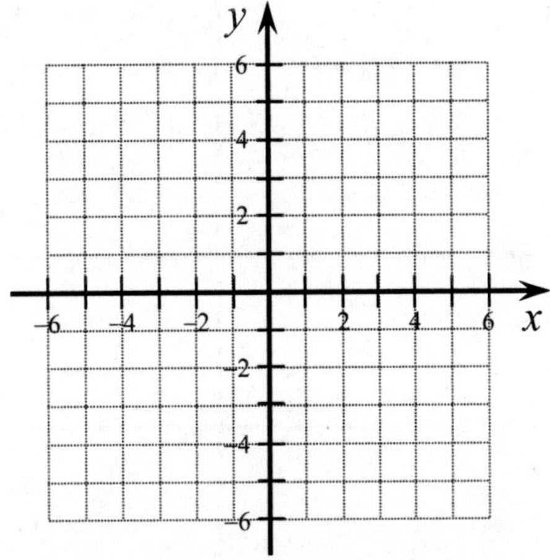

Function?

3. Graph: $x = |y|$

x	y
	-2
	-1
	0
	1
	2

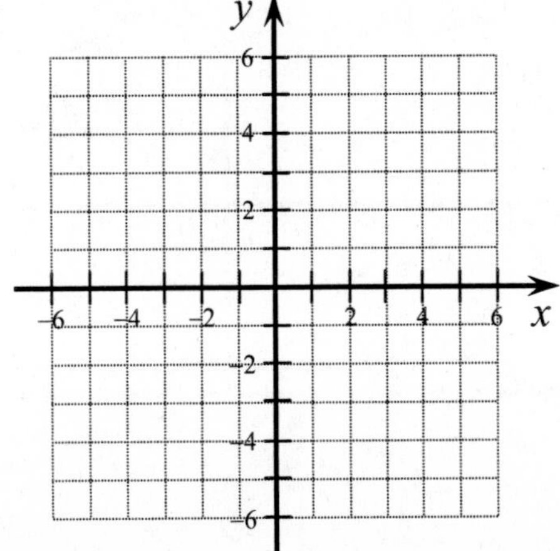

Function?

Assessment 3C

Chapter 3: End-of-Chapter Assessment for Understanding

For each of the following, describe the type of problem and the strategies and key steps to remember while doing the problem. You do **not** have to complete the problems.

	Type of Problem	Strategies and Key Steps
1. Graph the line that has slope 1 and passes through $(-2,-3)$.		
2. Graph $y=-x+4$ using the slope-intercept method.		
3. If $f(x)=-3x+1$, find $f(4)$.		
4. Find the equation of a vertical line that passes through $(3,7)$.		
5. Given the graph of a line, find the equation.		
6. Find the intercepts of $3x-5y=15$.		
7. Is the relation $\{(2,3),(4,1),(3,4)\}$ a function?		
8. Graph the solution of $y<2x-3$.		
9. Find the equation of the line that passes through $(0,3)$ and $(-1,5)$.		
10. Is $(6,1)$ a solution of $x-2y\leq 4$?		

Assessment 3D

Chapter 3: Metacognitive Skills Assessment

Metacognitive skills refer to the ability to judge how well you have learned something and to effectively direct your own learning and studying. This is a self evaluation tool designed to help you focus your studying and to improve your metacognitive skills with regards to this math class.

Fill the 1st column out before you begin studying.
Fill the 2nd column out after you study and before you take the test.
Go back to this page after your test and circle any of the ratings that you would now change – this identifies the "disconnects" between what you think you know well and what you actually know well.

Use the scale below to assign a number to each topic.

5 I am confident I can do any problems in this category correctly.
4 I am confident I can do most of the problems in this category correctly.
3 I understand how to do the problems in this category, but I still make a lot of mistakes.
2 I feel unsure about how to do these problems.
1 I know I don't understand how to do these problems.

Topic or Skill	**Before Studying**	**After Studying**
Plotting points or identifying points on a graph using an ordered pair.		
Identifying the quadrant or axis that a point is located in/on.		
Creating a line graph given a set of data.		
Deciding whether an ordered pair is a solution of a linear equation.		
Deciding whether an ordered pair is a solution of a linear inequality.		
Constructing a table of solutions for a linear equation.		
Graphing a linear equation by plotting points.		
Finding the x- and y-intercepts of a linear equation (if they exist).		
Graphing a linear equation by using the x- and y-intercepts.		
Interpreting a graph or answering questions based on the graph of a linear equation.		
Understanding the equations and graphs of vertical and horizontal lines.		
Finding the slope of a line by looking at its graph.		
Finding the slope of a line if you are given two points.		
Understanding the slopes of horizontal and vertical lines.		
Calculating a rate of change (including the units).		
Understanding the special properties of slope for parallel and perpendicular lines.		
Finding the slope and y-intercept of a line given in slope-intercept form.		
Rearranging a linear equation so that it is written in slope-intercept form.		

Continued on next page.

Use the scale below to assign a number to each topic.

5 *I am confident I can do any problems in this category correctly.*
4 *I am confident I can do most of the problems in this category correctly.*
3 *I understand how to do the problems in this category, but I still make a lot of mistakes.*
2 *I feel unsure about how to do these problems.*
1 *I know I don't understand how to do these problems.*

Topic or Skill	**Before Studying**	**After Studying**
Writing an equation of a line if you are given the slope and y-intercept.		
Using the slope and y-intercept to graph a linear equation.		
Using the slope and any point on the line to graph a linear equation.		
Using the point-slope form to write an equation of a line.		
Writing an equation for a line if you are given two points on the line.		
Writing an equation of a horizontal or a vertical line.		
Knowing the formulas for slope, slope-intercept form, and point-slope form.		
Applying the formulas for linear equations to write an equation to model data.		
Graphing a linear inequality using a check point.		
Understanding whether the boundary line in a linear inequality is dashed or solid and which side of the boundary line to shade.		
Solving for y in a linear inequality (including the special rule that applies to dividing or multiplying by a negative number when solving inequalities)		
Finding the domain and range of a relation.		
Understanding function notation.		
Evaluating a function written in function notation for a specified value.		
Identifying whether a relation is a function (including the vertical line test).		
Solving application problems that involve inequalities.		
Solving application problems that involve function notation.		

Student Workbook: Chapter 4

Table of Contents: *Solving Systems of Equations and Inequalities*

Assessment 4A

Pretest and Diagnostic Tool: Equations, Inequalities, and Problem Solving

Directions: Complete this assessment without looking back at your notes or your book. **Do not use a calculator on this assessment.**

1. If $x = -2$ and $y = 1$, is the equation $x - 2y = 0$ true or false? ________

2. If $x = 5$, solve the equation $x + 2y = 1$ for y.

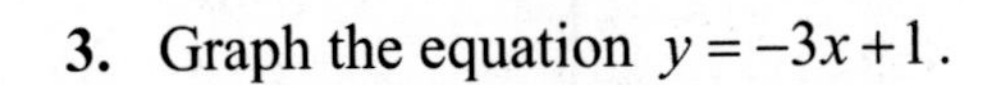

3. Graph the equation $y = -3x + 1$.

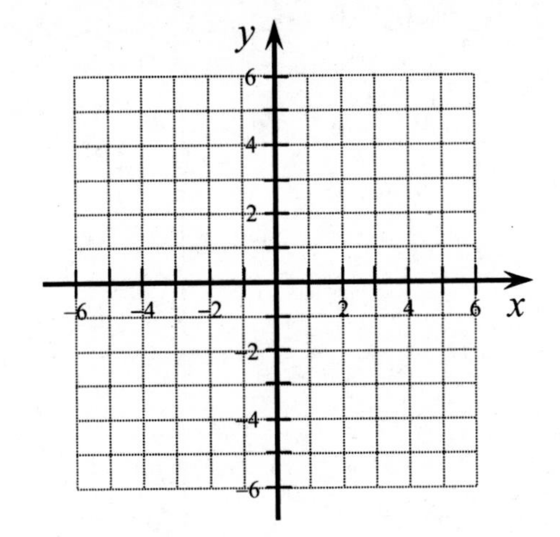

4. Write the equation $x - 2y = 8$ in slope-intercept form.

5. Solve for x: $3x - 2(x - 4) = 8$.

6. Solve for y: $y - (4y + 10) = -1$.

7. Solve for y: $12\left(\frac{3}{4}y\right) - 4 = y$.

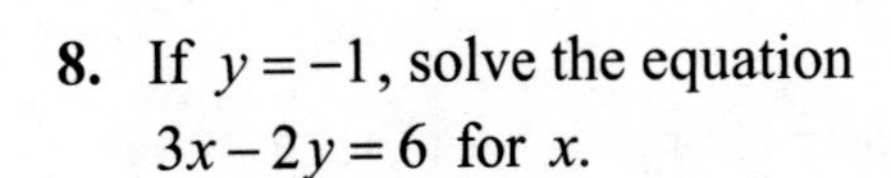

8. If $y = -1$, solve the equation $3x - 2y = 6$ for x.

9. Add the two expressions using the vertical method:

$$\begin{array}{r} 5x - 3y \\ +\ 2x + 3y \\ \hline \end{array}$$

10. Add the two expressions using the vertical method:

$$\begin{array}{r} -12x - 4y \\ +\ \ 8x - 4y \\ \hline \end{array}$$

11. Multiply both sides of the equation $6x + 5y = -1$ by 4.

12. Multiply both sides of the equation $x - 4y = 7$ by -2.

13. Multiply both sides of the equation $\frac{x}{3} - \frac{1}{2} = -2$ by 6.

14. Multiply both sides of the equation $0.05x + 0.06y = 210$ by 100.

15. Find an expression for the interest earned after one year if x dollars earns 4.5% annual interest.

16. Find an expression for the distance traveled by an airplane that travels for 4 hours at a rate of $x+15$ miles per hour.

17. Find an expression for the revenue made by a ticket office if x tickets are sold for \$12.00 each and y tickets are sold for \$6.00 each.

18. If $x=4$ and $y=-3$, is the inequality $2x+3y<-2$ true or false? _______

19. Is $(2,0)$ a solution of the inequality $y-x\le-2$? _______

20. Graph the inequality $y<2x-4$.

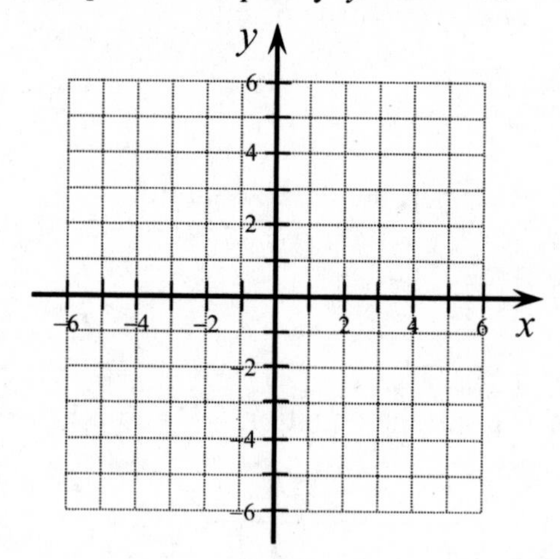

Student Activity A

Section 4.1: Targeting Solutions

Directions: In each problem below, test the ordered pairs that are provided to see if they are possible solutions to either of the equations in the system of equations. After testing, place the ordered pairs into the appropriate region of the "target" for each problem.

Outer target region	The ordered pair IS a solution of the first equation, but not the second equation.
Middle target region	The ordered pair is not a solution of the first equation, but IS a solution of the second equation.
Bulls-eye (center target region)	The ordered pair is a solution of both equations.

1. $\begin{cases} y < 5x + 2 \\ y > -x + 8 \end{cases}$

$(1, 6)$

$(4, 25)$

$(2, 10)$

$(-10, 20)$

$(-3, -20)$

$\left(\frac{1}{2}, 9\right)$

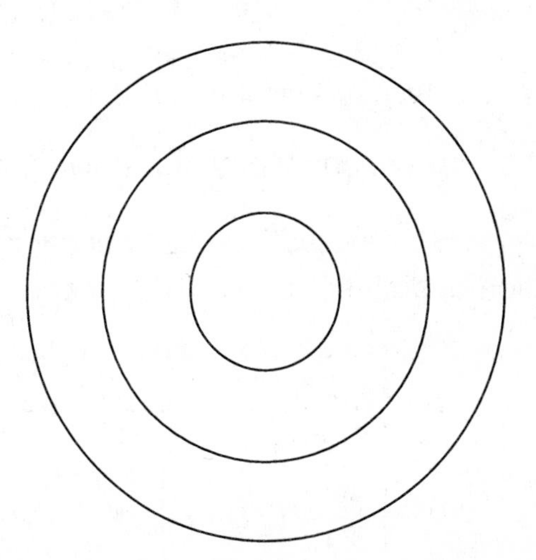

2. $\begin{cases} 4x + 2y = 8 \\ -2x + 4y = 1 \end{cases}$

$(5, -6)$

$\left(7, \frac{15}{4}\right)$

$\left(\frac{3}{2}, 1\right)$

$\left(-4, -\frac{7}{4}\right)$

$(0, 4)$

$(-3, 10)$

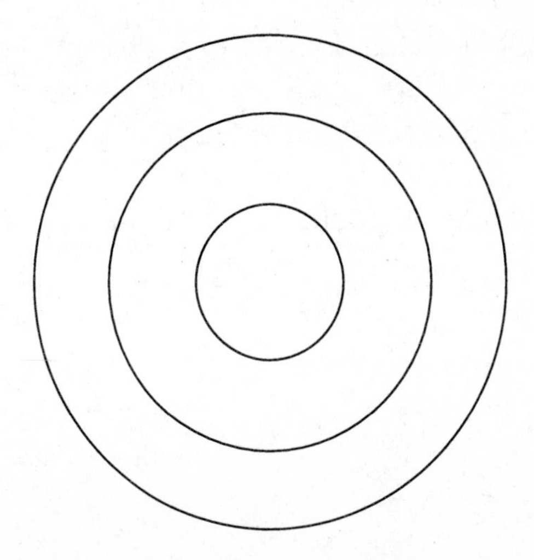

Guided Learning Activity

Section 4.1: Graphing to Solve Systems of Equations

Example 1:

Write some ordered-pair solutions of the equation $x+y=5$.

(,) (,) (,) (,) (,) (,)

Write some ordered-pair solutions of the equation $x-y=-3$.

(,) (,) (,) (,) (,) (,)

When we write the two equations like this: $\begin{cases} x+y=5 \\ x-y=-3 \end{cases}$

it is called a *system of equations*. When we *solve* a systems of two equations, we are looking for all of the ordered pairs (x,y) that satisfy *both* equations.

Maybe you found an ordered pair that was a solution of both equations? If so, circle it.

Example 2: What does it mean to be a solution to a system?

Consider the system of equations given by $\begin{cases} y=2x+2 \\ y=x-1 \end{cases}$ and the ordered pair $(-3,-4)$.

You can see by evaluating each equation for $x=-3$ and $y=-4$ that this ordered pair is a solution of both equations. This means it is a solution of the system of equations.

$y=2x+2$	$y=x-1$
$y=2(\ )+2$ $-4\stackrel{?}{=}2(-3)+2$ $-4=-4$	$y=(\ )-1$ $-4\stackrel{?}{=}(-3)-1$ $-4=-4$

So what does the solution of a system of equations look like? Graph $y=2x+2$ and $y=x-1$ on the graphing grid provided to the right.

What is special about the point $(-3,-4)$ on the graph?

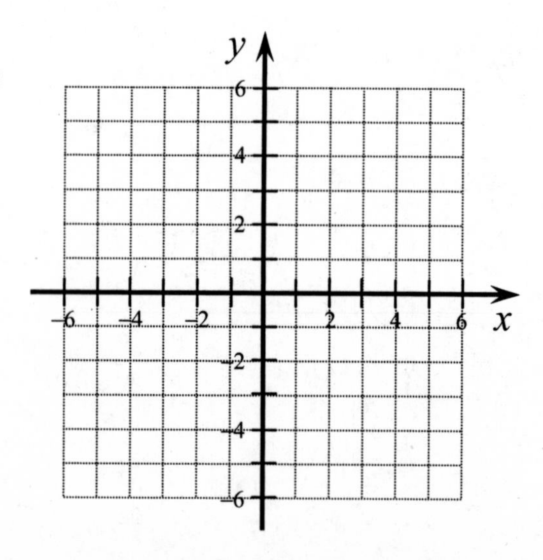

Now solve these systems of equations by graphing.

1. $\begin{cases} y = -3x + 6 \\ y = 2x - 4 \end{cases}$

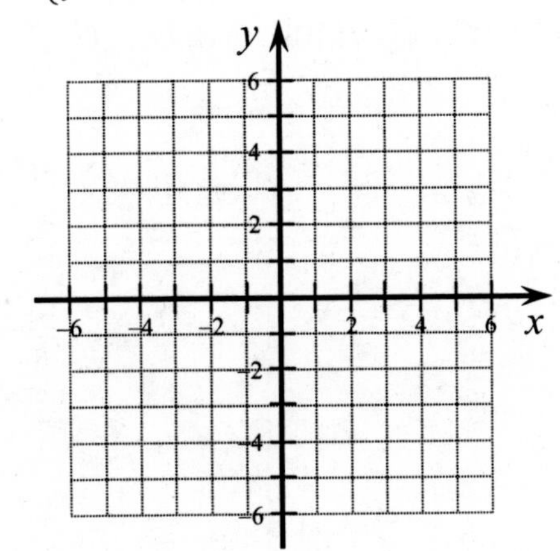

2. $\begin{cases} 5x - y = 6 \\ 3x - 3y = 6 \end{cases}$

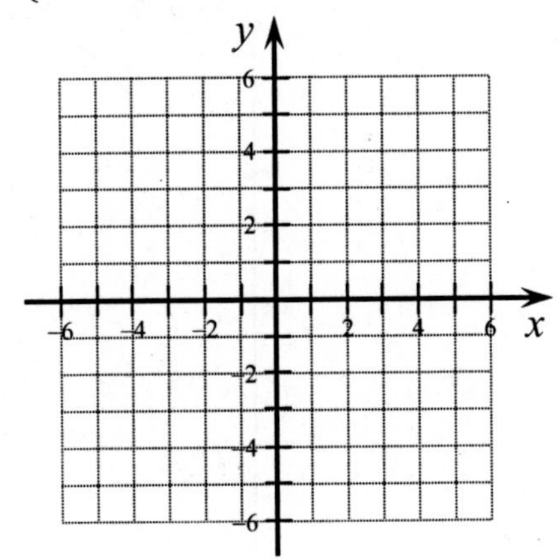

3. $\begin{cases} -4x + 2y = 8 \\ 6x - 3y = 3 \end{cases}$

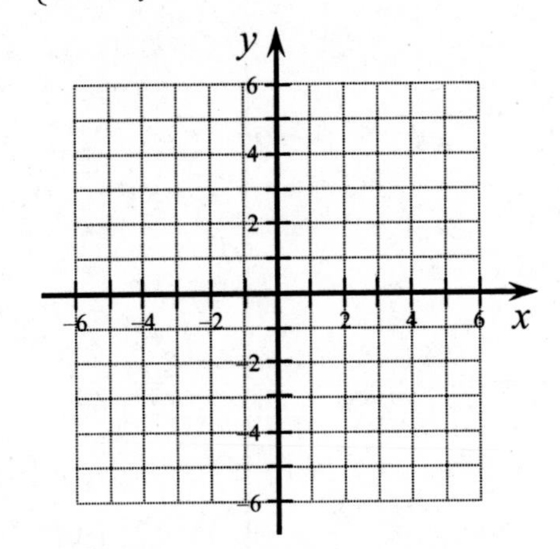

4. $\begin{cases} y = -3x - 2 \\ y = 2x + 3 \end{cases}$

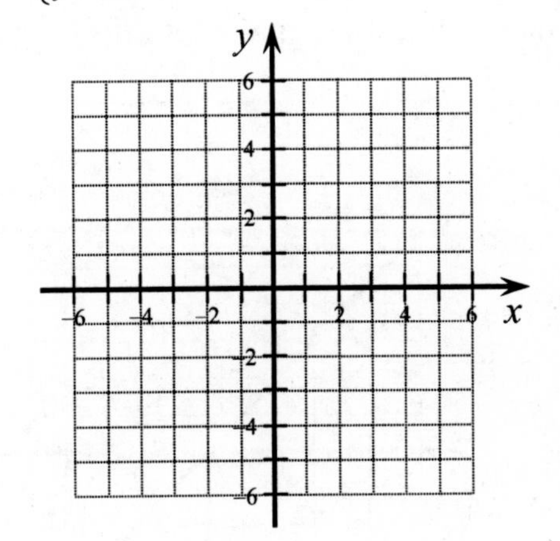

5. $\begin{cases} -2x - 4y = -2 \\ 4x + 8y = 4 \end{cases}$

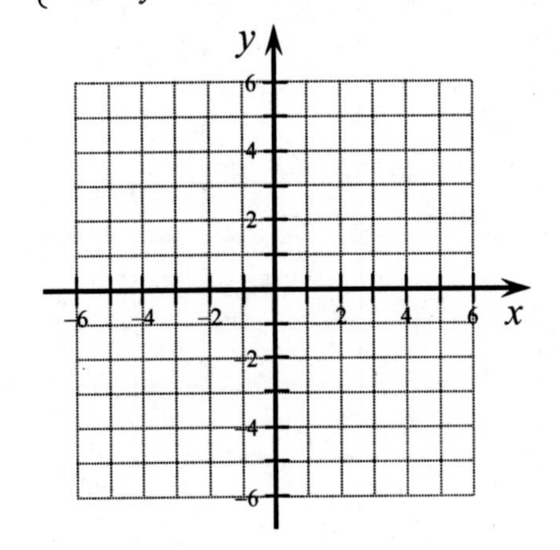

6. $\begin{cases} y = \frac{3}{4}x - 1 \\ y = -\frac{4}{3}x - 1 \end{cases}$

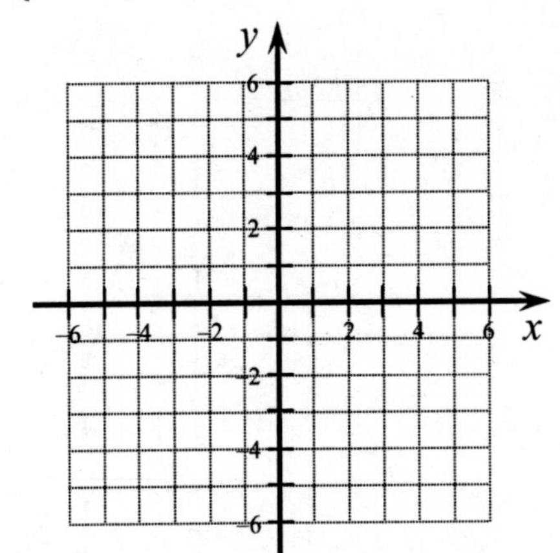

Student Activity B

Section 4.1: Following the Clues Back to the System of Equations

Directions: In each "crime-scene" below, you are shown the graph of a system of equations. Use your mathematical powers of reasoning (and detective skills) to determine what the system of equations must have been to result in this graph.

1.

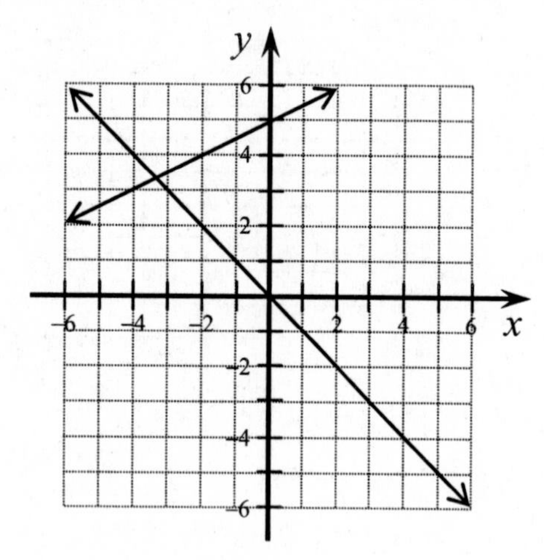

2.

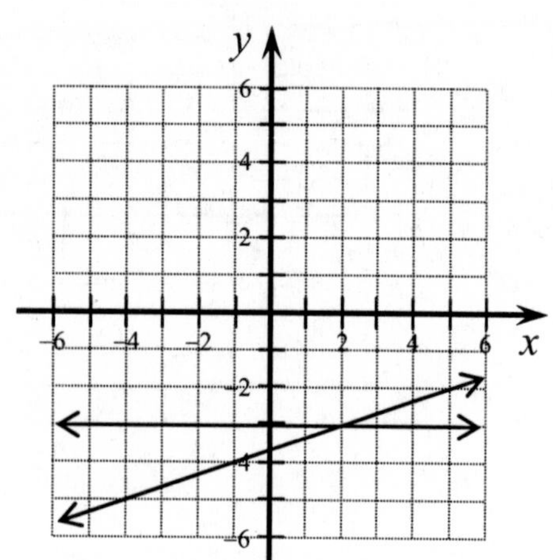

3.

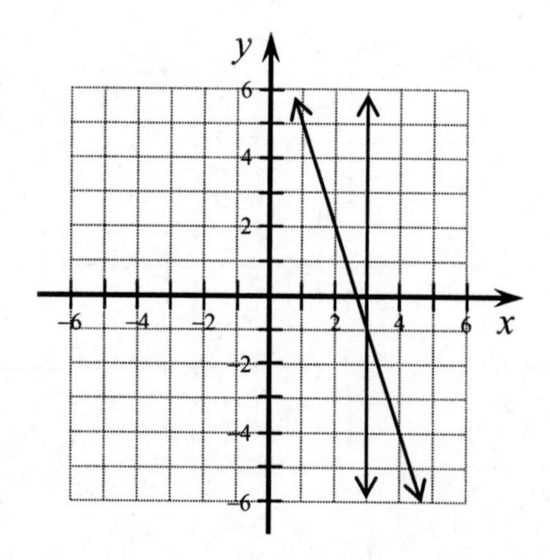

Student Activity C

Section 4.1: Graphing Systems of Equations with a Calculator

You can use your graphing calculator to solve systems of linear equations by graphing. Each model of calculator will use a different set of keystrokes to graph these equations. Before you start to graph any system of equations, it is a good idea to guess what your graph should look like (in case you type something in your calculator incorrectly). The standard zoom on most graphing calculators goes from -10 to 10 on the x-axis and from -10 to 10 on the y-axis.

1. Sketch a graph of the system of equations $\begin{cases} y = 5x - 5 \\ y = -3x + 3 \end{cases}$ on the axes below, and label the solution point.

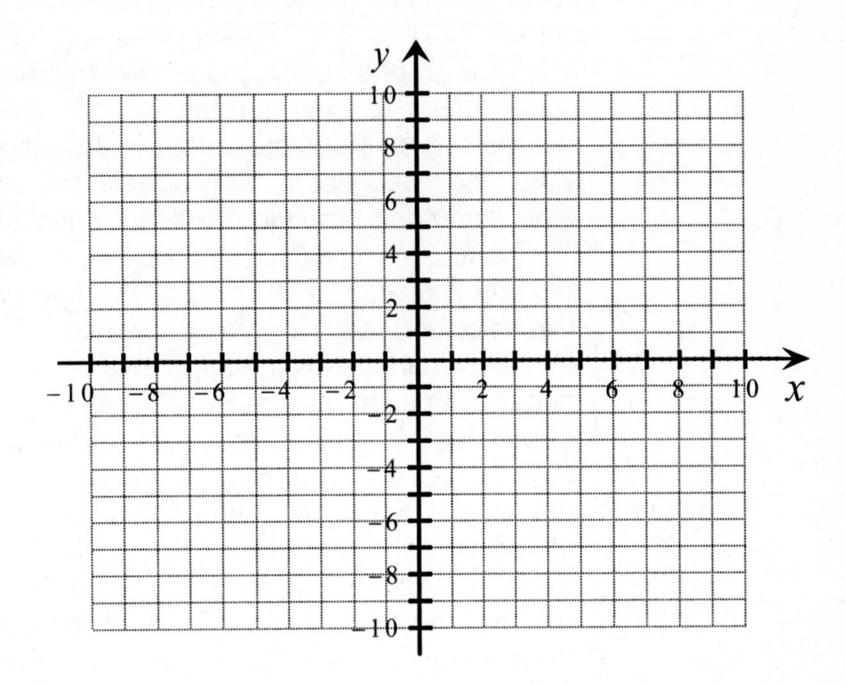

Note: In Chapter 3 you learned how to graph equations on your calculator (if you have forgotten how, look back at *Section 3.3 Student Activity C: Graphing Linear Equations with a Calculator*).

2. Find the $y =$ screen on your calculator. Place your cursor next to $Y_1 =$ and enter the first equation in your system. Then move your cursor to $Y_2 =$ and enter the second equation. Make sure you are in the standard viewing window and then view the graph. Hopefully this looks like the graph you produced above.

3. In order to find the solution to this system of equations on the calculator, we need to find the point of intersection. **From your graph screen**, find the *Math* or *Calculate* menu. You'll know you have found it when you see a menu that contains the functions *zero (or root)*, *max*, *min*, and *intersect* (it will also have a few others).

Draw the keys or write the steps that were necessary to find this menu.

4. Select *intersect* (or *intersection*) from the menu you just found. Your calculator will give you a series of prompts. At the end of all of the prompts, your calculator should display numerical values labeled $x=$ and $y=$. This should match the solution you found at the beginning of the activity.

Calculator Prompt	What does it mean? What should you do?	Possible Errors
First Curve?	The calculator is asking you to tell it which curve is the first curve in the system of equations. Place your cursor on one of the lines you graphed using the arrow keys and press *enter.* You will then be asked for the second curve. Move the cursor to the other curve and press *enter. Why does it ask this? Well, if you had graphed four lines, it would want to know which lines (curves) were the two that you want to find the intersection of.*	If you give the calculator the **same** curve for both the first and second prompts, you will likely get an error.
Lower Bound?	The calculator is asking you to give it a smaller region to look in to find the intersection point. It is asking you for a value to the left of the point of intersection (lower bound) first. Use your arrow keys to place your cursor to the left of the point of intersection and press *enter* or use the number keypad to enter a value that you know is to the left of the intersection point, and then press *enter.* You will then be asked for the Upper Bound. Repeat the process for a value to the right of the intersection point.	The upper bound must be **above** (to the right of) the lower bound. If you try to give the calculator an upper bound value that is below the lower bound value, you will get an error.
Guess?	It is easier for the calculator to find the exact intersection point if you give it a guess point near the intersection point. Just position your cursor near the intersection point and press enter. *Why does it ask for a guess? If you were graphing more complicated curves (for example polar curves), there might be more than one intersection point in the same lower and upper boundary. In this case, the calculator would want to know which point you wanted to find out exactly.*	You must give the calculator a guess that is **between** the lower and upper bounds. If your guess is outside the region where you told the calculator to look, it will be (understandably) unhappy with you.

Note: If you do not see the intersection values, it could be that you pressed enter one too many times when going through the prompts (inadvertently skipping the solutions). If this happens, repeat the steps a little more slowly, making sure to only press *enter* once for each prompt.

Student Activity A

Section 4.2: Which Egg is Easier to Crack?

When solving systems of equations by substitution, we begin by solving one of the equations for x or y; this is called the *substitution equation*. Sometimes one of the variables is easier to solve for than the other (that is, it will take less steps to solve for one than the other).

Directions: Look at the given equations and find the substitution equation where solving for the variable requires **the least number of steps**. Then place the substitution equation in the egg that corresponds to the variable that was solved for. For example, in the first equation, it would be much easier to solve for x, resulting in the equation $x = 2 - 3y$.

$x + 3y = 2$	$7x + y = 14$	$3y = 2x + 2$	$-2y + x = 18$
$x - 2y = 9$	$15 = -5y + x$	$16x = 16y + 16$	$\frac{2}{3}x + y = \frac{1}{3}$
$25x = 225y + 225$	$12x - y = 0$	$3y = 3x + 6$	$5x + y = 10$
$18x = 9y + 36$	$\frac{1}{10}x + y = 1$	$12x + y = 2$	$4x = 2y + 6$

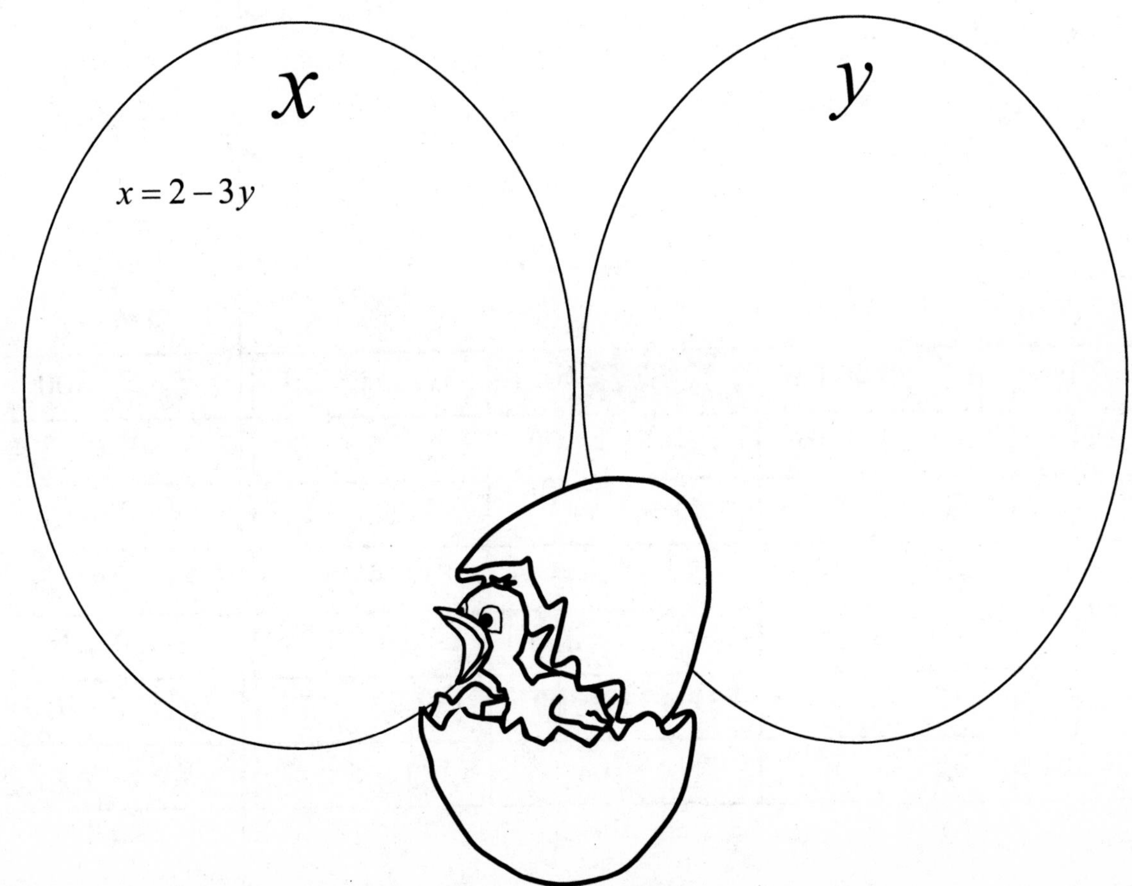

Student Activity B

Section 4.2: Clear the Way!

Directions: Clear each equation of decimals or fractions and shade in the corresponding square in the grid below. The first one has been done for you. There's a surprise when you're finished!

1. $\frac{1}{10}x - \frac{1}{5}y = 1$

   $10\left(\frac{1}{10}x - \frac{1}{5}y\right) = 10(1)$

   $\boxed{x - 2y = 10}$

2. $0.1x + 2 = 0.2y$

3. $\frac{1}{100}(3x - 5y) = 3$

4. $0.04(x - 50) = 0.02y$

5. $\frac{1}{10}x + \frac{1}{15}y = \frac{1}{30}$

6. $0.33(x + 1) = 7y$

7. $\frac{1}{8}(4x - 2y) = \frac{1}{16}$

8. $0.001x + 0.0002 = 0.003y$

9. $\frac{2}{3}y + \frac{4}{9} = \frac{1}{18}x$

10. $0.1x + 0.01y = 0.001$

$4x - 2 = 20y$	$4x - 50 = 20y$	$x - 2y = 10$	$3x - 5y = 30$	$x - y = 300$
$4x - 20 = 20y$	$4x - 200 = 2y$	$3x - 5y = 300$	$3x - 5y = 3$	$3x - 500y = 300$
$3x + 2y = 10$	$2x + 3y = 1$	$3x + 2y = 1$	$3x + y = 1$	$x + 2y = 1$
$4x - 2y = 1$	$8x - 2y = 1$	$8x - 4y = 1$	$8x - 4y = 2$	$4x - 2y = 2$
$x + 2 = 2y$	$x + 20 = y$	$x + 20 = 2y$	$x + 20 = 0.2y$	$x - 20 = 2y$
$x + 2 = 3y$	$10x + y = 1$	$33x + 33 = 700y$	$33x + 33 = 7y$	$33x + 1 = 70y$
$10x + 20 = 30y$	$10x + 2 = 30y$	$100x + 10y = 1$	$12y + 8 = x$	$2y + 4 = x$

Student Activity C

Section 4.2: Double Trouble on Substitution

Directions: When you solve a system of equations by substitution, there are *at least* two ways to tackle the problem. For example, in the first system, you could easily solve for y in the first equation, or for x in the second equation to find your substitution equation. **Some choices are easier than others!** For each of the systems below, solve the system of equations two ways as indicated. Your ordered pair solution should be the same for both methods. If they are not, you'll have to go back and look for a mistake. The first one has been started for you.

	System of Equations	Solve for a variable in the FIRST equation, then substitute and solve.	Solve for a variable in the SECOND equation, then substitute and solve.
1.	$\begin{cases} 2x + y = 5 \\ x - 4y = 7 \end{cases}$	Solve for y in the 1st equation: $y = -2x + 5$ Substitute into 2nd equation and solve: $x - 4(-2x + 5) = 7$ $x + 8x - 20 = 7$ $9x - 20 = 7$ $9x = 27$ $x = 3$ Find the missing coordinate of the ordered pair: $y = -2(3) + 5$ $= -6 + 5 = -1$ $(3, -1)$	
2.	$\begin{cases} x + 8y = 20 \\ 4x - y = 14 \end{cases}$		

	System of Equations	Solve for a variable in the FIRST equation, then substitute.	Solve for a variable in the SECOND equation, then substitute.
3.	$\begin{cases} 3x - y = -2 \\ -12x + 4y = 16 \end{cases}$		
4.	$\begin{cases} 3x + 4y = 0 \\ \frac{3}{4}x + \frac{8}{3}y = -\frac{5}{12} \end{cases}$		
5.	$\begin{cases} x - 7y = -1 \\ -\frac{1}{7}x + y = \frac{1}{7} \end{cases}$		

Student Activity A

Section 4.3: Forced Elimination

Directions: In order to develop some intuition about which variable is easier to eliminate, we will **start** a few problems by trying both of the possible eliminations. Use the elimination method to solve for a variable. **Then circle the elimination that was easier.** The first one has been started for you.

1.

Eliminate x	Eliminate y
$\begin{cases} x+3y=7 \\ 2x-3y=-4 \end{cases}$ Multiply the first equation by -2. $\begin{cases} -2x-6y=-14 \\ 2x-3y=-4 \end{cases}$ $-9y=-18$ $y=2$	$\begin{cases} x+3y=7 \\ 2x-3y=-4 \end{cases}$

2.

Eliminate x	Eliminate y
$\begin{cases} -10x+y=-5 \\ 10x-2y=0 \end{cases}$	$\begin{cases} -10x+y=-5 \\ 10x-2y=0 \end{cases}$

3.

Eliminate x	Eliminate y
$\begin{cases} 5x+7y=12 \\ 10x-3y=7 \end{cases}$	$\begin{cases} 5x+7y=12 \\ 10x-3y=7 \end{cases}$

4.

Eliminate x	**Eliminate y**
$\begin{cases} 3x + y = 2 \\ 4x - 2y = 6 \end{cases}$	$\begin{cases} 3x + y = 2 \\ 4x - 2y = 6 \end{cases}$

Now inspect these systems of equations. For each system, decide which variable is going to easier to eliminate and describe the process you will go through to achieve this elimination. Problem 5 has been done for you.

5. $\begin{cases} 3x + y = 1 \\ 2x + 3y = -11 \end{cases}$ Eliminate y.
Multiply the first equation by -3 and then add the equations.

6. $\begin{cases} 5x - y = -5 \\ x + y = 5 \end{cases}$

7. $\begin{cases} x + y = 12 \\ -x + y = 0 \end{cases}$

8. $\begin{cases} -4x + 2y = 2 \\ 4x + 6y = 22 \end{cases}$

9. $\begin{cases} 3x + 14y = -7 \\ 2x + 2y = 12 \end{cases}$

10. $\begin{cases} 3x + 8y = 2 \\ -6x - 7y = 5 \end{cases}$

Student Activity B

Section 4.3: Triple the Fun on Systems of Equations

Directions: For each of the problems below, solve the system of equations by graphing, by substitution, and by elimination – your solution should be the same for all three methods methods. If they are not, go back and look for a mistake in your work.

1.

System of Equations:

$$\begin{cases} 10x + 5y = 15 \\ x + 2y = -6 \end{cases}$$

Solve by Graphing:

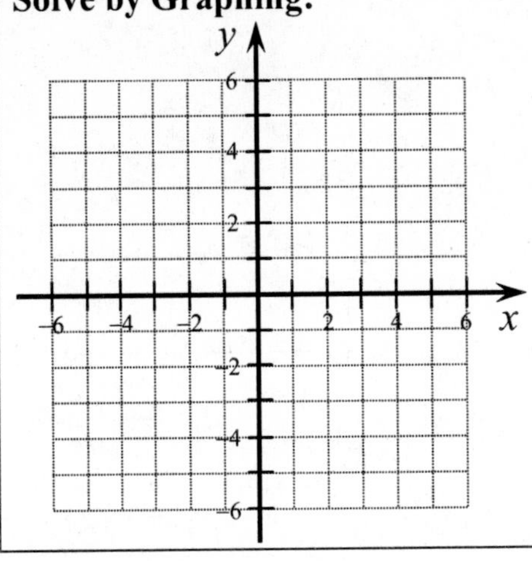

Solve by Substitution

Solve by Elimination

2.

System of Equations:

$$\begin{cases} \frac{1}{4}x - y = -\frac{1}{3} \\ 3x - 12y = -12 \end{cases}$$

Solve by Graphing:

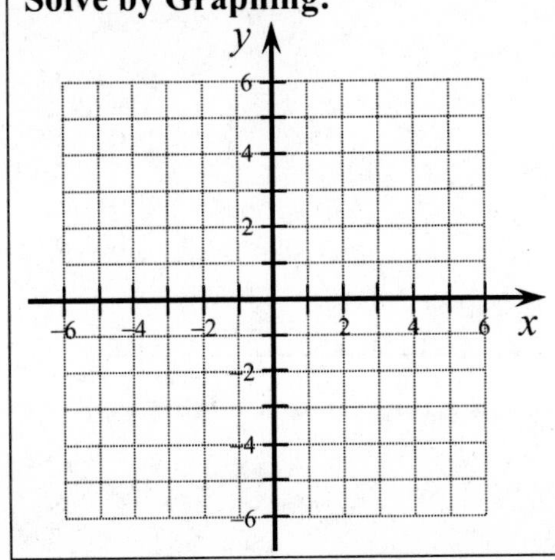

Solve by Substitution

Solve by Elimination

3.

System of Equations:

$$\begin{cases} 3y - 4x = 2 \\ 3x - 4y = -5 \end{cases}$$

Solve by Graphing:

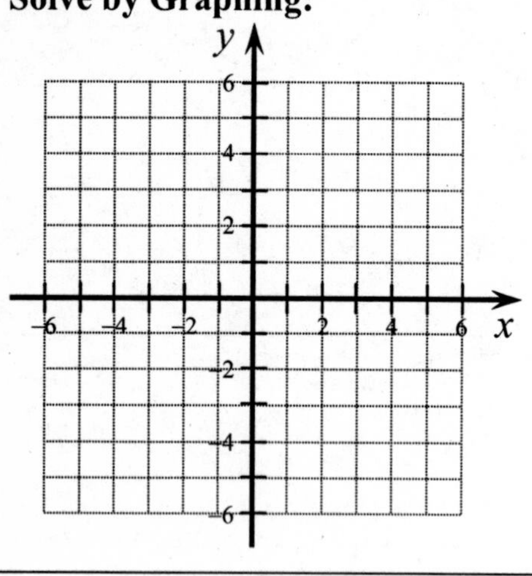

Solve by Substitution

Solve by Elimination

4.

System of Equations:

$$\begin{cases} -4x + y = -5 \\ 12x - 3y = 15 \end{cases}$$

Solve by Graphing:

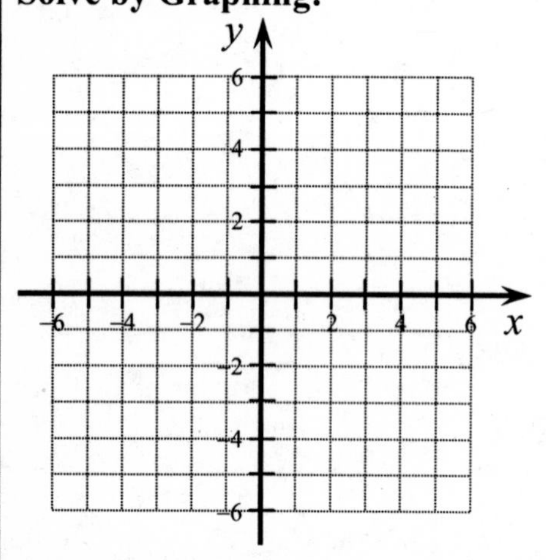

Solve by Substitution

Solve by Elimination

Student Activity C

Section 4.3: Choose Your Tactic

Directions: For each system of equations below, decide whether it will be easier to solve using substitution or elimination (choose your tactic).

	System of Equations	**Tactic** (substitution or elimination)	**Tactical Plan** **Substitution** – which variable in which equation will you solve for? **Elimination** – which variable will you eliminate and how?
1.	$\begin{cases} 5x+3y=-9 \\ y=2x+8 \end{cases}$		
2.	$\begin{cases} 2x+y=9 \\ 5x+3y=26 \end{cases}$		
3.	$\begin{cases} y=-x \\ 6x+6y=0 \end{cases}$		
4.	$\begin{cases} 4x+11y=7 \\ 4x+3y=-1 \end{cases}$		
5.	$\begin{cases} 16x-2y=16 \\ 4x=2y-8 \end{cases}$		
6.	$\begin{cases} 0.02x+0.01y=0.1 \\ -2x+3y=-18 \end{cases}$		
7.	$\begin{cases} -\frac{x}{5}-\frac{y}{3}=2 \\ -\frac{3x}{10}+\frac{2y}{10}=\frac{9}{10} \end{cases}$		
8.	$\begin{cases} x=12y-7 \\ 12y-x=12 \end{cases}$		

Assessment 4B

Chapter 4: Mid-chapter Assessment for Understanding

For each of the following, describe the type of problem and the strategies and key steps to remember while doing the problem. You do **not** have to complete the problems.

	Type of Problem	Strategies and Key Steps
1. Use substitution to solve the system. $\begin{cases} x-3y=-1 \\ 3x-4y=7 \end{cases}$		
2. Is the system below consistent or inconsistent? $\begin{cases} x-4y=6 \\ 2x=8y+10 \end{cases}$		
3. Is $(1,-2)$ a solution to the system? $\begin{cases} 3x+y=1 \\ x-y=-1 \end{cases}$		
4. Use elimination to solve the system. $\begin{cases} x+3y=-2 \\ x+4y=0 \end{cases}$		
5. Solve the system. $\begin{cases} x=y-7 \\ x-2y=14 \end{cases}$		
6. Is the pair of equations dependent or independent? $\begin{cases} x-4y=6 \\ 2x=8y+10 \end{cases}$		
7. Solve the system by graphing. $\begin{cases} y=-2x+4 \\ y=2x \end{cases}$		
8. Solve the system. $\begin{cases} 3x-4y=7 \\ 5x-8y=8 \end{cases}$		

Student Activity A

Section 4.4: The Last Sentence

Directions: What you see below is the last sentence from each of the example problems in this section of the textbook. Just from this last sentence, you can make a fairly good guess about what the two variables should be for the problem. The first one has been done for you.

Example 1: *Find the height of the figure and the height of the base.*

Let f = the height of the figure.
Let b = the height of the base.

Example 2: *Find the measure of each angle.*

Example 3: *Find the length and width of the poster.*

Example 4: *Find the cost of a class picture and the cost of an individual wallet-size picture.*

Example 5: *How much money was invested in each account?*

Example 6: *Find the speed of the boat in still water and the speed of the current.*

Example 7: *How many fluid ounces of each batch should she use?*

Example 8: *How many ounces of each ingredient should be used to create a 20-ounce box of raisin bran cereal that can be sold for $0.15 an ounce?*

Now go back and read all the example problems in this section. Can you add any detail to your variable declarations after reading the complete problems.

Student Activity B

Section 4.4: Working with the Smaller Pieces

Practice with Distance-Rate-Time Problems

Distance, rate and time can be related using the formula: $d = rt$

1. If you travel 8 miles from your house to visit a friend, what is your round-trip distance? ___________

2. Will you run faster when you run *with* the wind or *against* it? ___________________

3. How far can you drive in 90 minutes at a rate of 70 miles per hour? _____________

4. If you work 30 miles away from your home, how long will it take you to get there if your average rate is 45 miles per hour? _____________

5. If you drive for 3 hours and travel 210 miles, what was your average rate?

____

6. If you are in a boat that can travel 4 miles per hour in still water, will you be traveling *faster* or *slower* than 4 miles per hour when traveling upstream? ____________

Practice with Value Mixture and Percent Mixture Problems

7. If a store owner mixes peanuts that cost \$3 per pound with cashews that cost \$8 per pound, a pound of mixed nuts should cost

a. less than \$3 **b.** between \$3 and \$8 **c.** more than \$8

8. Simon has to mix two solutions in his chemistry class. One contains 3% sulfuric acid, and the other contains 10% sulfuric acid. How much sulfuric acid will the resulting mixture contain?

a. less than 3%
b. between 3% and 10%
c. between 10% and 13%
d. more than 13%

9. Lucy has to combine plain water with a solution that contains 15% ammonia. The resulting solution will be

a. less than 15% ammonia **b.** more than 15% ammonia

10. If you wanted to obtain 5 liters of a solution that is 4% nitric acid, which solution could you **not** add?

a. 3 liters of a 3% solution
b. 2 liters of a 5% solution
c. 6 liters of a 2% solution

11. At Meg's coffee shop, frozen coffees cost \$3.50 each. If Meg took in a total of \$42 on Tuesday, what is the maximum number of frozen coffees she could have sold?________

Practice with Investment Problems

Interest earned, principle invested, interest rate per year, and time invested (years) can be related using the formula: $I = Prt$

12. If you invest $2,500 for one year at an annual rate of 6%, how much interest does the investment earn? ________

13. If you invested a total of $15,000 in two different accounts, and you put $6,000 in the first account, how much did you put in the second account?________

14. If you invest $1,000 in an account that pays 5% annual interest, how much money will you have after one year?__________

15. Martin invests $2,000 in a bank CD. If the interest after one year is $160, what was the annual interest rate? __________

Guided Learning Activity

Section 3.3: Using the Problem Solving Strategy

Problem A: At a scrapbooking party, customers can purchase two special packages, the vacation scrapbook package and the baby album package. Shaude purchases 2 vacation packages and one baby album package for a total of $93.50. Stephan purchases one vacation package and three baby album packages for a total of $138. Find the cost of each of the special packages that were offered at the party.

1. Analyze the problem.

2. Define two variables. Write two equations.

3. Solve the system of equations.

4. State the conclusion.

5. Check the result.

Problem B: Jasper invests a total of 250,000 L$ (Linden dollars). He splits his investment between a 30-day Linden Bank CD that pays 7% interest at the end of the term and a 30-day Second Life real estate investment scheme organized by his brother that ends up losing 4% of its original value at the end of 30 days. At the end of the 30 days, Jasper makes 2,100 L$. How much did Jasper invest in the CD and how much did he invest in the real estate scheme?

1. Analyze the problem.

2. Define two variables. Write two equations.

3. Solve the system of equations.

4. State the conclusion.

5. Check the result.

Problem C: Technical grade hydrogen peroxide is 35% H_2O_2. Common household hydrogen peroxide is only 3% H_2O_2. Imogen wants to make a 10% H_2O_2 solution to use in removing a toxic mold in a city building. What amount of the technical grade solution and what amount of the household solution (to the nearest 0.1 mL) should she mix to create the 500 mL of solution she desires?

1. Analyze the problem.

2. Define two variables. Write two equations.

3. Solve the system of equations.

4. State the conclusion.

5. Check the result.

Student Activity C

Section 4.4: Setting Up the System

Directions: Read each problem carefully.
a) **Define the variables** you would use to solve the problem.
b) **Write a system** of two equations that could be used to solve each problem. You do not need to solve the application problems.

1. The elevation in Denver, Colorado, is 13 times greater than that of Waikapu, Hawaii. The sum of their elevations is 5,600 feet. Find the elevation of each city.

2. The owner of a pet food store wants to mix birdseed that costs $1.25 per pound with sunflower seeds that cost $0.75 per pound to make 50 pounds of a mixture that costs $1.00 per pound. How many pounds of each type of seed should he use?

3. An airplane can fly 500 miles in 2 hours when flying with the wind. The same journey takes 4 hours when flying against the wind. Find the speed of the wind and the speed of the plane.

4. Barb invested a total of $25,000 in two accounts. One account has an annual interest rate of 3.5%, while the other has an annual rate of 6%. She earned $1,257 in interest in one year. How much did she invest in each account?

5. Two angles are supplementary. The measure of the first angle is 10 degrees more than three times the second angle. Find the measure of each angle.

Student Activity D

Section 4.4: The Moving Walkway Problem

A moving walkway is like an escalator (transporting you from one place to another), only it is a flat belt that you can walk on. Moving walkways can commonly be found now in airports, museums, amusement parks, or zoos. When you walk onto a moving walkway going in your direction and continue to walk, your walking speed is increased by the speed of the walkway.

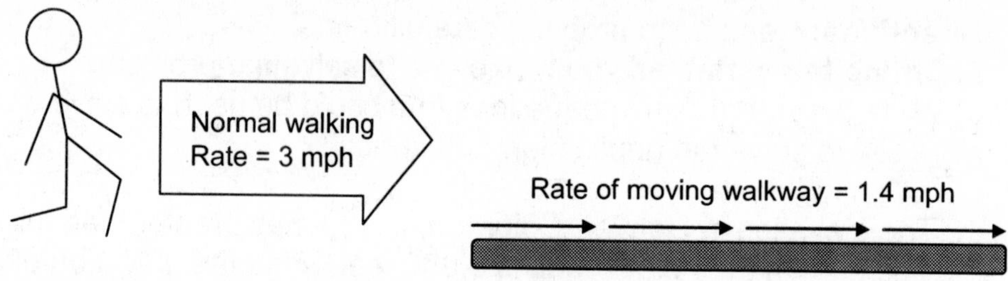

1. Using the information from the diagram above, fill in the missing rates.

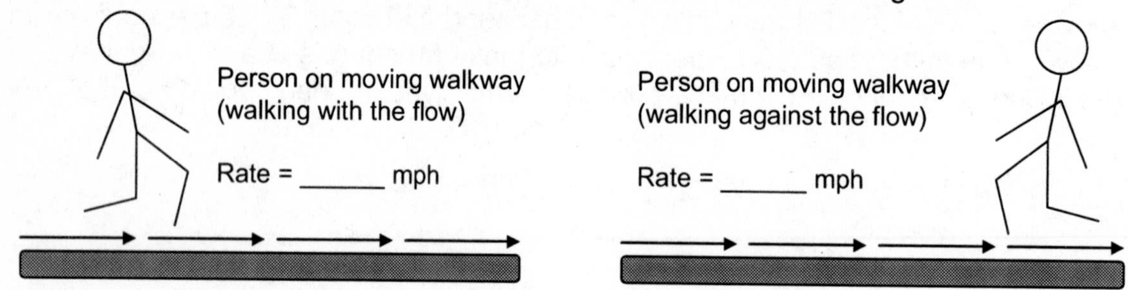

2. Often, you do not know one of the rates. In these cases, you have to express the rate as an algebraic expression. Assume the person is walking at a rate of 2.7 mph and the rate of the moving walkway is x mph. Fill in the missing rates.

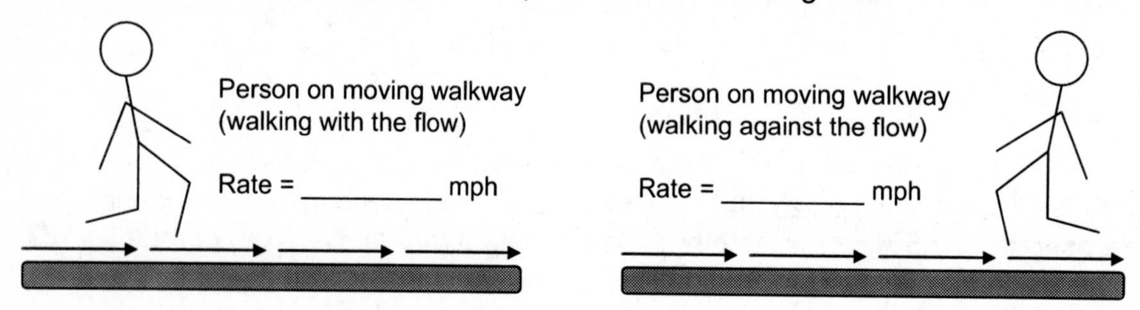

3. Now the rate of the person is R mph and the rate of the walkway is 2.3 mph. Fill in the missing rates.

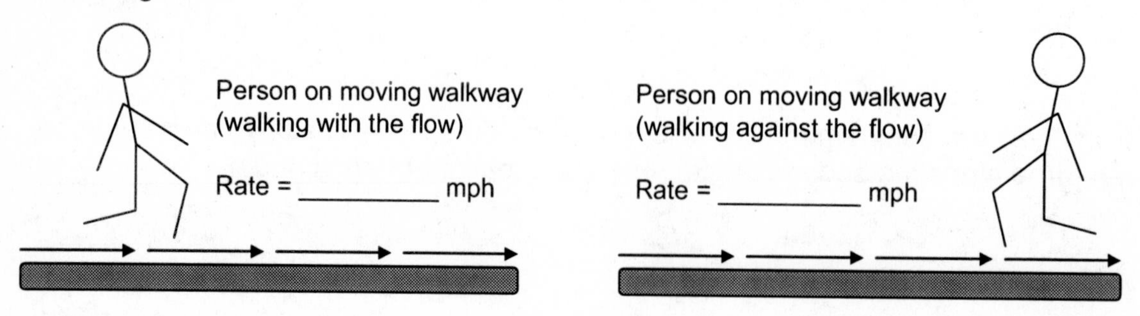

4. At the Montparnasse station in Paris, there is a high-speed moving walkway that was designed to help commuters get through the station faster. This moving walkway, which is 0.18 km long, moves at a speed of 9 km/hr (originally the walkway was designed to go 11 km/hr, but there were too many accidents). David walks at his normal pace in the direction of flow on the walkway – it takes 54 seconds to go the full distance of the walkway. How fast was David walking?

5. While on a layover at the Detroit International Airport, a bored math teacher decides to calculate the speed of one of the many moving walkways in the airport. She knows (from years of treadmill walking) that her normal walking speed is 2.5 mph (or 2.2 ft/sec). Walking with the flow of the walkway, it takes 43 seconds to go the distance of the walkway. Walking against the flow of the walkway, it takes 2 minutes and 47 seconds. What is the approximate rate of the moving walkway in feet per second and what is the length of the walkway in feet?

	Rate	Time	Distance
With Flow of Walkway			
Against Flow of Walkway			

6. The Central-Mid-Levels Escalator in Hong Kong is the world's longest covered escalator, stretching 800m (2,625 ft). A trip on the escalator, from top to bottom, takes about 20 minutes when a person is stands still. What is the approximate rate of the escalator in feet per second? How long would the trip take if the person walked downward at a rate of 2 ft/sec?

Student Activity A

Section 4.5: Tic-tac-toe on Inequalities

Directions for Tic-tac-toe #1: If the ordered pair in the square **IS** a solution of the system of inequalities shown below, then circle the ordered pair (thus putting an **O** on the square). If it **IS NOT** a solution, then put an **X** over the ordered pair on the square.

$$\begin{cases} y < 4x + 2 \\ 3x - 3y \geq 6 \end{cases}$$

$(-2,-6)$	$(1,1)$	$(4,1)$
$(6,2)$	$(8,6)$	$(4,4)$
$(2,0)$	$(0,2)$	$(0,-2)$

Directions for Tic-tac-toe #2: If the ordered pair in the square **IS** a solution of the graphed system of inequalities, then circle the ordered pair (thus putting an **O** on the square). If it **IS NOT** a solution, then put an **X** over the ordered pair on the square.

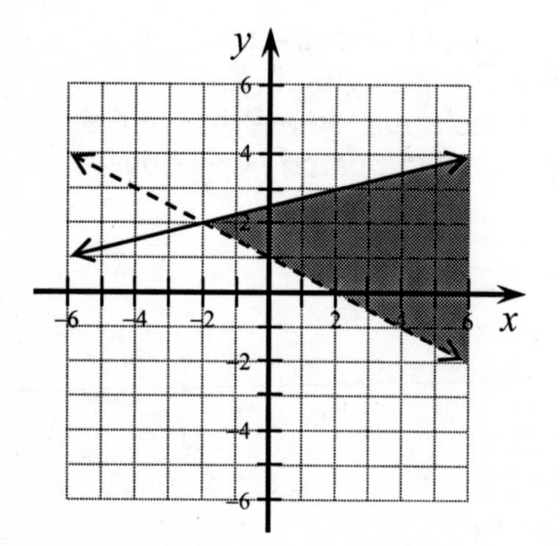

$(-2,2)$	$(2,0)$	$(2,3)$
$(8,3)$	$(3,4)$	$(0,4)$
$(6,4)$	$(6,-2)$	$(4,0)$

Student Activity B

Section 4.5: Testing Test Points

Directions: In each problem below you are given a system of inequalities, and the graphed inequality boundary lines. In each region separated by the boundary lines, a test point is provided. Test this point in both inequalities to determine if it is true (T) or false (F). Shade the region of the graph where the ordered pair **is** a solution to both inequalities. The first test point has been done for you.

1.

	$y<\frac{1}{2}x$	$y>2x-8$
$(0,3)$	F	T
$(2,-2)$		
$(-5,0)$		
$(-4,-4)$		

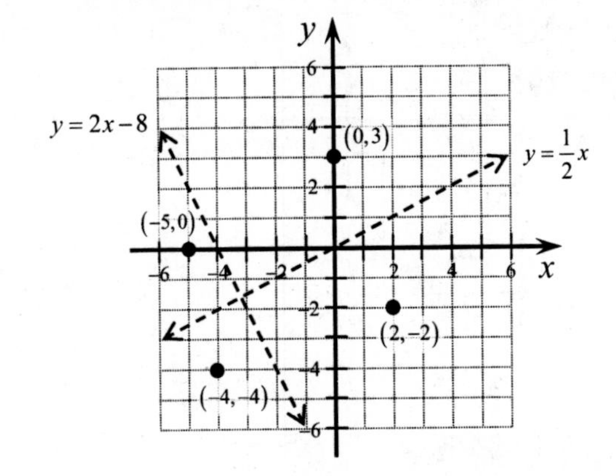

2.

	$y\le-\frac{1}{3}x+2$	$y\ge-\frac{3}{2}x-\frac{3}{2}$
$(-6,5)$		
$(-6,1)$		
$(6,4)$		
$(6,-2)$		

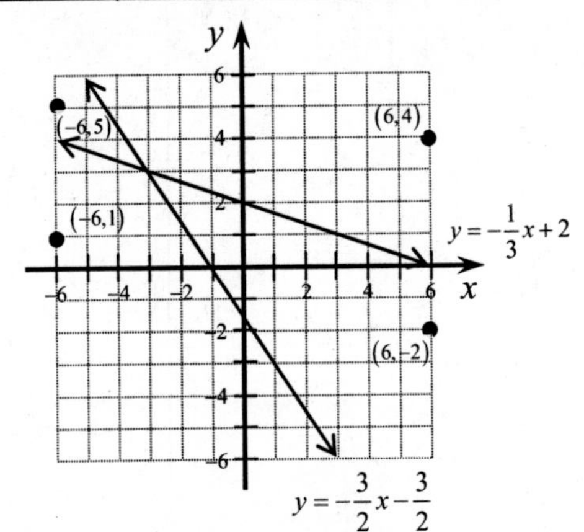

3.

	$y<\frac{2}{3}x+4$	$y>\frac{2}{3}x$
$(-6,3)$		
$(-3,1)$		
$(3,-2)$		

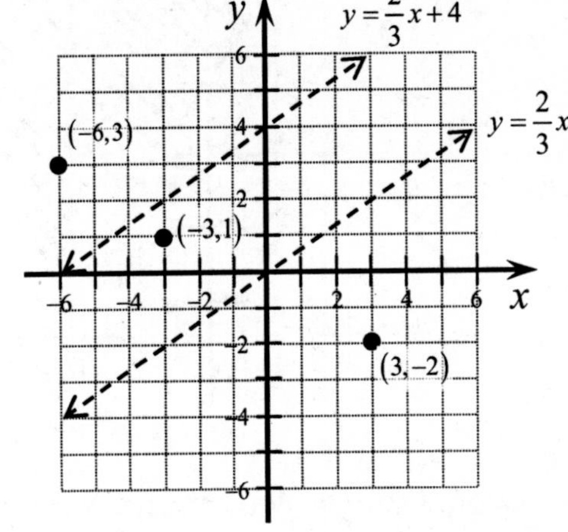

4.

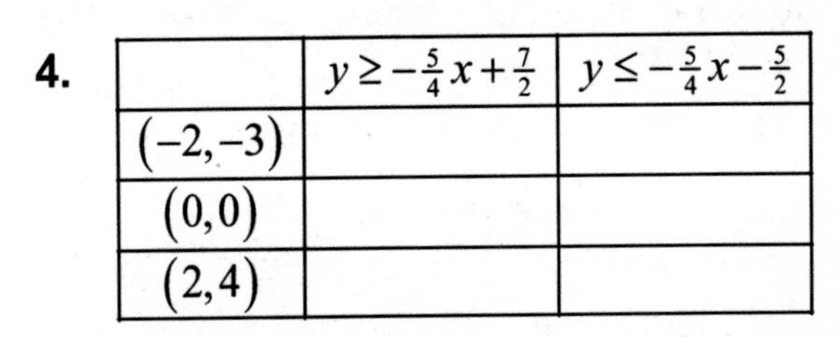

	$y\ge-\frac{5}{4}x+\frac{7}{2}$	$y\le-\frac{5}{4}x-\frac{5}{2}$
$(-2,-3)$		
$(0,0)$		
$(2,4)$		

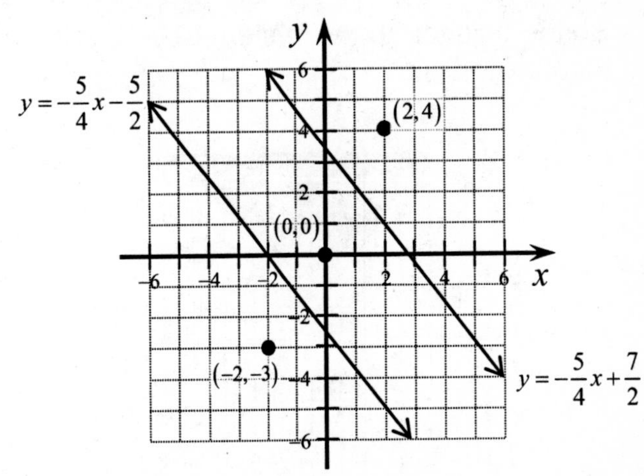

Guided Learning Activity

Section 4.5: Graphing Systems of Linear Inequalities

Example: Consider the system of inequalities given by $\begin{cases} y \geq -\dfrac{4}{3}x + 2 \\ y > \dfrac{5}{6}x + 1 \end{cases}$

What does the solution of a system of inequalities look like? When you graph the inequalities on the same set of axes, the solution is the region where the two graphs intersect. Separately, the graphs of our inequalities look like this:

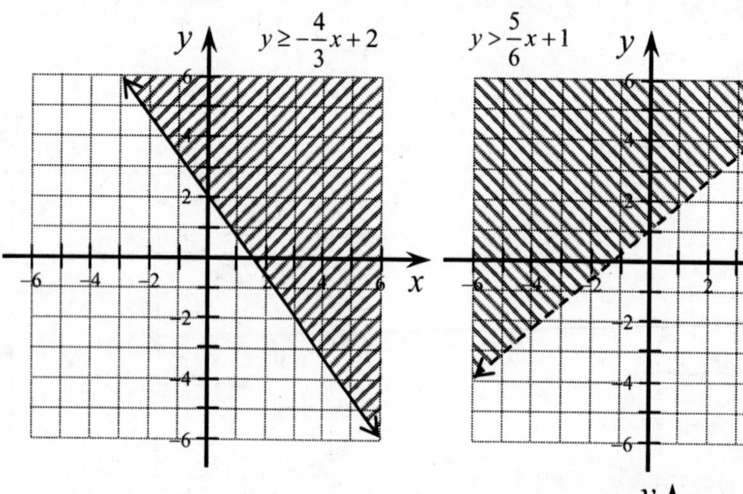

If we overlap these two graphs, we can find the area of intersection (shown here in solid gray).

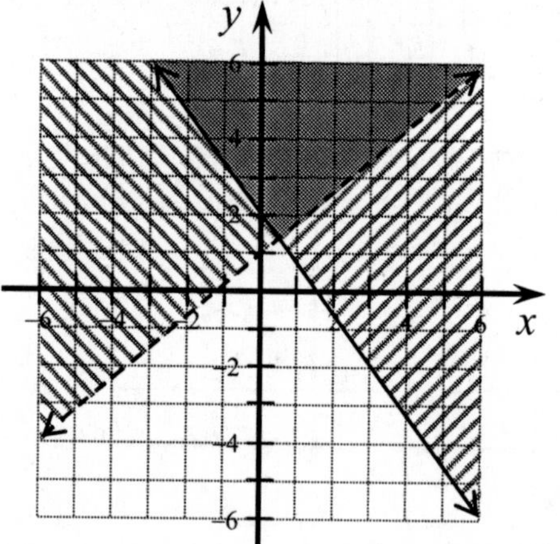

So the solution to the system of inequalities is:

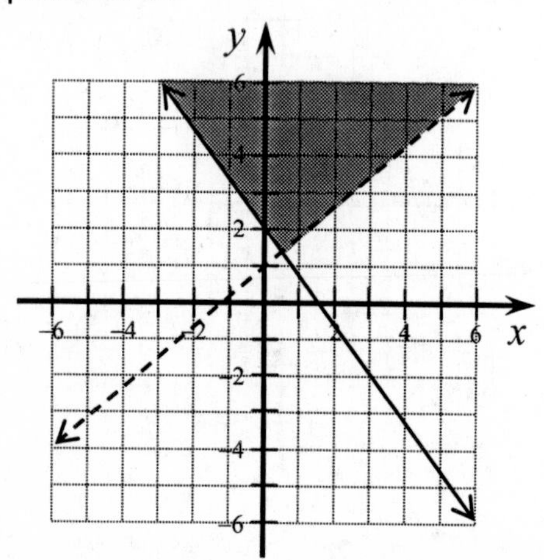

Now we'll solve these systems of inequalities by graphing.

1. $\begin{cases} y < -2x + 3 \\ y < 2x - 3 \end{cases}$

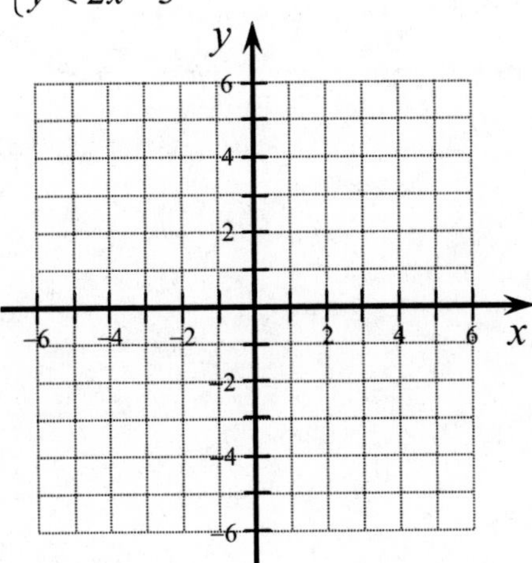

2. $\begin{cases} 2 - y \le 0 \\ y - \frac{1}{2}x \le 4 \end{cases}$

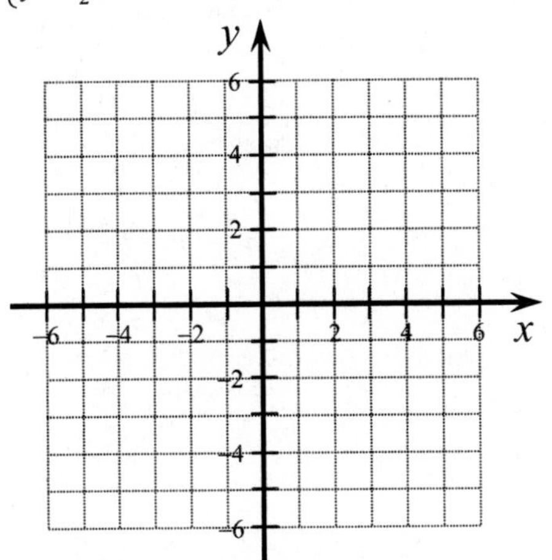

3. $\begin{cases} y > \frac{1}{3}x + 1 \\ y < \frac{1}{3}x - 3 \end{cases}$

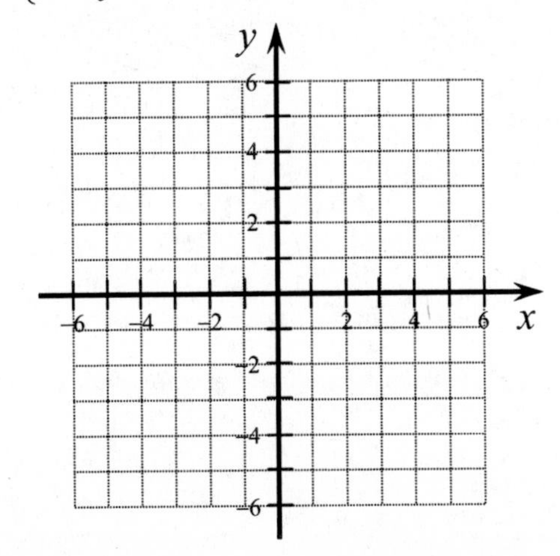

4. $\begin{cases} 3x + y \ge -5 \\ -3x - y \ge -5 \end{cases}$

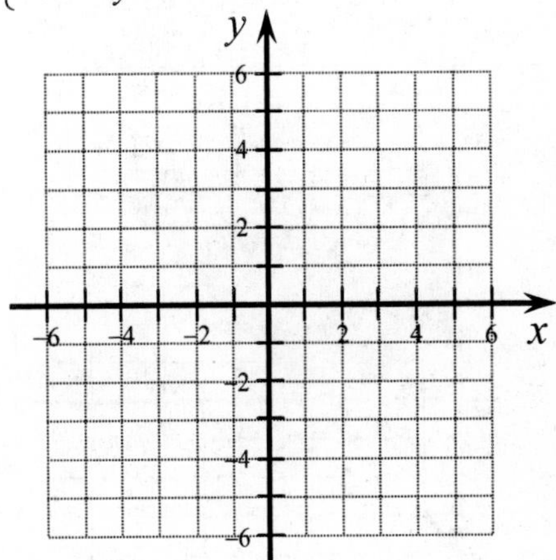

5. $\begin{cases} y \le 0 \\ x \le 0 \end{cases}$

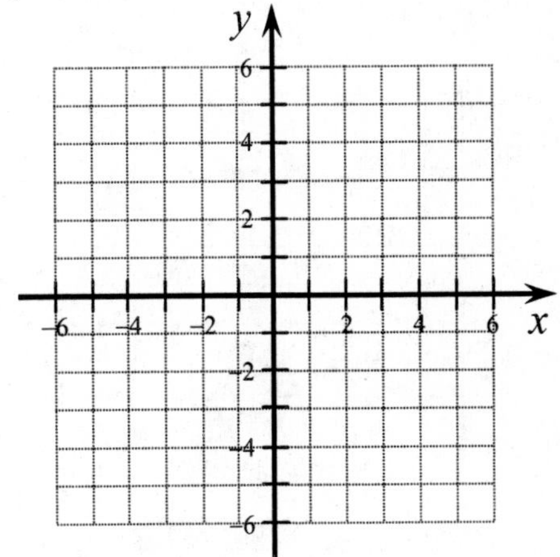

6. $\begin{cases} y \le -\frac{1}{2}x + 2 \\ y > 2x + 2 \end{cases}$

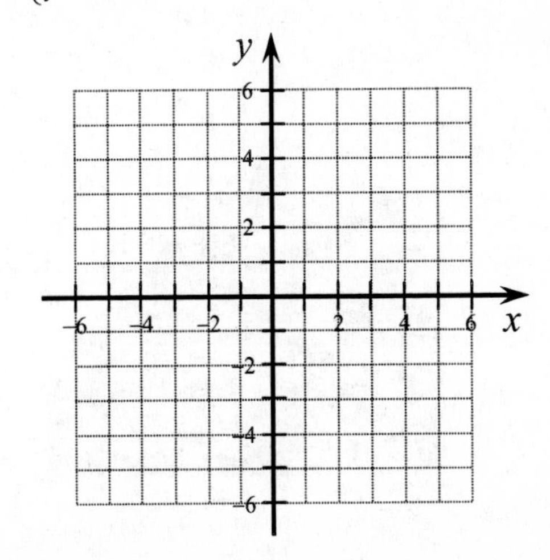

Student Activity C

Section 4.5: Following the Clues Back to the System of Inequalities

Directions: In each "crime-scene" below, you are shown the graph of a system of inequalities. Use your mathematical powers of reasoning (and detective skills) to determine what the inequality must have been to result in this graph.

1.

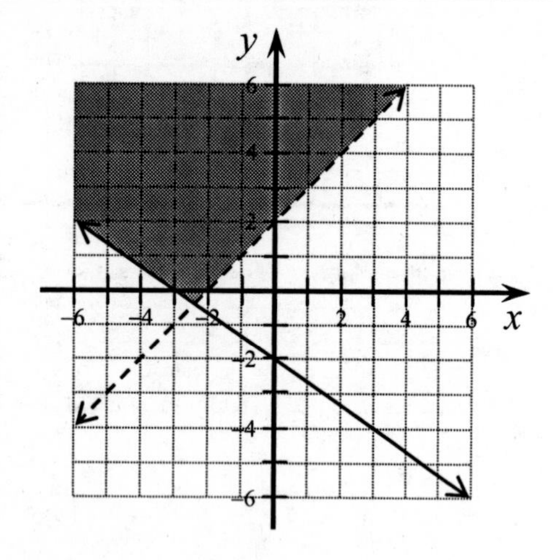

2.

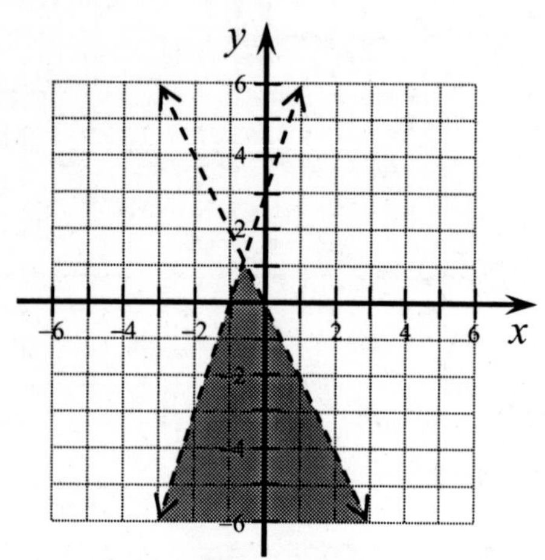

3.

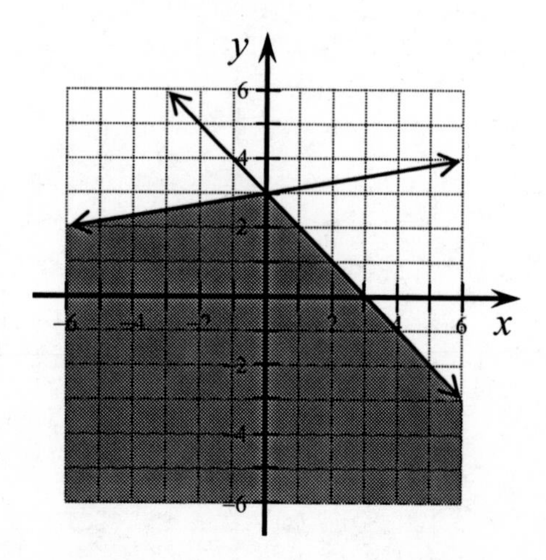

Assessment 4C

Chapter 4: End-of-chapter Assessment for Understanding

For each of the following, describe the type of problem and the strategies and key steps to remember while doing the problem. You do **not** have to complete the problems.

	Type of Problem	Strategies and Key Steps
1. Rafi paddles his canoe upstream 24 miles in 4 hours. If he can make the downstream trip in 3 hours, find the rate that Rafi can paddle and the rate of the current.		
2. Is $(1,-3)$ a solution to the system? $\begin{cases} 3x+y<1 \\ x-y>-1 \end{cases}$		
3. Two angles are supplementary and one angle measures 30° more than the other angle. Find the measure of each angle.		
4. Solve the system by graphing. $\begin{cases} y-3x=2 \\ 4x-y=-1 \end{cases}$		
5. Graph the solution to the system. $\begin{cases} x \le 3 \\ y \le \frac{1}{2}x-1 \end{cases}$		
6. Is the pair of equations dependent or independent? $\begin{cases} 3x+y=8 \\ 6x+2y=8 \end{cases}$		
7. Graph the solution to the system. $\begin{cases} 3x+y<4 \\ 2y-x>-2 \end{cases}$		

Assessment 4D

Chapter 4: Metacognitive Skills Assessment

Metacognitive skills refer to the ability to judge how well you have learned something and to effectively direct your own learning and studying. This is a self evaluation tool designed to help you focus your studying and to improve your metacognitive skills with regards to this math class.

Fill the 1st column out before you begin studying.
Fill the 2nd column out after you study and before you take the test.
Go back to this page after your test and circle any of the ratings that you would now change – this identifies the "disconnects" between what you think you know well and what you actually know well.

Use the scale below to assign a number to each topic.
5 *I am confident I can do any problems in this category correctly.*
4 *I am confident I can do most of the problems in this category correctly.*
3 *I understand how to do the problems in this category, but I still make a lot of mistakes.*
2 *I feel unsure about how to do these problems.*
1 *I know I don't understand how to do these problems.*

Topic or Skill	Before Studying	After Studying
Checking whether an ordered pair is a solution of a system of equations.		
Classifying a system of equations as consistent or inconsistent.		
Classifying a system of equations as dependent or independent.		
Graphing a system of equations to find the solution.		
Understanding what the "solution" to a system of equations is.		
Finding or identifying a substitution equation for a system of equations.		
Solving a system of equations by substitution.		
Clearing the fractions from an equation.		
Identifying a variable to eliminate and performing the steps to set up that elimination in a system of equations.		
Solving a system of equations by elimination.		
Deciding whether substitution or elimination would be an easier method for solving a system of equations.		
Clearing the decimals from an equation.		
Declaring the variables for an application involving a system of equations.		
Writing the system of equations for a problem involving $d = rt$.		
Writing the system of equations for a value-mixture problem or a percent-solution problem.		
Writing the system of equations for an investment problem.		
Solving a system of equations from an application problem.		
Writing the conclusion to an application problem involving a system of equations.		
Checking whether an ordered pair is a solution of a system of inequalities.		
Determining whether the boundary lines for a system of inequalities are dashed or solid.		
Graphing the boundary lines for a system of inequalities.		
Determining the proper region to shade in a system of inequalities.		

Student Workbook: Chapter 5

Table of Contents: *Exponents and Polynomials*

Assessment 5A

Pretest and Diagnostic Tool: Exponents and Polynomials

Directions: Complete this assessment without looking back at your notes or your book. **Do not use a calculator for this assessment.**

1. Multiply: $3 \cdot 3 \cdot 3$. _____

2. Multiply: $(-2)(-2)$. _____

3. Simplify: -4^2. _____

4. Write 2^3 using multiplication.

____

5. Simplify: $\dfrac{2 \cdot 2 \cdot 2 \cdot 2 \cdot 3}{2 \cdot 2 \cdot 3}$.

6. Write $(4 \cdot 4)(4 \cdot 4 \cdot 4)$ as 4 to some power. ______

7. Fill in the blank to complete this pattern: 8, 4, 2, ___, $\frac{1}{2}$, $\frac{1}{4}$.

8. Fill in the blanks to complete this pattern: 27, 9, 3, 1, ___, ___.

9. -3 is the same as $\dfrac{1}{3}$. True or false?

____

10. Evaluate: $3-(-4)$. ______

11. $47{,}000 = 4.7(1000)$ True or false?

____

12. $0.00058 = 5.8(0.001)$ True or false?

____

13. $\dfrac{1}{10^2} = \dfrac{1}{100}$ True or false? _______

14. Identify the exponent in 4^2. ____

15. Write $3, 10, 5$ in descending order.

____

16. Add: $(5x+2)+(3x-6)$.

17. Simplify: $-(4x-8)$.

18. Subtract: $(2x+3y)-(4x-2y)$.

19. Multiply.

$$\begin{array}{r} 425 \\ \times\ 36 \\ \hline \end{array}$$

20. Multiply: $(2x)(-5)$.

21. If $M = 2AB$ and we insert $A = 4x$ and $B = -3$, what is the new formula for M? __________

22. Evaluate: $(2+5)^2$.

23. Simplify: $x^2 - x - x + 1$.

24. Divide. $12\overline{)1632}$

25. Evaluate: $\dfrac{12+36}{12}$.

Student Activity A

Section 5.1: Exponent Duos

Directions: In each of the "duos" below, place two equivalent exponential expressions of the following format:

Expanded Expression using multiplication
Compact exponential expression

The first two duos have been done for you.

Expanded	Compact
$(3x)(3x)(3x)$	$(3x)^3$
$3\cdot x\cdot x\cdot x$	$3x^3$
$(-5x)(-5x)$	
$-(5x)(5x)$	
	$(2a^2)^3$
	$2a^5$
$(x+1)(x+1)$	
	$(a+b)^2$
$\left(\frac{1}{2}\right)\left(\frac{1}{2}\right)\left(\frac{1}{2}\right)\left(\frac{1}{2}\right)$	
	$\left(-\frac{2}{3}\right)^2$
$(10ab)(10ab)$	
$6\cdot x\cdot x\cdot y\cdot y$	
$7\cdot m\cdot m\cdot m\cdot p$	
	$\left(\frac{a}{b}\right)^3$
$\left(\frac{1}{z}\right)\left(\frac{1}{z}\right)$	
$(a-2)(a-2)$	
$-(4x)(4x)(4x)$	
	$(-2.1)^3$
$\pi\cdot r\cdot r$	
	$\frac{4}{3}\pi r^3$

Student Activity B

Section 5.1: Match Up on Basic Exponent Rules

Match-up: Match each of the expressions in the squares of the grid below with an equivalent simplified expression from the top. If an equivalent expression is not found among the choices A through D, then choose E (none of these).

A x^5 **B** x^6 **C** $9x^4$ **D** $12x^6$ **E** None of these

$(6x^2)(2x^4)$	$4x^4-(-5x^4)$	$x^3 \cdot x^2$	$(-3x^3)(-3x)$
$6x^3+6x^3$	$-4x^5+5x^5$	$12(x^2)^3$	$(-3x^3)(-4x^3)$
$(x^2)(x^2)(x)$	$(27x^3)\left(\frac{1}{3}x\right)$	$(x^3)^2$	$(12x^3)^2$
$(3x^2)^2$	$3(-2x^3)^2$	$3x^3+6x$	$(5x^2)\left(\frac{1}{5}x^3\right)$
x^4+x	$x^4 \cdot x$	$(-3x^2)^2$	$8x^6+4x^6$

Guided Learning Activity

Section 5.2: Zero and Negative Exponents

To develop the concepts of zero and negative exponents, we will first look at several patterns of values.

1. Fill in the blanks to complete each sequence below:

$81, 27, 9, \_\_\_, \_\_\_, \_\_\_, \_\_\_$ How do you get the next term? ____________

$16, 8, 4, \_\_\_, \_\_\_, \_\_\_, \_\_\_$ How do you get the next term? ____________

$x^4, x^3, x^2, \_\_\_, \_\_\_, \_\_\_$ How do you get the next term? ____________

$4^4, 4^3, 4^2, \_\_\_, \_\_\_, \_\_\_$ How do you get the next term? ____________

$\frac{1}{125}, \frac{1}{25}, \frac{1}{5}, \_\_\_, \_\_\_, \_\_\_$ How do you get the next term? ____________

2. Now complete each table below by finishing the patterns:

81	3^4
27	3^3
9	3^2

16	2^4
8	2^3
4	2^2

$\frac{1}{125}$	
$\frac{1}{25}$	
$\frac{1}{5}$	
	5^2
	5^3

64	4^3
	4^2
	4^1

3. Based on your observations in the patterns above, finish the exponent rules below:

Rule for Zero Exponents: $x^0 = \_\_\_$ (for $x \neq 0$)

Rule for Negative Exponents: $x^{-1} = \_\_\_$ (for $x \neq 0$)

$x^{-2} = \_\_\_$

$x^{-3} = \_\_\_$

$x^{-n} = \_\_\_$

4. Apply the rules you have just developed to write each of these exponential expressions without zero or negative exponents.

$6^{-2} =$ $\quad 10^0 =$ $\quad y^{-4} =$ $\quad 2^{-5} =$ $\quad z^{-1} =$ $\quad (4a)^0$

Student Activity A

Section 5.2: Match Up on Trickier Exponent Rules

Match-up: Match each of the expressions in the squares of the grid below with an equivalent simplified expression from the top. If an equivalent expression is not found among the choices A through D, then choose E (none of these).

A 1 **B** $\frac{4}{x^2}$ **C** $9x^2y^3$ **D** $\frac{-9x^4}{y^3}$ **E** None of these

$(4x)^{-2}$	$(4x^{-2})^0$	$4x^{-2}$	$4x^0$
$\frac{-(3x^2y)^2}{y^5}$	$\frac{(-3xy)^2}{y^{-1}}$	$\frac{3^{-2}y^{-3}}{x^{-4}}$	$\left(\frac{x}{2}\right)^{-2}$
$8x^2\left(\frac{x^{-2}}{8}\right)$	$(9x^2y^3)^0$	$\left(\frac{x}{2}\right)^2$	$y^7\left(\frac{y^2}{3x}\right)^{-2}$
$4\left(\frac{1}{x^2}\right)^0$	$(2x)^{-2}$	$3(x^2y^2)(3x^2y^2)^{-1}$	$\frac{(2x^{-1}z^2)^2}{z^4}$
$\left(-\frac{y}{a^4b^4}\right)\left(\frac{3xab}{y}\right)^4$	$\left(\frac{100x^{27}y^{35}}{a^4b^5}\right)^0$	$(2yz)^2(xyz)^{-2}$	$\frac{-12x^4}{5}\left(\frac{5}{-12x^4}\right)$

Student Activity B

Section 5.2: Double the Fun on Exponent Rules

Directions: For many problems involving the simplification of exponents, there are at least two ways to tackle the problem. For each of the problems below, try the problem both ways – your answers should be the same for both methods. If they are not, you'll have to go back and look for a mistake. The first one has been done for you.

Simplify this expression	Apply exponent rules directly	First move factors to avoid working with negative exponents
$\dfrac{x^{-2}y^5}{x^{-6}y^{-1}}$	$\dfrac{x^{-2}y^5}{x^{-6}y^{-1}} = x^{-2-(-6)}y^{5-(-1)}$ $= x^{-2+6}y^{5+1} = x^4y^6$	$\dfrac{x^{-2}y^5}{x^{-6}y^{-1}} = \dfrac{x^6y^5y^1}{x^2} = x^{6-2}y^{5+1} = x^4y^6$
$\dfrac{16a^3b^{-2}}{8a^{-1}b^5}$		
$3x^{-2}x^3x^{-1}$		
$(2x)^{-2}(-6x^3)$		
$\left(\dfrac{a^4b}{ab^2}\right)^{-2}$		
$\left(\dfrac{x^{-2}y^5}{x^{-1}y^6}\right)^3$		
$\dfrac{(xy^3)^{-1}}{2x^4y^{-1}}$		

Student Activity C

Section 5.2: Mathematical Heteronyms

Directions: In writing, there are words that are spelled the same but have different pronunciations and different definitions; these are called heteronyms. Many mathematical expressions look similar but are really very different (almost like mathematical heteronyms). In each set of expressions below, pay close attention to the use of parentheses and the mathematical operations and notation.

1.

$3-1$	$3(-1)$	3^{-1}	$3-(-1)$	-3^{-1}	$3\div(-1)$

2.

$2-5$	$2(-5)$	2^{-5}	$2^{-1}-5$	$2^{-1}-5^{-1}$	$2-5^{-1}$

3.

x^2x^{-3}	$x^2(-x^3)$	x^2x^3	$x^{-2}x^3$	x^2-x^3	$(x^2)^{-3}$

4.

$2a^{-3}$	$(-2a)^3$	$2^3(-a)$	$2-a^3$	$(2a)^{-3}$	$(-2a)\cdot 3$

5.

$\left(\frac{3}{4}\right)^{-2}$	$\frac{3^{-2}}{4}$	$\left(\frac{3}{4}\right)(-2)$	$\frac{3}{4}-2$	$\frac{3}{4^{-2}}$	$\left(-\frac{3}{4}\right)^2$

Student Activity D

Section 5.2: Exponents Using a Calculator

When you input exponents into a calculator, you must be careful to tell the calculator which part is the base and which part is the exponent. Each calculator requires a specific set of keystrokes to evaluate exponential expressions and this can become quite complicated when the expression also involves negatives. We will fill out the table below, first performing each calculation by hand, then finding the proper keystrokes to do the evaluation on your calculator.

First you need to locate your exponent button. It may look like $\boxed{\wedge}$ or $\boxed{x^y}$.
Reminders: There is a difference between the minus key and the negative key on your calculator. Fractions need to be placed inside parentheses for proper evaluation.

	Expression	Evaluate by hand	Calculator Keystrokes to get equivalent result
a.	2^3		
b.	2^{-1}		
c.	5^{-2}		
d.	$\left(\frac{1}{2}\right)^3$		
e.	$\left(\frac{2}{3}\right)^{-2}$		
f.	-2^4		
g.	$(-2)^4$		
h.	-4^{-2}		
i.	$(-4)^{-2}$		
j.	$2^{-1}+3^{-1}$		
k.	$(2+3)^{-1}$		
l.	$5-2^{-1}$		

Guided Learning Activity

Section 5.3: Charting Scientific Notation

Scientific notation is based on using powers of 10. Fill in the Powers of 10 with their decimal notation in the table below. 10^3 has been done for you.

Power of 10	10^0	10^1	10^2	10^3	10^4	10^5	10^6
Value				1,000			

Power of 10	10^{-6}	10^{-5}	10^{-4}	10^{-3}	10^{-2}	10^{-1}	10^0
Value							

Scientific notation: $N \times 10^n$ where $1 \le N < 10$, and n is an integer.
With your class, fill out the table below. The first one has been done for you.

Number in standard notation (decimal notation)	Put together the $N \times 10^n$ part in pieces			
	N	2nd factor in decimal notation	2nd factor as a power of 10	$N \times 10^n$
0.00495	4.95	0.001	10^{-3}	4.95×10^{-3}
8,000,000,000				
4,002,000				
0.00000034				
				9.2×10^5
	6.843	0.0000001		
				8.04×10^{-7}
	2	10,000,000		

Rather than remembering whether the decimal place moves left or right, it is easier to remember that 10 raised to negative powers means you are looking at a number that is small (between zero and 1) and 10 raised to positive powers means you are looking at a number that is large (>1). Simply move the decimal point the correct number of digits in the direction that creates the appropriate bigger or smaller number!

Decimal Value	**Is it a *small* or *large* number?**	**Is the power on the 10 *positive* or *negative?***	**Scientific Notation**
0.0000725			
			1.2×10^{7}
			4.534×10^{-6}
84,000,000,000			
48			
			9×10^{-4}
674,000			
			2.99×10^{8}
			6.67×10^{-11}
0.000000056704			

Student Activity A

Section 5.3: Escape the Matrix

Directions: Begin at the box marked START. By shading in pairs of adjacent squares that represent equal numbers, you will eventually find the path to "escape" this matrix of boxes. The first "step" in the path and two of the middle steps in the path have been shaded for you.

START 0.00043	4.3×10^{-4}	5×10^{6}	500,000	7.6×10^{3}	760,000
4.3×10^{3}	4.3×10^{-3}	5,000,000	5.0×10^{7}	50×10^{6}	76,000
43×10^{3}	789×10^{-2}	7.89	0.023	2.3×10^{-2}	0.0076
39×10^{8}	390,000,000	3.90×10^{7}	2.3×10^{2}	23,000	7.6×10^{-4}
16×10^{4}	3.90×10^{8}	1,000,000	10×10^{5}	5.555×10^{-5}	0.00005555
1.6×10^{2}	0.00016	1×10^{5}	10×10^{3}	5.555×10^{5}	555,500
32,300	32.30×10^{5}	3.230×10^{6}	0.003230×10^{6}	3230	5.555×10^{4}
3.230×10^{4}	6.89×10^{4}	689	689×10^{-1}	323×10^{0}	2.02×10^{2}
323.0×10^{-2}	3.230	32.30	3.20×10^{1}	**ESCAPE** the Matrix	200

Student Activity B

Section 5.3: Scientific Notation Using a Calculator

There is a special button on most calculators that is used for scientific notation. Sometimes it is denoted with [EE] or [EXP]. This button stands for "$\times 10$ to the," so, for example we could enter 4×10^3 as [4] [EE] [3]. When using the scientific notation button, you **do not need to** multiply by 10 too.

1. First we need to find the scientific notation button on your calculator. Try entering the expression: 4×10^3 then press [ENTER] or [=]. You should see either 4000 or 4e3 or $4\ ^{03}$ or something like this. Ask your instructor if you are unsure about whether you have correctly located the scientific notation button.

Draw the series of keystrokes you used to enter 4×10^3 here:

2. Try another one: 2.5×10^{12} and press [ENTER] or [=]. This number is too big to display on most calculator screens, so you should see the calculator give the number to you in it's version of scientific notation, like 2.5e12, 2.5E12, or $2.5\ ^{12}$.

Draw how you entered 2.5×10^{12} **and** the way the number appears on the screen:

3. Now we try a number with a negative exponent. Key in the number3.4×10^{-3} followed by [ENTER] or [=]. This should display as 0.0034, 3.4e-3, or $3.4\ ^{-03}$. Remember to use the negative key and not the minus key to enter the negative exponent.

Draw the series of keystrokes you used to enter 3.4×10^{-3} here:

4. Try another one: 4.62×10^{-14} and press [ENTER] or [=]. This number is too big to display on most calculator screens, so you should see the calculator give the number to you in it's version of scientific notation, like 4.62e-14, 4.62E-14, or $4.62\ ^{-14}$.

Draw how you entered 4.62×10^{-14} **and** the way it appears on the screen:

The real usefulness of the scientific notation button becomes obvious when we begin to perform calculations that involve scientific notation.

For example, if we wanted to calculate $\frac{2.4\times10^{-3}}{8\times10^{-12}}$ **without** the scientific notation button, here's what I'd have to enter on *my* calculator:

$$\boxed{(}\ 2.4\ \boxed{\times}\ 10\ \boxed{\wedge}\ \boxed{(-)}\ 3\ \boxed{)}\ \boxed{\div}\ \boxed{(}\ 8\ \boxed{\times}\ 10\ \boxed{\wedge}\ \boxed{(-)}\ 12\ \boxed{)}$$

Using the scientific notation button, this is much easier:

$$2.4\ \boxed{EE}\ \boxed{(-)}\ 3\ \boxed{\div}\ 8\ \boxed{EE}\ \boxed{(-)}\ 12$$

5. Try the calculation $\frac{2.4\times10^{-3}}{8\times10^{-12}}$ on your calculator. (the answer is 3×10^{8})

Draw how you entered $\frac{2.4\times10^{-3}}{8\times10^{-12}}$ here:

For practice, try these calculations and write down the answers using the proper scientific notation. In other words, don't write an answer like $8e3$; write the answer as 8×10^{3} instead.

6. $\left(8.2\times10^{9}\right)\left(2\times10^{4}\right)$

7. $\frac{1.5\times10^{-4}}{2.5\times10^{3}}$

8. $\left(1.1\times10^{-5}\right)^{2}$

9. $\frac{\left(5.4\times10^{-6}\right)\left(2\times10^{3}\right)}{9\times10^{8}}$

10. $\frac{3.6\times10^{-4}}{\left(1.2\times10^{3}\right)\left(1.5\times10^{-10}\right)}$

Guided Learning Activity
Section 5.4: Language of Polynomials

A **polynomial** is a single term or a sum of terms in which all variables have whole-number exponents and no variable appears in a denominator. Recall that the **terms** of an algebraic expression are separated by addition (remember that subtraction can be rewritten as addition of a negative term). A polynomial with exactly one term is called a **monomial**; exactly two terms, a **binomial**; and exactly three terms, a **trinomial**. A polynomial can be written with one or more variables.

With your class, fill out the table below. The first one has been done for you.

Expression (rewrite with addition symbols if necessary)	**How many variables?**	**Classification**			
		Monomial	**Binomial**	**Trinomial**	**Polynomial**
$5x^2 - 2x + 7$ $5x^2 + (-2x) + 7$	1			X	X
$y - 2.5$					
$-5x^2y$					
$4x^2 - \frac{2}{x^3}$					
$b^3 + \frac{1}{3}b^2 - \frac{1}{2}b + 12$					
$x^2 - 3xy - 10y^2$					

Polynomials are often written in **descending powers** of the variable (the variable exponents decrease from left to right). When a polynomial is written in descending powers, the first term is called the **lead term**. The coefficient of the lead term is called the **lead coefficient**. Recall that a term that consists of a single number is called the **constant term**.

With your class, fill out the table below. The first one has been done for you.

Expression	Rewrite the expression in descending powers of x	Lead term	Lead coefficient	Constant term (if there is one)
$4x-x^2$	$-x^2+4x$	$-x^2$	-1	None
$3x+5-2x^2$				
$x^3+200x+300x^2$				
$7x^3-2-3x^4$				

The **degree of a term** of a polynomial in one variable is the value of the exponent on the variable. Thus, the degree of $7x^4$ is equal to 4. If the polynomial has more than one variable, the degree of a term is the *sum* of the exponents on the variables in that term. Thus, the degree of $7x^2y^3$ is 5. For a constant term, we can imagine an unwritten variable with a zero power, consider that 7 could be written as $7x^0$. Thus, the **degree of a nonzero constant** is zero. We can also discuss the **degree of a polynomial** which is the same as the degree of the highest degree term of the polynomial.

With your class, fill in the table below. The first one has been done for you.

Expression	Degree of...				Degree of polynomial
	1st term	2nd term	3rd term	4th term	
$5x^2-3x+7$	2	1	0	None	2
$y-2.5$					
$-5x^4y+2x^2y^2$					
$b^3+\frac{1}{3}b^2-\frac{1}{2}b+12$					
$x^2-3xy-10y^2$					
$x^3+300x^2+200x-1000$					
$-3v^4+7v^3-2$					

Student Activity A

Section 5.4: Match Up on Polynomial Evaluation

Remember that it may be helpful to first create a parentheses skeleton for each polynomial before you substitute the designated values. For example, $x^2 - 3x$ would be first rewritten as $(\quad)^2 - 3(\quad)$.

Match-up: Match each of the answers in the squares of the grid below with its value in choices A through D. If you do not see the value in choices A through D, then choose choice E (none of these).

A 9 **B** 1 **C** −2 **D** 0 **E** None of these

Evaluate $x^2 + 9x + \frac{23}{9}$ for $x = \frac{2}{3}$.	$y = 2.25x^3 + 7.75x - 1$ Find y when $x = 1$.	Evaluate $5x^2 - 5x + \frac{9}{4}$ for $x = \frac{1}{2}$.	Simplify: $3x^2 + 5 - 2x^2 - 7 - x^2$
Evaluate $\frac{-(x^3 - 3x)}{2}$ for $x = 3$.	Evaluate $5x^2 + 10x$ for $x = -2$.	Evaluate x^2y for $x = -1$ and $y = -2$.	Evaluate $x^2 + 2x - 1$ for $x = -1$.
$y = x^2 - 2$ Find y when $x = 0$.	$y = 3x^2 - 2x - 7$ Find y when $x = 2$.	Evaluate $9x^4 - 9x^3 + 9x^2 - 9x + 9$ for $x = 1$.	Evaluate $x^3 + x^2 + x - 2$ for $x = -2$
Simplify: $\frac{x^5}{2} + 3 - 2x^5 + \frac{3x^5}{2} + 6$	$y = x^3 - 3.3x^2 + 1$ Find y when $x = 3.3$.	Evaluate $x^3 - 1$ for $x = 1$.	Evaluate $x^3 - 1$ for $x = -1$.

Student Activity B

Section 5.4: Caught in the Net

Directions: On each of the graphing grids below, there are several fish. For each problem there are two equations to graph on the grid (these represent the "nets"). See which of the fish get "trapped" (enclosed) by the nets and which escape. Give the letters of the fish that get trapped.

Problem 1: Graph $y = x^2 - 4$ and $y = \frac{1}{2}x + 5$

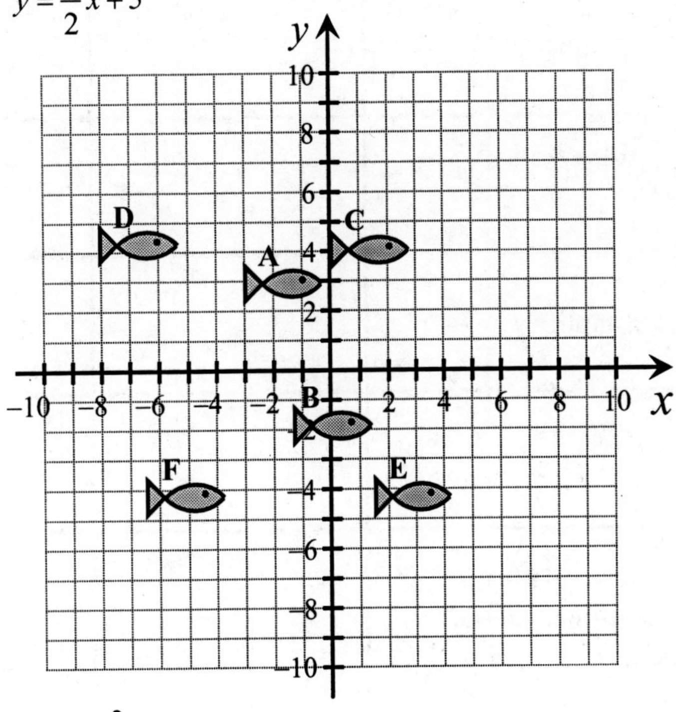

Problem 2: Graph $y = \frac{1}{9}x^3 + 1$ and $y = -\frac{3}{4}x^2 + 10$

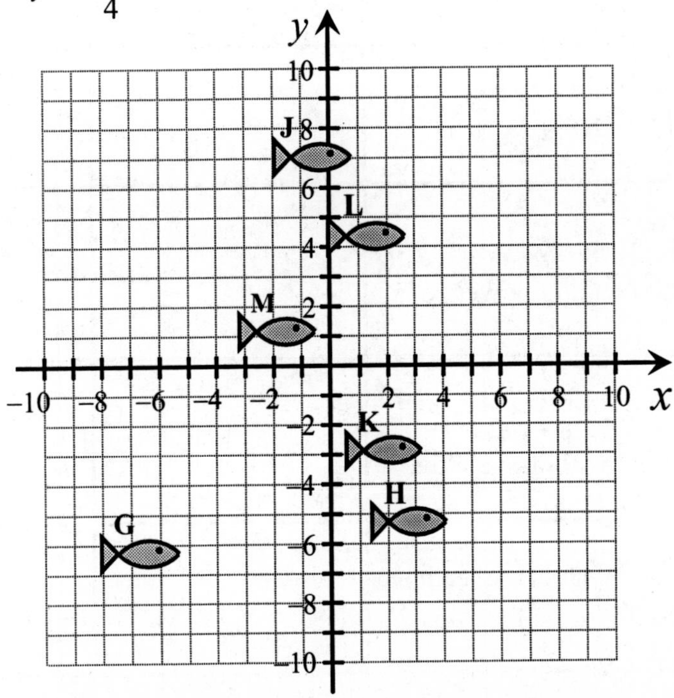

Assessment 5B

Chapter 5: Mid-chapter Assessment for Understanding

For each of the following, describe the type of problem and the strategies and key steps to remember while doing the problem. You do **not** have to complete the problems.

	Type of Problem	Strategies and Key Steps
1. Evaluate: 3^{-2}.		
2. What is the degree of $12x^2y$?		
3. Simplify: $\frac{b^2 \cdot b^5}{b}$.		
4. Simplify: $(3x^{-2})^3(2x^4)$.		
5. Write 0.000027 in scientific notation.		
6. Simplify: $5x^0$.		
7. Multiply: $(3.1\times10^5)(2\times10^{-3})$.		
8. Write $6-5x^2+3x$ in descending powers of x.		
9. Simplify $\frac{6x^{-3}}{-2x^2}$ and write without negative exponents.		
10. What is the leading coefficient of x^2+5x?		

Guided Learning Activity

Section 5.5: Vertical Form of Polynomial Addition and Subtraction

The expressions $342+605$ and $\left(3x^4+4x+2\right)+\left(6x^2+5\right)$ are simplified in almost the same way: For both expressions, we line up the like terms vertically; then add.

Example 1:

$$\begin{array}{rrr} & 3 & 4 & 2 \\ + & 6 & 0 & 5 \\ \hline & 9 & 4 & 7 \end{array} \qquad \begin{array}{rlll} & \text{3 hundreds} & \text{4 tens} & \text{2 ones} \\ + & \text{6 hundreds} & \text{0 tens} & \text{5 ones} \\ \hline & \text{9 hundreds} & \text{4 tens} & \text{7 ones} \end{array} \qquad \begin{array}{rrrr} & 3x^2 & +4x & +2 \\ + & 6x^2 & +0x & +5 \\ \hline & 9x^2 & +4x & +7 \end{array}$$

One of the differences between the vertical addition of the expressions is that we cannot "carry" coefficients greater than 9 to the next column like we carry numbers greater than 9. Another difference is that we can have negative coefficients if that is part of the original expression.

Example 2: $\left(4x^2+5x+9\right)+\left(-2x^2+5x-6\right)$ becomes

$$\begin{array}{rrrr} & 4x^2 & +5x & +9 \\ + & -2x^2 & +5x & -6 \\ \hline & 2x^2 & +10x & +5 \end{array}$$

Now try these. Make sure to **line up the like terms** in the same columns and keep the signed coefficients with the terms in the columns.

a. $\left(y^3+3y^2-4y+2\right)+\left(9y^2+4y-7\right)$ **b.** $\left(a^2+4ab-2b^2\right)+\left(a^2+8ab-3b^2\right)$

In subtraction, we have to remember that the second row of each column is being subtracted from the first row **for every entry**. Below we look at the similar expressions $749-615$ and $\left(7x^2+4x+9\right)-\left(6x^2+x+5\right)$.

Example 3:

$$\begin{array}{rrrr} & 7 & 4 & 9 \\ - & 6 & 1 & 5 \\ \hline & 1 & 3 & 4 \end{array} \qquad \begin{array}{rlll} & \text{7 hundreds} & \text{4 tens} & \text{9 ones} \\ - & \text{6 hundreds} & \text{1 ten} & \text{5 ones} \\ \hline & \text{1 hundred} & \text{3 tens} & \text{4 ones} \end{array} \qquad \begin{array}{rrrr} & 7x^2 & +4x & +9 \\ - & 6x^2 & +1x & +5 \\ \hline & 1x^2 & +3x & +4 \end{array}$$

We've learned on many occasions that subtraction of grouped terms can be tricky as the -1 must be distributed to the grouped terms. Because some of the terms in the second row have a + in front of the coefficients or constants, it is confusing to think of this as a subtraction problem. To be clear, we should write the vertical format of subtraction like this:

$$\begin{array}{rrrr} & 7x^2 & +4x & +9 \\ - & (6x^2 & +x & +5) \\ \hline \end{array}$$

And since polynomial subtraction can be rewritten without the grouping as polynomial addition (after the distribution), we can further alter the problem:

$$\begin{array}{rrr} 7x^2 & +4x & +9 \\ -\ (6x^2 & +x & +5) \\ \hline \end{array} \quad \text{becomes} \quad \begin{array}{rrr} 7x^2 & +4x & +9 \\ +\ -6x^2 & -x & -5 \\ \hline 1x^2 & +3x & +4 \end{array}$$

Example 4: Let's go through this process with $(4x^2+3x-6)-(3x^2-2)$

First we line up the like terms, still using the grouping symbols for the subtraction:

$$\begin{array}{rrr} 4x^2 & +3x & -6 \\ -\ (3x^2 & +0x & -2) \\ \hline \end{array} \quad \text{becomes} \quad \begin{array}{rrr} 4x^2 & +3x & -6 \\ +\ -3x^2 & -0x & +2 \\ \hline 1x^2 & +3x & -4 \end{array}$$

Now try these. Write the polynomial subtraction in the vertical format, first with grouping symbols as subtraction, then performing the distribution by -1 and rewriting as addition.

c. $(y^3+9y^2-4y+2)-(6y^2+2y-7)$

d. $(a^2+4ab-2b^2)-(a^2-8ab+3b^2)$

e. $(6x^3-5x-4)-(2x^2-8x+3)$

Student Activity A

Section 5.5: When Does Order Matter?

Directions: Add or subtract in each of the expressions below and then combine the like terms. Write the simplified expression in descending powers of x.

1. Add: $(x^2+3x-7)+(4x^2+5x-2)$

2. Add: $(4x^2+5x-2)+(x^2+3x-7)$

3. What lesson can we learn from problems **1** and **2**?

---

4. Subtract: $(x^2+3x-7)-(4x^2+5x-2)$

5. Subtract: $(4x^2+5x-2)-(x^2+3x-7)$

6. What lesson can we learn from problems **4** and **5**?

---

7. Subtract: $(x^2+3x-7)-(4x^2+5x-2)$

8. Simplify: $x^2+3x-7-4x^2+5x-2$

9. What lesson can we learn from problems **7** and **8**?

---

10. Add: $(x^2+3x-7)+(4x^2+5x-2)$

11. Simplify: $x^2+3x-7+4x^2+5x-2$

12. What lesson can we learn from problems **10** and **11**?

Guided Learning Activity

Section 5.6: Vertical Form of Polynomial Multiplication

The expressions $304 \cdot 21$ and $(3x^2+4)+(2x+1)$ are simplified in almost the same way: For both expressions, we line up the like terms vertically; then multiply.

Example 1:

		3	0	4
×			2	1
		3	0	4
	6	0	8	
	6	3	8	4

		3 hundreds	0 tens	4 ones
×			2 tens	1 one
		3 hundreds	0 tens	4 ones
	6 thousands	0 hundreds	8 tens	
	6 thousands	3 hundreds	8 tens	4 ones

		$3x^2$	$+0x$	$+4$
×			$+2x$	$+1$
		$+3x^2$	$+0x$	$+4$
	$6x^3$	$+0x^2$	$+8x$	
	$6x^3$	$+3x^2$	$+8x$	$+4$

One of the differences between the vertical multiplications of the expressions is that we cannot "carry" coefficients greater than 9 to the next column like we carry numbers greater than 9. Another difference is that we can have negative coefficients if that is part of the original expression.

Example 2: $(4x+5)(2x-6)$ becomes

		$4x$	$+5$
×		$2x$	-6
		$-24x$	-30
	$8x^2$	$+10x$	
	$8x^2$	$-14x$	-30

Now try these. Make sure to **line up the like terms** in the same columns and keep the signed coefficients with the terms in the columns.

a. $(7y-6)(y+5)$

b. $(x^2-3x+6)(x-9)$

c. $(a^2+8a+6)(a^2+5)$

d. $(2x^2+5x-4)(x^2-6x+7)$

Student Activity A

Section 5.6: Monomial Addition and Multiplication Tables

Here are simple addition and multiplication tables with monomial inputs.

Addition:

$+$	x	$2x$	x^2	$2x^2$
x	$2x$	$3x$		
$2x$	$3x$	$4x$		
x^2			$2x^2$	$3x^2$
$2x^2$			$3x^2$	$4x^2$

Multiplication:

$\bullet$	x	$2x$	x^2	$2x^2$
x	x^2	$2x^2$	x^3	$2x^3$
$2x$	$2x^2$	$4x^2$	$2x^3$	$4x^3$
x^2	x^3	$2x^3$	x^4	$2x^4$
$2x^2$	$2x^3$	$4x^3$	$2x^4$	$4x^4$

Notice that there are shaded spaces left in the addition table where unlike terms can not be combined.

Directions: The following tables are addition or multiplication tables involving monomial inputs. Fill in the missing squares with the appropriate monomials. Write NL (for "not like") or shade the grid spaces where the terms *cannot* be combined.

$+$	$-3x^2$	$-x^2$	$-3x$	$-x$	0	x	$3x$	x^2	$3x^2$
$-3x^2$									
$-x^2$									
$-3x$									
$-x$									
0									
x									
$3x$									
x^2									
$3x^2$									

•	$-3x^2$	$-x^2$	$-3x$	$-x$	0	x	$3x$	x^2	$3x^2$
$-3x^2$									
$-x^2$									
$-3x$									
$-x$									
0									
x									
$3x$									
x^2									
$3x^2$									

Directions: The following tables are addition or multiplication tables with missing information. Fill in the missing squares with the appropriate monomials. Write NL for "not like" in the squares where unlike terms cannot be added.

+	$7x$				x^2
$5x$				$-2x$	
	$7x$	$-2x^2$	0		x^2
		$-6x^2$	$-4x^2$		$-3x^2$
$-4x$				$-11x$	
			$6x^2$		$7x^2$

•		x^6		
x^4	$-5x^6$	x^{10}		
	$-15x^7$	$3x^{11}$		$24x^8$
$-7x$	$35x^3$			
		x^6		$8x^3$
		x^7	$2x^2$	

Student Activity B

Section 5.6: Multiplying Binomials with FOIL

When we multiply two **binomials**, we can use a shortcut method called the FOIL method. FOIL stands for **F**irst **O**uter **I**nner **L**ast. The first one has been done for you.

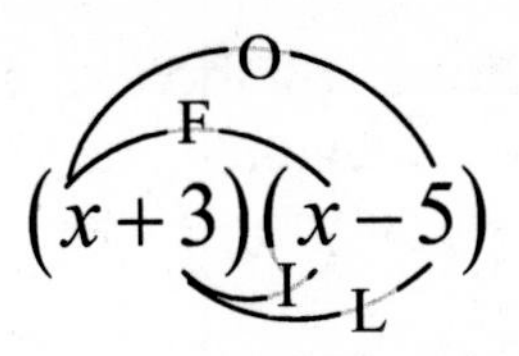

Expression	Find these terms of the multiplication...				Combine like terms and simplify
	First	Outer	Inner	Last	
$(x+3)(x-5)$	x^2	$-5x$	$3x$	-15	$x^2-2x-15$
$(u-4)(u-8)$					
$(y+3)(y+3)$					
$(x+7)(x-7)$					
$\left(a+\frac{1}{3}\right)\left(a+\frac{2}{3}\right)$					
$(2x+5)(3x-4)$					
$(4x+3)(6x+7)$					
$(x^2+4)(x^2+6)$					
$\left(2w+\frac{1}{2}\right)\left(w-\frac{3}{2}\right)$					
$(s^3-2)(s^3-8)$					
$(3.6x+1)(2x-4.5)$					

Student Activity C

Section 5.6: Not All Multiples are the Same

Directions: All the expressions below are written in triples. First multiply and simplify each expression. Then circle the pairs from each problem that are really equivalent (if there are any).

1. $(x+3)(x-5)$ $\qquad$ $(x-5)(x+3)$ $\qquad$ $(x-3)(x+5)$

2. $(x-2)(x+6)$ $\qquad$ $(x+2)(x-6)$ $\qquad$ $(x-6)(x+2)$

3. $(x-2)(x-2)$ $\qquad$ $(2-x)(2-x)$ $\qquad$ $(x-2)(2-x)$

4. $(x+3)(x+4)$ $\qquad$ $(3+x)(4+x)$ $\qquad$ $(x-3)(x-4)$

5. $(x+2)(x^2-2x+4)$ $\qquad$ $(x-2)(x^2+2x-4)$ $\qquad$ $(x-2)(x-2)(x-2)$

6. $(x+5)(x-5)$ $\qquad$ $(x-5)^2$ $\qquad$ $(5+x)(5-x)$

Student Activity A

Section 5.7: Charting Squared Binomials

When we **square a binomial**, we are really just multiplying two binomials and can use the FOIL method to do it. For example, $(x-6)^2$ is really the same as $(x-6)(x-6)$. Fill in the empty boxes in the table below. The first one has been done for you.

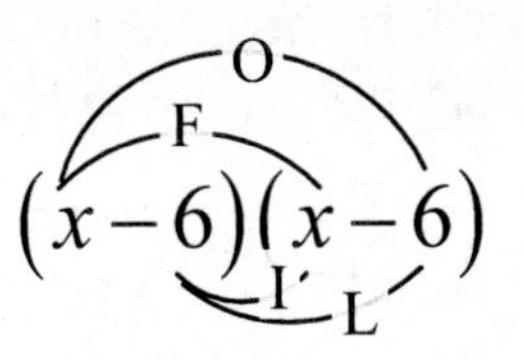

Expression	Find these terms of the multiplication...				Combine like terms and simplify
	First	Outer	Inner	Last	
$(x-6)^2$ $(x-6)(x-6)$	x^2	$-6x$	$-6x$	36	$x^2-12x+36$
$(u+4)^2$ $(u+4)(u+4)$					
$(a-5)^2$ $(a-5)(a-5)$					
$(x+\frac{1}{2})^2$ $(x+\frac{1}{2})(x+\frac{1}{2})$					
$(A+B)^2$ $(A+B)(A+B)$					
$(A-B)^2$ $(A-B)(A-B)$					

Notice that in the first row, $-6x-6x=-12x$ is the same as $2(-6x)=-12x$. If you look back through the rows in the table above, you should see a similar pattern in every row.

We can thus use the formula below to determine the product of a binomial squared.

$$(A+B)^2 = A^2+2\cdot A\cdot B+B^2$$

The middle term is found by taking twice the product of A and B. Make sure that you use parentheses when you are squaring the A-term and the B-term! For example, the term like $2x^2$ is not the same as $(2x)^2$.

Directions: Using the binomial squared formulas, fill in the empty boxes in the table below. The first one has been done for you.

Expression	Find these terms of binomial squared.			Final expression
	The square of the first term: A^2	Twice the product of the first and second terms: $2 \cdot A \cdot B$	The square of the last term: B^2	$A^2 + 2AB + B^2$
$(x-6)^2$	$(x)^2$	$2(x)(-6)$	$(-6)^2$	$x^2 - 12x + 36$
$(u+4)^2$				
$(a-5)^2$				
$\left(x+\frac{1}{2}\right)^2$				
$(2y+5)^2$				
$(3a-1)^2$				
$(5x+4y)^2$				
$(9-u)^2$				
$(v^3+2)^2$				
$\left(6z-\frac{4}{3}\right)^2$				
$(x^4-y^3)^2$				

Student Activity B

Section 5.7: Charting a Sum-Difference

There's one more special product to look at, but again, we'll start by going back to the FOIL method to figure it out. Fill in the empty boxes in the table below. The first one has been done for you.

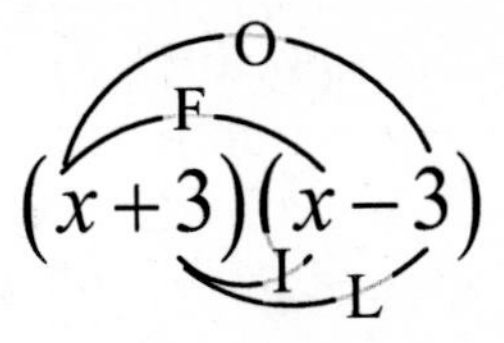

Expression	Find these terms of the multiplication...				Combine like terms and simplify
	First	Outer	Inner	Last	
$(x+3)(x-3)$	x^2	$-3x$	$3x$	-9	x^2-9
$(x-3)(x+3)$					
$(x-3)(x-3)$					
$(x+3)(x+3)$					
$(a-5)(a+5)$					
$(a+5)(a-5)$					
$(a-5)(a-5)$					
$(a+5)(a+5)$					
$(x+\frac{1}{2})(x-\frac{1}{2})$					
$(A+B)(A-B)$					

Circle all the rows where the inner and outer terms were additive inverses (where their sum was zero). Did it happen in every row? ____ What is special about the beginning expressions where the middle terms **were** additive inverses?

____

Student Activity C

5.7 Visual Representation for Multiplying Binomials

Multiplication can be represented using a diagram called an area model. Find an expression for the area of each of the squares and rectangles below. Remember that the formula for finding the area of a rectangle is Area = (length)(width). Write the expression for the area inside each square or rectangle, and then write it inside the same square or rectangle in the composite figure below. **Thus each area will be written in two places.**

Problem 1:

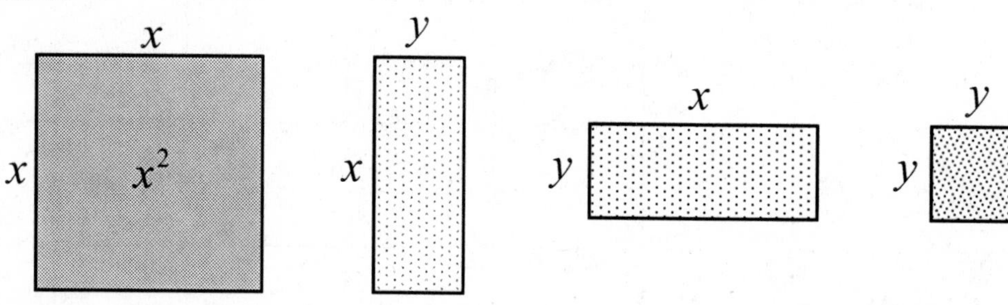

To find the area of the composite rectangle, we could either add the areas of all the smaller pieces, which would give this area:

____

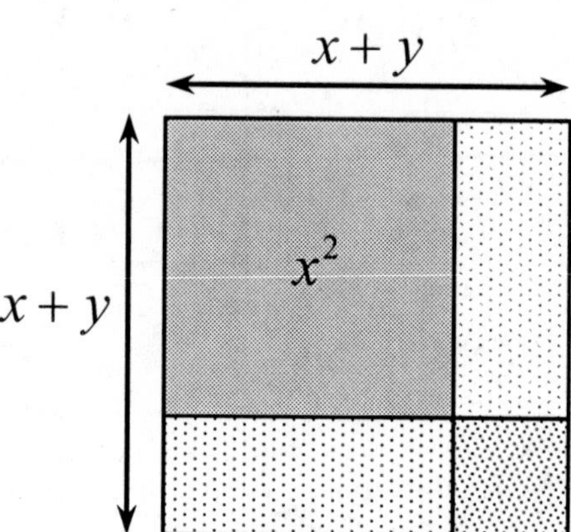

Or we could find the area by using the area formula, which would give

Area $=(x+y)(x+y)=(x+y)^2$

This tells us that $(x+y)^2 =$ ____________________.

Problem 2: Do this one the same way.

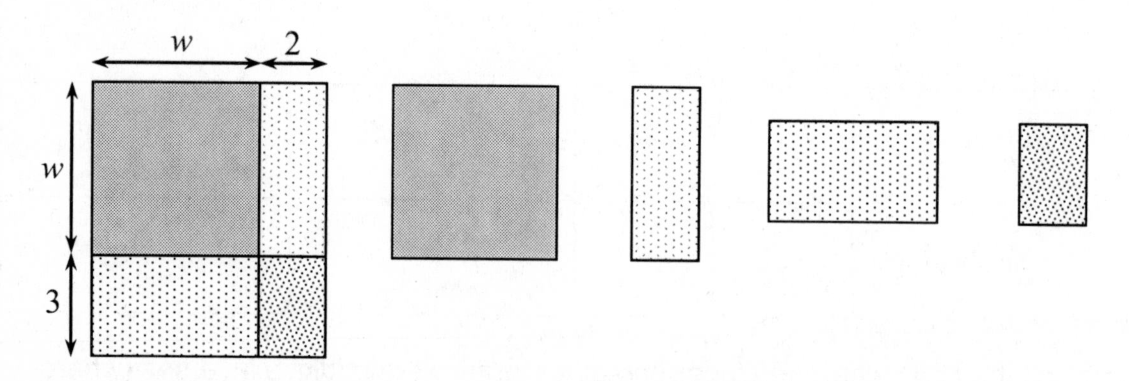

What mathematical equation does this set of figures tell us?

________________ = ______________________

Student Activity D

Section 5.7: Exponential and Polynomial Heteronyms

Directions: In writing, there are words that are spelled the same but have different pronunciations and different definitions; these are called heteronyms. Many polynomial and exponential expressions look similar but are really very different (almost like mathematical heteronyms). In each set of expressions below, pay close attention to the use of parentheses and the mathematical operations and notation.

1.

$(3x)^2$	$2(3x)$	$2(x+3)$	$(x+3)^2$	$(2x)^3$

2.

$(x-4)^2$	$2(x-4)$	$2(-4x)$	$(-2x)^4$	$(4x)^2$

3.

$(x+3)(x-2)$	$(3x)(-2x)$	$3x(x-2)$	$-2x(x+3)$	$x^3(-2x)$

4.

$(x+5)(x-5)$	$(5x)(x-5)$	$5x(-5x)$	$x^5(x-5)$	$x^5 \cdot x^5$

5.

$5x^2(4x^3)$	$(5+2x)(4x+3)$	$5x^2(4x+3)$	$(5x)^2(4x^3)$	$(5x)^2(4x+3)$

Guided Learning Activity

Section 5.8: Polynomial Long Division

The expressions $168 \div 14$ and $\left(1x^2+6x+8\right)+\left(1x+4\right)$ are simplified in almost the same way using long division. For both expressions, we line up the like terms vertically; then divide.

Example 1:

$$
\begin{array}{r}
12 \\
14\overline{)\,168} \\
-14 \\
\hline
28 \\
-28 \\
\hline
0
\end{array}
\qquad\qquad
\begin{array}{r}
1x \quad +2 \\
1x+4\overline{)\,1x^2 \quad +6x \quad +8} \\
-(1x^2 \quad +4x) \quad\quad \\
\hline
2x \quad +8 \\
-(2x \quad +8) \\
\hline
0
\end{array}
$$

Notice that in both long division problems, the like terms line up in the same columns.

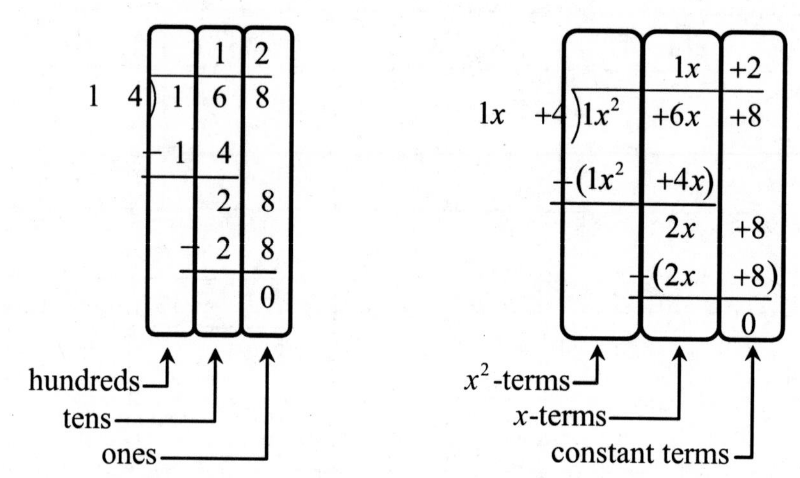

As you perform the operations in polynomial long division, you will want to make sure that you line up the like terms as well.

If you are missing a column of terms, you will need to insert a placeholder. For example, $12\overline{)1008}$ would not give the same quotient as $12\overline{)18}$, the two zeros in 1008 act as placeholders for the tens place and the hundred place. Also, we would not change the order of the digits in the columns, for example $12\overline{)1008}$ is not the same as $12\overline{)8010}$.

In a similar fashion, we would not write $x+2\overline{)x^3+8}$ because the x-term and the x^2-term columns are missing. Instead we write $x+2\overline{)x^3+0x^2+0x+8}$ to hold the places of the necessary columns.

In polynomial long division, the terms of the dividend and the divisor must be written in descending order with no missing terms.

1. Rewrite these long division problems so that all the columns are accounted for and the terms are in the proper order. Write in any unwritten ones. For example, write x^2 as $1x^2$.

a. $x-2\overline{)x^3-4x+5}$ **b.** $x-1\overline{)4-5x+x^2}$ **c.** $2+a\overline{)a^2-4}$

Here is a polynomial long division example shown step by step. Your instructor will walk you through the steps. You should add your own notes so that you will remember what has happened on each step. At the very least, highlight the new part of each step in color so that it is easier to see later on when you go back through your notes.

Example 2:

$$\begin{array}{r} x \\ x+8\overline{)x^2+5x-24} \end{array}$$

$$\begin{array}{r} x \\ x+8\overline{)x^2+5x-24} \\ \underline{x^2+8x} \end{array}$$

$$\begin{array}{r} x \\ x+8\overline{)\ x^2+5x-24} \\ \underline{-\left(x^2+8x\right)} \end{array}$$

$$\begin{array}{r} x \\ x+8\overline{)\ x^2+5x-24} \\ \underline{-x^2-8x} \end{array}$$

$$\begin{array}{r} x \\ x+8\overline{)\ x^2+5x-24} \\ \underline{-x^2-8x} \\ -3x \end{array}$$

$$\begin{array}{r} x \\ x+8\overline{)\ x^2+5x-24} \\ \underline{-x^2-8x} \\ -3x-24 \end{array}$$

$$\begin{array}{r} x\ \ -3 \\ x+8\overline{)\ x^2+5x-24} \\ \underline{-x^2-8x} \\ -3x-24 \end{array}$$

$$\begin{array}{r} x\ \ -3 \\ x+8\overline{)\ x^2+5x-24} \\ \underline{-x^2-8x} \\ -3x-24 \\ \underline{-3x-24} \end{array}$$

$$\begin{array}{r} x\ \ -3 \\ x+8\overline{)\ x^2+5x-24} \\ \underline{-x^2-8x} \\ -3x-24 \\ \underline{-(-3x-24)} \end{array}$$

$$\begin{array}{r} x\ \ -3 \\ x+8\overline{)\ x^2+5x-24} \\ \underline{-x^2-8x} \\ -3x-24 \\ \underline{+3x+24} \end{array}$$

$$\begin{array}{r} x\ \ -3 \\ x+8\overline{)\ x^2+5x-24} \\ \underline{-x^2-8x} \\ -3x-24 \\ \underline{+3x+24} \\ 0 \end{array}$$

Usually, in practice, we do not write this step:

$$x+8\overline{)\;x^2+5x-24}\quad \text{quotient } x,\quad \underline{-(x^2+8x)}$$

We skip directly from

$$\begin{array}{r} x \\ x+8\overline{)x^2+5x-24} \\ \underline{x^2+8x} \end{array}$$

to

$$\begin{array}{r} x \\ x+8\overline{)\;x^2+5x-24} \\ \underline{-x^2-8x} \\ -3x \end{array}$$

by changing the signs.

Change the signs using a colored pencil, so that you can see that you have done this step *after* writing the terms for the original row.

Directions: Try each of the problems below. Stop after each problem and wait for your instructor to go over the answer. If you make any mistakes, correct your mistakes using a colored pencil so that you will see the corrections (and original mistakes) when you go back to study later on.

2. $(2x^2-5x-12)\div(x-4)$

3. $(7x^2+x^3-28+5x)\div(x+4)$

If there is a remainder after the long division is done, we write this as a ratio with the divisor. Consider the following example that shows you why we do it this way.

Example 3:

$\frac{25}{4}=6\frac{1}{4}$ or we could write it like this

$$\begin{array}{r} 6 \\ 4\overline{)\;25} \\ \underline{-24} \\ 1 \end{array}$$

, which gives the result $6 \text{ R } 1$.

If we want to write $6\text{R}1$ mathematically, it should be $6\frac{1}{4}$, which is the same as $6+\frac{1}{4}$.

Example 4:

Likewise, when we do the following division, we get a remainder:

$$\begin{array}{r} x\ +3 \\ x-5\overline{)\ x^2-2x-16} \\ \underline{-x^2+5x}\quad\;\; \\ 3x-16 \\ \underline{-3x+15} \\ -1 \end{array}$$

So we write the answer: $x+3+\dfrac{-1}{x-5}$ $\begin{array}{l}\leftarrow \text{remainder} \\ \leftarrow \text{divisor}\end{array}$

Directions: Try each of the problems below. Stop after each problem and wait for your instructor to go over the answer. If you make any mistakes, correct your mistakes using a colored pencil so that you will see the corrections (and original mistakes) when you go back to study later on.

4. $\left(x^2+4x-8\right)\div\left(x+6\right)$

5. $\left(x^3+4x^2-5x-22\right)\div\left(x+4\right)$

Student Activity A

Section 5.8: Match Up on Polynomial Division

Match-up: Match each of the expressions in the squares of the grid below with an equivalent simplified expression from the top. If an equivalent expression is not found among the choices A through D, then choose E (none of these). Be careful – you may have to simplify rational expression terms.

A $x+4$ **B** $2x-3$ **C** $x+4+\dfrac{3}{x-1}$ **D** $2x-3-\dfrac{4}{x+3}$ **E** None of these

$x-1\overline{)x^2+3x-1}$	$x+7\overline{)x^2+11x+28}$	$x+3\overline{)x^2+7x+8}$
$x+3\overline{)2x^2+3x-9}$	$(4x^2+6x-26)\div(2x+6)$	$(x^2+7x+16)\div(x+3)$
$3x+9\overline{)6x^2+9x-39}$	$(10x^2+30x-10)\div(10x-10)$	$18x+9\overline{)18x^2+81x+36}$
$(x^2-16)\div(x-4)$	$\dfrac{6x^2y+2x^3y}{2x^2y}$	$\dfrac{36a^2b^2cx-54a^2b^2c}{18a^2b^2c}$

Student Activity B

Section 5.8: Pay Attention!

Directions: Many problems in this chapter look very similar. You **must** pay attention to the operation and directions for the problems. Try the problems below, but please, **pay attention** to what you're doing!

1. Multiply: $(x^2+5x-14)(x+7)$

2. Add: $(x^2+5x-14)+(x+7)$

3. Divide using long division: $(x^2+5x-14)\div(x+7)$

4. Subtract: $(x^2+5x-14)-(x+7)$

5. Add: $(27x^3+12x^2)+(3x^2)$

6. Divide: $(27x^3+12x^2)\div(3x^2)$

7. Multiply: $(27x^3+12x^2)(3x^2)$

8. Subtract: $(27x^3+12x^2)-(3x^2)$

Student Activity C

Section 5.8: Language of Polynomial Operations

Directions: We have worked with the language of algebra before. Now that we've made it through the whole chapter, let's practice the same language using polynomials. You do not have to carry out the mathematics in each problem; simply translate the sentence into mathematical notation. The first one has been done for you.

	Directions in words	**Write the expression with mathematical notation, using the proper grouping terms, fraction bars, or long division notation where necessary.**
1.	Find the sum of $x+4$ and $x-1$.	$(x+4)+(x-1)$
2.	Divide x^2-3x+2 by $x-1$.	
3.	Find the difference of x^2-2x-7 and x^2-4.	
4.	Find the quotient of $-6x^2y^3$ and $2x^2y$.	
5.	Find the product of x^2+4x+3 and $x+5$.	
6.	Find the product of $x+6$ and $x-9$.	
7.	Find the total of $x+3$, $x+4$, and $x-2$.	
8.	Find twice the sum of x^2 and x^2-4.	
9.	Find the quotient of $10a^3b-25a^4$ and $5a^2$.	
10.	What is $x^2-10x+16$ less $2x+4$?	

Assessment 5C

Chapter 5: End-of-Chapter Assessment for Understanding

For each of the following, describe the type of problem and the strategies and key steps to remember while doing the problem. You do **not** have to complete the problems.

	Type of Problem	Strategies, Rules and Key steps
1. Multiply: $(3x-5)^2$.		
2. Divide: $(4x^2-9)\div(2x+3)$.		
3. Multiply: $(x-5)(x-6)$.		
4. Simplify: $\left(\frac{6x^0}{x^3}\right)^{-2}$.		
5. Subtract: $(2x^2-4x)-(3x-7)$.		
6. Multiply: $(12+a)(12-a)$.		
7. Divide: $(14x^3-21x^2+28x)\div(7x)$.		
8. Simplify $\frac{(-4x)^2}{(3x)(2x^3)}$.		
9. Multiply: $(x+3)(x^2-6x+8)$.		
10. Evaluate $y=x^3-4x^2+12$ for $x=-3$.		

Assessment 5D

Chapter 5: Metacognitive Skills Assessment

Metacognitive skills refer to the ability to judge how well you have learned something and to effectively direct your own learning and studying. This is a self evaluation tool designed to help you focus your studying and to improve your metacognitive skills with regards to this math class.

Fill the 1st column out before you begin studying.
Fill the 2nd column out after you study and before you take the test.
Go back to this page after your test and circle any of the ratings that you would now change – this identifies the "disconnects" between what you think you know well and what you actually know well.

Use the scale below to assign a number to each topic.
5 *I am confident I can do any problems in this category correctly.*
4 *I am confident I can do most of the problems in this category correctly.*
3 *I understand how to do the problems in this category, but I still make a lot of mistakes.*
2 *I feel unsure about how to do these problems.*
1 *I know I don't understand how to do these problems.*

Topic or Skill	Before Studying	After Studying
Understanding what an exponent represents (what does 2^5 mean?).		
Understanding how expressions like $(-3)^2$ and -3^2 OR $2x^2$ and $(2x)^2$ are different.		
Knowing the product, quotient and power rules for exponents.		
Correctly applying the product, quotient, and power rules for exponents.		
Knowing the exponent rules for zero and negative exponents.		
Rewriting an expression to eliminate the negative exponents.		
Simplifying exponential expression involving negative exponents.		
Simplifying exponential expressions that have zero exponents.		
Simplifying expressions involving many exponent rules all mixed up (terms in parentheses, fractions, negative exponents, zero exponents, etc.).		
Converting back and forth between decimal notation and scientific notation.		
Performing operations (like multiplication and division) using scientific notation		
Using the scientific notation button on a calculator to perform calculations involving scientific notation.		
Identifying the number of terms in a polynomial.		
Writing a polynomial in descending order.		
Identifying the degree or coefficient of a term or of a polynomial.		

Continued on next page.

Use the scale below to assign a number to each topic.

5 *I am confident I can do any problems in this category correctly.*

4 *I am confident I can do most of the problems in this category correctly.*

3 *I understand how to do the problems in this category, but I still make a lot of mistakes.*

2 *I feel unsure about how to do these problems.*

1 *I know I don't understand how to do these problems.*

Topic or Skill	**Before Studying**	**After Studying**
Evaluating a polynomial for a given value (especially negative values).		
Graphing an equation that involves x^2 or x^3 by making a table of values and plotting points.		
Adding or subtracting polynomials.		
Multiplying a monomial by a polynomial. Example: $3x^2(x^2-2x+4)$.		
Multiplying two binomials using FOIL. Example: $(x+3)(x+7)$		
Multiplying any two polynomials. Example: $(x+3)(x^2-4x+7)$		
Multiplying a special sum-difference product. Example: $(a+7)(a-7)$		
Squaring a binomial. Example: $(x-4)^2$ (be careful, the answer's **not** x^2-16)		
Distinguishing between addition and multiplication, because the rules are not the same! For example, $(x+5)^2$ vs. $(5x)^2$ OR $(x+2)(x+3)$ vs. $(x+2)+(x+3)$.		
Dividing a polynomial by a monomial. Like $(6x^3+3x^2-9x)\div 3x$		
Using polynomial long division. Including what to do with a remainder! Example: $(x^2-6x+8)\div(x-2)$.		
Knowing how to make adjustments in polynomial long division when there's a term missing.		

Student Workbook: Chapter 6

Table of Contents: *Factoring and Quadratic Equations*

Assessment 6A

Pretest and Diagnostic Tool: Factoring and Quadratic Equations

Directions: Complete this assessment without looking back at your notes or your book. **Do not use a calculator for this assessment.**

For problems 1-2, find the largest integer that divides both numbers exactly (with no remainder).

1. 16 and 24? _____

2. 27 and 36? _____

For problems 3-5, true or false?

3. $4x+20=4(x+5)$ _______

4. $-3x+15=-3(x+5)$ _______

5. $4(-x+6)=-4(x-6)$ _______

For problems 6-8, write at least three factor pairs for each of the numbers below. For example, three factor pairs for 12 are $1\cdot 12$, $2\cdot 6$, and $3\cdot 4$.

6. 18: _______ _______ _______

7. 48: _______ _______ _______

8. 50: _______ _______ _______

9. Multiply: $(x+4)^2$.

____

10. Multiply: $(x+6)(x-6)$.

____

11. Circle the numbers below that are perfect squares:

16 27 36 44 78 81

12. Make this equation true: $x^8=(\_\_\_)^2$.

13. Simplify: $(3xy)^2$. _______

14. Simplify: $(5y^2)^3$. _______

15. Multiply: $(x-4)(x^2+4x+16)$.

16. Circle the numbers below that are perfect cubes:

8 16 27 36 49 64 125

17. Make this equation true: $x^6=(\_\_\_)^3$.

Use the expression $4x^2+5x-7$ to answer questions 18-20.

18. How many terms are in the expression? _____

19. What is the lead coefficient in the expression? _____

20. What is the constant term? _____

21. $x^2+16=x^2-16$ True or false? ____

22. $9-x^2=x^2-9$ True or false? ______

23. Rearrange the equation and combine like terms so that $x^2+5x=3x+4$ so that $=0$ is on the right-hand side.

24. Is $x=-3$ a solution of the equation $x^2-2x-15=0$? ______

25. Is $x=\frac{1}{2}$ a solution of the equation $(2x-1)(x+3)=0$? ______

Guided Learning Activity

Section 6.1: Charting Factors and GCFs

Directions Part I: For each expression, decide if the terms in the top row would be factors of that expression. Place an **X** in any column for which the term **would** be a factor, then use those decisions to help you find the GCF.

		2	3	4	6	8	12	x	x^2	x^3	x^4	y	y^2	y^3	y^4	GCF
a.	$6x^3-3x^2$		X					X	X							$3x^2$
b.	$18y^2+24y$															
c.	$19x^4y-3xy^2$															
d.	$16+40x^3$															
e.	$24x^4y^4+36x^2y^2$															
f.	x^4+y^4															
g.	$36x+2x^2$															
h.	$9y^4+3xy^4$															
i.	$x^3y^2w+x^2y^3z$															

Directions Part II: Given each expression, find the numerical GCF, the x GCF, and the y GCF. Then use the product to find the overall GCF for the expression. The first one has been done for you.

	Expression	Constant GCF	x GCF	y GCF	Overall GCF	Factored Expression
a.	$15x^4-25x^3$	5	x^3		$5x^3$	$5x^3(3x-5)$
b.	$7y^3+49y^2$					
c.	$16xy^2+32y^4$					
d.	$24x^8-46x^7$					
e.	$4x^6+4y^6$					
f.	$12x^2+144$					
g.	$33xy-44x^2y^2$					
h.	$9x^{10}y^5+3x^7y^4$					

Student Activity A

Section 6.1: Hatch the Missing Factors

Directions: Fill in the blank in each equation to make each statement true. All of the solutions are provided inside the "egg". When you have correctly placed all of the factors, you have "hatched" the egg.

$8x^2 \cdot \_\_\_\_\_\_ = 24x^3$

$-4x \cdot \_\_\_\_\_\_ = 36x^4$

$9y^2 \cdot \_\_\_\_\_\_ = 54y^6$

$\_\_\_\_\_\_ \cdot 6x^3 = -36x^3$

$2x^2y \cdot \_\_\_\_\_\_ = -8x^2y^3$

$5x^2 \cdot \_\_\_\_\_\_ = 5x^2$

$5x^2 \cdot \_\_\_\_\_\_ = -5x^2$

$\_\_\_ \cdot (-3z) = 3z^2$

$7w^3 \cdot \_\_\_\_\_\_ = -49w^5$

$5uv \cdot \_\_\_ = 5u^4v$

$16ab^2 \cdot \_\_\_ = 32a^2b^2$

$9r^2 \cdot \_\_\_ = 36r^3s^2$

$t^7u \cdot \_\_\_ = 16t^8u^2v$

$\_\_\_ \cdot (-7x) = 14x$

$c \cdot \_\_\_\_\_\_ = mc^2$

$12x^4 \cdot \_\_\_ = -144x^{12}$

$18x^4 \cdot \_\_\_ = -36x^5y^3$

$-10x^3y^2z \cdot \_\_\_\_\_\_ = -120x^5y^3z^4$

Egg:

$6y^4$ $-z$ $-2xy^3$ $16tuv$ -1 $-4y^2$ -6 $-7w^2$ $12x^2yz^3$ $-12x^8$ $-9x^3$ mc u^3 $2a$ -2 1 $3x$ $4rs^2$

Student Activity B

Section 6.1: Match Up on Factoring (GCFs and Grouping)

Match-up: In each box of the grid, you will find an expression that needs to be factored. Once you have factored each expression, look to see whether the factor appears in the list at the top. If it does, list that letter, if none of the factors are listed, then choose F (none of these are factors). The first one has been done for you. Some boxes may have more than one answer.

A $x+3$ **B** $x-2$ **C** $x+2y$ **D** $y-5$ **E** $x+4$ **F** None of these are factors

$x^2-3x+2xy-6y$ $x(x-3)+2y(x-3)$ Ans: $(x+2y)(x-3)$ C	$3x^2y^2+9xy^2$	$4xy^4+8y^4$	$xy+3y-5x-15$
$x^2+4x+2xy+8y$	$3x^2y^2z^3-6xy^2z^3$	$xy+4y-5x-20$	$7xy-14y-3x+6$
$6xy^2+10y^2+3xy+5y$	$x^2+3x+2xy+6y$	$3a^2b^3x+12a^2b^3$	$10x^2y-50x^2+9xy-45x$
$xy+3y-2x-6$	$x^2+2xy+2xy+4y^2$	$xy+2y^2-5x-10y$	$xy-2y-5x+10$

Student Activity A

Section 6.2: Revenge of Factor Pairings

Directions: In each diagram, there is a number in the top box and exactly enough spaces beneath it to write all the possible factor-pairs involving **integers**. The number -12 has been done for you. See if you can find all the missing factor-pairs.

−12	
$1(-12)$	$-1(12)$
$2(-6)$	$-2(6)$
$3(-4)$	$-3(4)$

−32	

−45	

50	

63	

−18	

28	

−30	

100	

−36	
	X

56	

−42	

Directions: In the tables below, you are given a factor pair. If you **ADD** the two factors instead of multiplying them, what is the result? Complete the tables to find the sum of each factor pair. The first one has been started for you.

−12	Sum
$-1(12)$	11
$1(-12)$	−11
$-2(6)$	
$2(-6)$	
$-3(4)$	
$3(-4)$	

32	Sum
$1(32)$	
$-1(-32)$	
$2(16)$	
$-2(-16)$	
$4(8)$	
$-4(-8)$	

−64	Sum
$-1(64)$	
$1(-64)$	
$-2(32)$	
$2(-32)$	
$-4(16)$	
$4(-16)$	
$-8(8)$	

30	Sum
$1(30)$	
$-1(-30)$	
$2(15)$	
$-2(-15)$	
$3(10)$	
$-3(-10)$	
$5(6)$	
$-5(-6)$	

Guided Learning Activity

Section 6.2: Factoring Trinomials of the Form $x^2 + bx + c$

Factoring a trinomial of the form $x^2 + bx + c$ is really a special case of factoring a trinomial of the form $ax^2 + bx + c$ (where $a = 1$).

Key Product: The key product is the product $a \cdot c$. In this section, we are only looking at expressions where the lead coefficient is 1, so $a \cdot c$ becomes $1 \cdot c = c$. In this section, the key product is just c.

Key Sum: The key sum is the coefficient on the middle term, the b value.

Example 1: Factor $x^2 + 8x + 12$

The Key Product is last term of the trinomial: 12

The Key Sum is the coefficient of the middle term: 8

Key Product 12		Key Sum 8
1	12	13
−1	−12	−13
2	6	8
−2	−6	−8
3	4	7
−3	−4	−7

Once we find a row that works, we can construct the factors with the factors of the key product: $(x+2)(x+6)$

Notice that all the factor pairs that were negative numbers, resulted in a similarity in the sums. What was the similarity? ______________

Since we knew that the Key Sum was positive 12, was it necessary to consider the negative factor pairs for this expression? _____

Example 2: Factor $a^2 - 7a - 30$

The Key Product is the last term of the trinomial: -30

The Key Sum is the coefficient of the middle term: -7

Key Product −30		Key Sum −7
1	−30	−29
−1	30	29
2	−15	−13
−2	15	13
−3	10	7
3	−10	−7
−5	6	1
5	−6	−1

Once we find a row that works, we can construct the factors with the factors of the key product: $(a+3)(a-10)$

Once you find the row with the proper factor pair, you do not need to continue trying more possibilities. It is best to try factor pairs in an orderly fashion so that you do not accidentally forget any.

Now Try These! Fill in the empty charts below to help you factor the trinomials. Write the factored form of the trinomial in the "Answer" blank. It is possible that you may not have to fill in **all** the empty places to finish the problems or you may run out of boxes.

1. Factor: $x^2 - 12x + 20$

Key Product		Key Sum

Answer: ________________

2. Factor: $a^2 - 10a + 24$

Key Product		Key Sum

Answer: ________________

3. Factor: $w^2 + 14w + 48$

Key Product		Key Sum

Answer: ________________

4. Factor: $y^2 - 3y - 18$

Key Product		Key Sum

Answer: ________________

Student Activity B

Section 6.2: Find the Prime Trinomial

Directions: In each row there are three trinomials; two are factorable and one is not. If the trinomial is factorable, factor it. If it is not factorable, write "PRIME."

1.	x^2-5x-6	x^2-x+6	x^2-5x+6
2.	$x^2-5x+24$	$x^2-5x-24$	$x^2+5x-24$
3.	$x^2-8x+12$	x^2-x-12	$x^2+4x+12$
4.	$x^2-11x-12$	$x^2+11+12$	$x^2+13x+12$
5.	x^2+3x+2	$x^2+3x+10$	$x^2+3x-28$
6.	$x^2-19x+60$	$x^2-19x-60$	$x^2+19x+60$
7.	$x^2-4x-60$	$x^2+4x-60$	$x^2-4x+60$
8.	$x^2-16x+60$	$x^2+16x+60$	$x^2-16x-60$

Student Activity C

Section 6.2: Match Up on Factoring Trinomials

Match-up: In each box of the grid, you will find an expression that needs to be factored. Once you have factored each expression, look to see whether the factor appears in the list at the top. If it does, list that letter, if none of the factors are listed, then choose F (none of these are factors). The first one has been done for you. Some boxes may have more than one answer.

A $x+4$ **B** $x-2$ **C** $x-1$ **D** $x+5$ **E** $x-7$ **F** None of these are factors

x^2-3x+2 $(x-2)(x-1)$ Ans: B, C	$x^2+12x+35$	$x^2-2x-35$	$x^2+8x+16$
x^2-3x-4	x^2-4x+4	x^2+x-2	x^2-x-42
$x^2+13x+42$	x^2-6x-7	x^2+2x-8	x^2+x-20
$3x^2y-18xy+24y$	$5x^2+10x-15$	$5x^2+20x+15$	$10x^2y^2-30xy^2-280y^2$

Guided Learning Activity

Section 6.3: Factoring Trinomials of the Form ax^2+bx+c

Here we look at factoring a trinomial of the form ax^2+bx+c where $a \neq 1$.
Key Product: The key product is the product $a \cdot c$.
Key Sum: The key sum is the coefficient on the middle term, the b value.

Example 1: Factor $6x^2-23x+20$

The Key Product is $a \cdot c = 6 \cdot 20 = 120$.
The Key Sum is the coefficient of the middle trinomial term: -23

Remember that you can stop checking factor pairs when you find a row that works!

Now we break up the middle term into two terms using the factor pair as coeffients.

Key Product 120		Key Sum −23
−1	−120	−121
−2	−60	−62
−3	−40	−43
−4	−30	−34
−5	−24	−29
−6	−20	−26
−8	−15	−23

$6x^2-23x+20$

$6x^2 \underbrace{-8x-15x}_{\text{from the successful factor pair}} +20$

And then factor by grouping:

$6x^2-8x-15x+20$

$2x(3x-4)-5(3x-4)$

$(3x-4)(2x-5)$

Example 2: Factor $12a^2+11a+2$

The Key Product is $a \cdot c = 12 \cdot 2 = 24$
The Key Sum is the coefficient of the middle trinomial term: 11

Remember that you can stop checking factor pairs when you find a row that works!
(Can you see why we only need to try positive factors of the Key Product?)

Key Product 24		Key Sum 11
1	24	25
2	12	14
3	8	11
4	6	10

Now we break up the middle term into two terms using the factor pair as coefficients.

$12a^2+11a+2$

$12a^2 \underbrace{+3a+8a}_{\text{from the successful factor pair}} +2$

And then factor by grouping:

$12a^2+3a+8a+2$

$3a(4a+1)+2(4a+1)$

$(3a+2)(4a+1)$

Now Try These! Fill in the empty charts below to help you factor the trinomials. Write the factored form of the trinomial in the "Answer" blank. It is possible that you may not have to fill in **all** the empty places to finish the problems or you may run out of boxes.

1. Factor: $8x^2 + 26x + 15$

Key Product		Key Sum

Answer: ____________________

2. Factor: $4w^2 - 4w - 15$

Key Product		Key Sum

Answer: ____________________

3. Factor: $-10x^2 + 29x - 21$

Key Product		Key Sum

Answer: ____________________

4. Factor: $18x^2 + 13x + 2$

Key Product		Key Sum

Answer: ____________________

Student Activity A

Section 6.3: Life After the GCF, Part I

Directions: For each expression below, find and factor out the GCF. Then categorize the resulting expression in the table as one of the following:

- GCF only: if no other factoring can be done within the parentheses
- GCF and form x^2+bx+c: expression within parentheses is of the form x^2+bx+c
- GCF and form ax^2+bx+c expression within parentheses is of the form ax^2+bx+c

The first one has been done for you.

	Expression (and factor out the GCF)	GCF only	GCF and form x^2+bx+c	GCF and form ax^2+bx+c
1.	$3x^2-12x-15$ $3(x^2-4x-5)$		X	
2.	$8x^3+16x^2-24x$			
3.	$20x^2yz+15x^2$			
4.	$9x^3y+15x^2y+6xy$			
5.	$60t^2x^2+78t^2x+24t^2$			
6.	$27x^2+27x+6$			
7.	$150a^2b^2c^2-125ab^2c$			
8.	$25x^5+75x^4+50x^3$			
9.	$26y^2+13y$			
10.	$36x^2+99x+54$			

Student Activity B

Section 6.3: Is it Completely Factored?

Is $(8x-2)(x+6)$ the completed factoring for the expression $8x^2+46x-12$?

1. Multiply $(8x-2)(x+6)$: ____________________

2. Is the result of the multiplication equivalent to $8x^2+46x-12$? _____

3. From the expression $8x^2+46x-12$, first factor out the 2: ______________

4. Then continue to factor by examining the trinomial within the parentheses:

5. Was the original expression, $(8x-2)(x+6)$, completely factored? _____

6. How can you make $(8x-2)(x+6)$ a completely factored expression?

7. Tic-tac-toe Directions: If the expression in the square **IS** completely factored, then circle it (thus putting an **O** on the square). If the expression **IS NOT** completely factored, then put an **X** on the square and finish the factoring.

$(x-2)(3x-12)$	$(7y-35)(2y+3)$	$(9x-28)(x+2)$
$(4x-9)(2x-5)$	$(3x+4)(4x-9)$	$(2a+8)(3a+12)$
$(2x+6)(2x-6)$	$(3w-16)(6w+21)$	$(5x-9)(6x+25)$

Teaching Guide

Section 6.4: Factoring Perfect Square Trinomials and the Difference or Difference of Two Squares

Preparing for Your Class

Suggested Class Time: 30-45 minutes

Materials Needed: Student activities

Vocabulary

- Perfect-square trinomials, difference of two squares

Instruction Tips

- Technically, you can still use the key number/key sum method to factor a perfect square trinomial and a difference of two squares. But it is useful for the student to begin to make the abstraction to the *form* of the expression and how to correctly put the pieces of a formula involving expressions together.

Factor: $x^2 - 49$

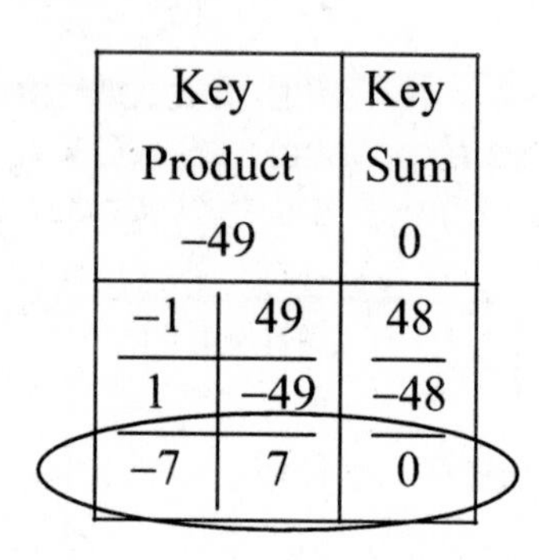

Key Product −49		Key Sum 0
−1	49	48
1	−49	−48
−7	7	0

Factored form: $(x+7)(x-7)$

Factor: $x^2 - 6x + 9$

Key Product 9		Key Sum −6
1	9	10
−1	−9	−10
3	3	6
−3	−3	−6

Factored form: $(x-3)^2$

- Some expressions that look unfactorable using a sum of squares, like $x^6 + 64$ or $a^6 + 1$, *are* factorable using a sum of cubes. So be careful in choosing your class examples.
- Students often forget to look for a GCF first. When this happens, they conclude that an expression like $4x^4 + 36x^2$ is prime, since it can be written as $(2x^2)^2 + (6x)^2$. They *must* still look for a GCF to factor out first.
- It is possible to squeeze Sections 5.4 and 5.5 into one class period if your class has good exponent skills.

Teaching Your Class

Student Activity A: *Recognizing Perfect Squares and Cubes.* In this activity, students practice writing an expression as its corresponding squared or cubed expression. This is a good warm up activity.

Recognizing a perfect-square trinomial:

- When we square a binomial, we get a special expression.
- $(A+B)^2 = A^2 + 2AB + B^2$ or $(A-B)^2 = A^2 - 2AB + B^2$
- If we recognize a trinomial as a perfect-square trinomial, then it is easy to factor.
- Which of these are perfect-square trinomials?
  - $x^2 + 16x + 64$
  - $a^2 - 10a + 25$
  - $z^2 - 4z - 4$ (no)
  - $x^2 + 2x + 1$
  - $x^2 + 12x - 36$ (no)
  - $x^2 - 8x + 9$ (no)

> **Student Activity B:** *Skeleton of a Perfect-Square Trinomial.* Students read through the examples, and then try to apply the formula for perfect-square trinomials on their own.

Recognizing a difference or sum of two squares:

- Sum of squares: $A^2 + B^2$
- Difference of squares: $A^2 - B^2$
- If we recognize a binomial as a difference of squares, then it is easy to factor.
- If we recognize a binomial as a sum of squares (after factoring out a GCF), then it is prime.
- Which of these are a either a sum of squares or a difference of squares?
  - $x^2 - 25$
  - $x^2 + 16$
  - $a^2 - 1$
  - $w^2 + 54$ (neither)
  - $4x^2 - 9$
  - $64y^2 + 1$

> **Student Activity C:** *Skeleton of the Difference or Sum of Two Squares.* Students read through the examples, and then try to apply the formulas for the difference or sum of two squares on their own.

More involved examples: (factor completely)

- $x^4 - 16 = (x^2+4)(x+2)(x-2)$ (difference of squares within a difference of squares)
- $2x^4 - 18x^2 = 2x^2(x+3)(x-3)$ (GCF, then difference of squares)
- $9x^2 + 54x + 81 = 9(x+3)^2$ (GCF, then perfect-square trinomial)
- $4x^4 + 36x^2 = 4x^2(x^2+9)$ (GCF, this is NOT prime)

Student Activity A

Section 6.4: Recognizing Perfect Squares and Cubes

Directions: For each expression below, determine whether it might be a perfect square or a perfect cube. If it is, write it in the squared $(\ \ )^2$ or cubed $(\ \ )^3$ form and drop it in the appropriate place (the square, the cube, or the recycle bin if it is neither). The first one has been done for you. Be careful, a couple of these will need to be filed in *both* squares and cubes! And some of the numbers have two $(\ \ )^2$ forms.

-27 $(-3)^3$	125	9	512	$216x^3y^3$
$16b^2$	-64	1	100	$1000x^3$
729	$49a^4b^4$	-1	$8x^3$	$81h^6$
144	$8x^2$	$25z^2$	$-25z^2$	64
$27a^3$	0	169	$216x^3y^9$	$100x^3$

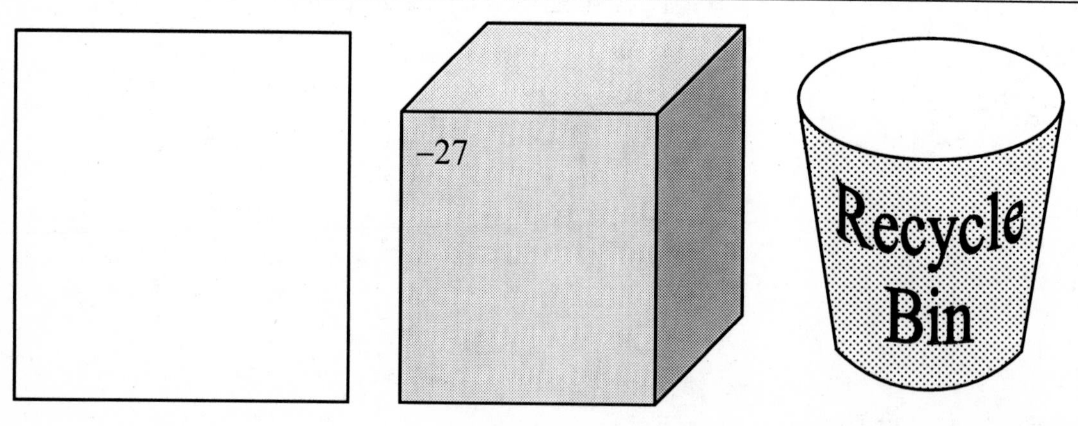

Student Activity B

Section 6.4: Skeleton of a Perfect Square Trinomial

In Chapter 5, we worked with squaring a binomial to achieve a perfect square trinomial. In this section, we take this process backwards. Here are two examples, with the description of each step.

Example 1: Factor: $x^2 - 12x + 36$

Step 1: Notice that both the first and last terms are squared terms: $(x)^2$ and $(6)^2$. This is the first clue that the trinomial might be a perfect square trinomial.

Step 2: In a perfect square trinomial, the middle term is supposed to be twice the product of the bases, in this case, x and 6. $2(x)(6)$ is $12x$, so the trinomial is a perfect square trinomial.

Step 3: Remember the factoring formula for a perfect square trinomial:

$$A^2 + 2AB + B^2 = (A+B)^2 \qquad A^2 - 2AB + B^2 = (A-B)^2$$

Step 4: Write the correct factored form: $(x)^2 - 2(x)(6) + (6)^2 = (x-6)^2$.

Example 2: Factor: $4a^2 + 12ab + 9b^2$

Step 1: Notice that both the first and last terms are squared terms: $(2a)^2$ and $(3b)^2$.

Step 2: Is the middle term twice the product of the bases? $2(2a)(3b) = 12ab$, so yes.

Step 3: Factoring formula for a perfect square trinomial: $A^2 + 2AB + B^2 = (A+B)^2$

Step 4: Write the correct factored form: $(2a)^2 + 2(2a)(3b) + (3b)^2 = (2a+3b)^2$

Now you try these! For each problem, fill in the skeleton for the sum or difference of squares, then factor appropriately.

1. Factor: $x^2 - 10x + 25$ $\quad = (\quad)^2 - 2(\quad)(\quad) + (\quad)^2 = (\qquad)^2$
2. Factor: $y^2 + 2y + 1$ $\quad = (\quad)^2 + 2(\quad)(\quad) + (\quad)^2 = (\qquad)^2$
3. Factor: $25x^2 - 60x + 36$ $\quad = (\quad)^2 + 2(\quad)(\quad) + (\quad)^2 = (\qquad)^2$
4. Factor: $v^4 + 6v^2 + 9$ $\quad = (\quad)^2 + 2(\quad)(\quad) + (\quad)^2 = (\qquad)^2$
5. Factor: $16 - 8x + x^2$ $\quad = (\quad)^2 + 2(\quad)(\quad) + (\quad)^2 = (\qquad)^2$
6. Factor: $a^2b^2 + 14abc + 49c^2$ $\quad = (\quad)^2 + 2(\quad)(\quad) + (\quad)^2 = (\qquad)^2$

Student Activity C

Section 6.4: Skeleton of the Difference or Sum of Two Squares

The best way to learn the sum and difference of squares formulas is to learn a procedure for constructing the formulas rather than memorizing them. Here are three examples, with the description of each step:

Example 1: Factor: $x^2 - 25$

Step 1: Construct a skeleton of the difference of squares: $(\quad)^2 - (\quad)^2$

Step 2: Fill in the skeleton with the appropriate bases: $(x)^2 - (5)^2$

Step 3: Construct the skeleton of the difference of squares formula: $(\quad + \quad)(\quad - \quad)$

Step 4: Insert the bases into both factors: $(x+5)(x-5)$

Example 2: Factor: $64y^2 - 81z^2$

Step 1: Construct a skeleton of the difference of squares: $(\quad)^2 - (\quad)^2$

Step 2: Fill in the skeleton with the appropriate bases: $(8y)^2 - (9z)^2$

Step 3: Construct the skeleton of the difference of squares formula: $(\quad + \quad)(\quad - \quad)$

Step 4: Insert the bases into both factors: $(8y+9z)(8y-9z)$

Example 3: Factor: $4x^2 + 9$

Step 1: Construct a skeleton of the sum of squares: $(\quad)^2 + (\quad)^2$

Step 2: Fill in the skeleton with the appropriate bases: $(2x)^2 + (3)^2$

Step 3: A sum of squares is prime. This binomial does not factor.

Now you try these! For each problem, fill in the skeleton for the sum or difference of squares, then factor appropriately.

1. Factor: $36 - a^2$

Step 1: Construct a skeleton of the difference of squares: $(\quad)^2 - (\quad)^2$

Step 2: Fill in the skeleton above with the appropriate bases.

Step 3: Construct the skeleton of the difference of squares formula: $(\quad + \quad)(\quad - \quad)$

Step 4: Insert the bases in the skeleton above.

2. Factor: $49y^2 + x^2 = (\quad)^2 + (\quad)^2 =$ ______________________

3. Factor: $a^2b^2 - 4 = (\quad)^2 - (\quad)^2 =$ ______________________

4. Factor: $w^4 - 49v^2 = (\quad)^2 - (\quad)^2 =$ ______________________

5. Factor: $81 + u^2 = (\quad)^2 + (\quad)^2 =$ ______________________

6. Factor: $9w^2 - z^6 = (\quad)^2 - (\quad)^2 =$ ______________________

Assessment 6B

Chapter 6: Mid-chapter Assessment for Understanding

For each of the following, describe the type of problem and the strategies and key steps to remember while doing the problem. You do **not** have to complete the problems.

	Type of Problem	Strategies and Key Steps
1. Factor: $y^2 + 7y - 18$.		
2. Factor: $16w^2 - 25$.		
3. Find the GCF of $8x^2$, $24x$, and $16x^3$.		
4. Factor: $3x^2 - 14x - 24$.		
5. Factor: $25a - 5$.		
6. Factor: $x^2 - 12x + 20$.		
7. Factor: $9y^2 + 4$.		
8. Factor: $xy - 2y + 3x - 6$.		
9. Factor: $2x^2 + 8x - 42$.		
10. Factor: $3x^2 - 12$.		

Guided Learning Activity

Section 6.5: Learning the Cubes Procedure

The best way to learn the sum and difference of cubes formulas is to learn a procedure for constructing the formulas rather than memorizing the formulas. Here are two examples, with the description of each step:

Example 1: Factor: $x^3 - 125$

Step 1: Construct a skeleton of the difference of cubes: $(\quad)^3 - (\quad)^3$

Step 2: Fill in the skeleton with the appropriate bases: $(x)^3 - (5)^3$

Step 3: Construct the skeleton of the cubes formula: $(\quad - \quad)(\quad + \quad + \quad)$

same sign as original expression (first sign, $-$)

opposite of first sign (second sign, $+$)

always positive (third sign, $+$)

Step 4: Insert the bases: $(x-5)(\quad + \quad + \quad)$

Step 5: Insert the squares of those bases: $(x-5)(x^2 + \quad + 25)$

$(x)^2$ → x^2; $(5)^2$ → 25

Step 6: Insert the product of the bases: $(x-5)(x^2 + 5x + 25)$

$(x)(5)$ → $5x$

Example 2: Factor: $8w^3 + 27x^6$

Step 1: Construct a skeleton of the sum of cubes: $(\quad)^3 + (\quad)^3$

Step 2: Fill in the skeleton with the appropriate bases: $(2w)^3 + (3x^2)^3$

Step 3: Construct the skeleton of the cubes formula: $(\quad + \quad)(\quad - \quad + \quad)$

same sign as original expression (first sign, $+$)

opposite of first sign (second sign, $-$)

always positive (third sign, $+$)

Step 4: Insert the bases: $(2w+3x^2)(\quad - \quad + \quad)$

Step 5: Insert the squares of those bases: $(2w-3x^2)(4w^2 + \quad + 9x^4)$

$(2w)^2$ → $4w^2$; $(3x^2)^2$ → $9x^4$

Step 6: Insert the product of the bases: $(2w-3x^2)(4w^2 + 6x^2w + 9x^4)$

$(2w)(3x^2)$ → $6x^2w$

Now you try these!

1. Factor: $x^3 - 8$

Step 1: Construct a skeleton of the sum or difference of cubes: $(\quad)^3 - (\quad)^3$

Step 2: Fill in the skeleton above with the appropriate bases.

Step 3: Construct the skeleton of the cubes formula below (put in the signs):

$$(\qquad)(\qquad\qquad\qquad)$$

$$\uparrow \qquad \uparrow \qquad \uparrow$$

$$(\quad)^2 \quad (\quad)(\quad) \quad (\quad)^2$$

Step 4: Insert the bases in the skeleton above.

Step 5: Insert the squares of those bases into the skeleton above.

Step 6: Insert the product of the bases into the skeleton above.

2. Factor: $64y^3 + 1$

$$64y^3 + 1 = (\quad)^3 + (\quad)^3$$

$$= (\qquad)(\qquad\qquad\qquad)$$

$$\uparrow \qquad \uparrow \qquad \uparrow$$

$$(\quad)^2 \quad (\quad)(\quad) \quad (\quad)^2$$

3. Factor: $a^9 + 8b^6$

$$a^9 + 8b^6 = (\quad)^3 + (\quad)^3$$

$$= (\qquad)(\qquad\qquad\qquad)$$

$$\uparrow \qquad \uparrow \qquad \uparrow$$

$$(\quad)^2 \quad (\quad)(\quad) \quad (\quad)^2$$

4. Factor: $512R^3 - 343r^3$

$$512R^3 - 343r^3 = (\quad)^3 - (\quad)^3$$

$$= (\qquad)(\qquad\qquad\qquad)$$

$$\uparrow \qquad \uparrow \qquad \uparrow$$

$$(\quad)^2 \quad (\quad)(\quad) \quad (\quad)^2$$

Student Activity A

Section 6.5: Match Up on Factoring Binomials

Match-up: In each box of the grid, you will find an expression. Decide how to classify the expression and look to see whether the classification appears in the list at the top. If it does, list that letter, if the classification is not listed, then choose E (none of these). The first one has been done for you. **Some boxes may have more than one answer.**

Also, draw the appropriate skeleton and insert the appropriate bases.

A Sum of Squares: $(\ )^2+(\ )^2$

B Difference of Squares: $(\ )^2-(\ )^2$

C Sum of Cubes: $(\ )^3+(\ )^3$

D Difference of Cubes: $(\ )^3-(\ )^3$

E None of these

x^2-16	$4z^4-9$	$9x^2+27$	y^2+8
w^4+4z^2	$8x^3-27$	x^6-64	$8x^9+125$
$729-x^6$	$4a^2+144b^6$	$169x^4-121y^2$	$343x^3+64$
x^2-3	x^3-1000	w^6-27x^3	w^6+1

Student Activity B

Section 6.5: Life After the GCF, Part II

Directions: For each expression below, find and factor out the GCF. Then categorize the resulting expression in the table by examining the expression that is within the parentheses. Some expressions may have more than one categorization. The first one has been done for you.

	Expression (factor out the GCF)	GCF only	GCF and form x^2+bx+c	GCF and form ax^2+bx+c	GCF and perfect-square trinomial	GCF and sum or difference of squares	GCF and sum or difference of cubes
1.	$12x^3+12x^2-72x$ $12x\left(x^2+x-6\right)$		X				
2.	$125x^5-625x^3$						
3.	$12x^3y^3+42x^2y^3+18xy^3$						
4.	$3t^2x^3+24t^2$						
5.	$40x^2-250$						
6.	$2x^2y^2+4xy^2+2y^2$						
7.	$24y^3+27$						
8.	$27x^4y^6-8x^4$						
9.	$4x^4-16x^3-16x^2$						
10.	$32t^2x^3-2t^2x$						

Guided Learning Activity

Section 6.6: How Recognizing a GCF Can Make Your Life Easier

Often it is tempting to start factoring a polynomial expression by immediately setting up a factor table or two sets of parentheses without first looking for a GCF. While you can still factor out a GCF at the end (as long as you remember to look for it), consider how much easier some factoring problems could be if you look for the GCF first.

Example 1: Factor $60x^2+120x+60$

Method 1: Finding GCF Last

First we find the key product and key sum.

Key Product 3600		Key Sum 120
1	3,600	3,601
2	1,800	1,802
3	1,200	1,203
⋮	⋮	⋮
48	75	123
50	72	122
60	60	120

There are 23 factor pairs for the key product before we get to the correct one.

Then we rewrite the middle term using the two terms from the table, and factor by grouping.

$$60x^2+120x+60$$
$$60x^2+60x+60x+60$$
$$60x(x+1)+60(x+1)$$
$$(60x+60)(x+1) \leftarrow$$
$$60(x+1)(x+1)$$
$$60(x+1)^2$$

Notice that if we stop here, the expression is not fully factored.

Method 2: Finding the GCF First

$$60x^2+120x+60$$
$$60(x^2+2x+1)$$

You may notice that the polynomial portion is a perfect square and go directly to the skeleton:

$$60\left[(x)^2+2(1)(x)+(1)^2\right]=60(x+1)^2.$$

Or, you could use the factor table:

Key Product 1	Key Sum 2
1 | 1	2

In this case, the factoring is significantly easier:

$$60(x^2+2x+1)$$
$$60(x+1)(x+1)$$
$$60(x+1)^2$$

You can see that the second method is much shorter and leaves much less room for arithmetic error. When factoring, it is always best to check for a GCF first!

Match-up: In each box of the grid, you will find an expression that needs to be factored. Find and factor out the GCF first. If you don't, you'll regret it! Once you have factored the expression, look to see whether any of the factors appear in the list at the top. If none of the factors are listed, then choose E (none of these are factors). Some boxes may have more than one answer. The first one has been done for you.

A $x+1$ **B** $2x+3$ **C** $4x-1$ **D** $x-5$ **E** None of these are factors

$40x^2+100x+60$ $20(2x^2+5x+3)$ $20(2x^2+2x+3x+3)$ $20[2x(x+1)+3(x+1)]$ $20(2x+3)(x+1)$ **A, B**	$120x^2+90x-30$	$70-70x^2$	$-30x^2+15x+90$
$-60x^2+315x-75$	$40x^2+80x+40$	$8x^2+10x-3$	$-600x^2-750x+225$
$60x^2-260x-200$	$-45x^2+195x+150$	$-15x^2-45x-30$	x^2-4x-5
$11x^2-44x-55$	$40x^2-140x-300$	$20x^2-40x+20$	$320x^2-160x+20$

Student Activity A

Section 6.6: Factoring Strategizing

For each type of factoring problem, describe the classification of the factoring problem and the strategies and key steps (like the parentheses skeleton) to remember while doing the problem. You do not have to complete the problems.

Choose from these classifications (more than one may apply):

Factor out a GCF
Factor by Grouping
Trinomial $x^2 + bx + c$
Trinomial $ax^2 + bx + c$
Sum of Squares
Difference of Squares
Sum of Cubes
Difference of Cubes
Perfect Square Trinomial

Expression to be factored	Classification	Strategies and Key Steps
1. $9x^2 - 4$		
2. $4u^4 + 32u$		
3. $t^2 - 12t + 35$		
4. $x^2 + 49$		
5. $x^3 + 5x^2 - 3x - 15$		
6. $15y^2 - 36y + 12$		
7. $x^2 + 14x + 49$		
8. $64z^3 - 27$		
9. $a^2 + 15ab + 54b^2$		
10. $4x^4 + 4x^3 - 80x^2$		

Student Activity B

Section 6.6: Factoring Variations on a Number

Directions: In each set there are six expressions to be factored, all ending with the same constant (positive or negative). If the expression is factorable, factor it. If it is not factorable, write "PRIME."

Variations on 1			
	$x^2 - 1$	$x^3 - 1$	$x^2 - 2x + 1$
	$x^2 + 1$	$x^3 + 1$	$x^2 + 2x + 1$

Variations on 64			
	$x^2 - 64$	$x^2 + 64$	$x^2 - 16x + 64$
	$x^3 - 64$	$x^3 + 64$	$x^2 + 16x + 64$

Variations on 36			
	$x^2 - 36$	$9x^2 + 36$	$x^2 + 12x + 36$
	$4x^2 - 36$	$9x^2 - 36$	$x^2 - 12x + 36$

Student Activity C

Section 6.6: Checking the Factoring with a Calculator

Directions: One way you can check your factoring is to evaluate both the original expression and the factored expressions for the same value of the variable. It is not wise to use the values of 0, 1, or 2 for these types of checks since these three numbers have some quirky properties:

$0+0=0$ and $0 \cdot 0=0$ $\qquad$ $1 \cdot 1=1$ $\qquad$ $2+2=4$ and $2 \cdot 2=4$

If a check by evaluation results in different answers, then you should reexamine the factoring. Note that getting the same result from evaluation does not **ensure** that your factors are correct, but it is likely to find a mistake in your work if there is one.

In each table below, check the student's factoring by evaluating for the given values with a calculator. It will help to write out the skeleton with the value substituted first. Then input the numerical expression into your calculator to get the result. If you find a factoring mistake, correct it! The first one has been started for you.

1.

	Evaluate for $x=3$.	**Evaluate for $x=2.4$.**
$3x^2+7x+2$	$3(3)^2+7(3)+2=50$	
$(x+2)(3x+1)$	$(3+2)(3(3)+1)=50$	

2.

	Evaluate for $x=4$.	**Evaluate for $x=1/3$.**
$x^2+4x-21$		
$(x-7)(x+3)$		

3.

	Evaluate for $x=1/2$.	**Evaluate for $x=-3$.**
$4x^2-25$		
$(2x+5)(2x-5)$		

4.

	Evaluate for $x=3.5$.	**Evaluate for $x=-5/4$.**
$4x^2+20x+25$		
$(4x+5)^2$		

5.

	Evaluate for $x=5$.	**Evaluate for $x=3$.**
x^3-8		
$(x-2)(x^2+4x+4)$		

Student Activity A

Section 6.7: Escape the Matrix by Solving Quadratic Equations

Directions: In each box of the grid, you will find a quadratic equation that needs to be solved. Once you have solved the equations, use pairs of matching solution numbers to navigate out your way out of the matrix. The first one has been done for you. For example, the repeated solution of -3 leads you to the box on the right, with solutions of -3 and 4. To move the next step on the escape route, you need to find an adjacent box with a solution of 4 and something else.

START HERE Solve: $(x+3)^2=0$ Solutions: -3 and -3	Solve: $x^2-x-12=0$ $(x-4)(x+3)=0$ Solutions 4 and -3	Solve: $x^2-8x-48=0$	Solve: $(x+12)^2=0$
Solve: $x^2+2x-8=0$	Solve: $(x-4)(x-\frac{1}{2})=0$	Solve: $x^2-\frac{5}{6}x+\frac{1}{6}=0$	Solve: $3x^2-19x+6=0$
Solve: $x^2-8x+15=0$	Solve: $x^2-x-42=0$	Solve: $x^2+x=2$	Solve: $(x-6)(x+2)=0$
ESCAPE the Matrix Solve: $x^2+8x+16=0$	Solve: $x^2-16=0$	Solve: $x^2-5x+4=0$	Solve: $x^2-11x=-24$

Student Activity B

Section 6.7: Multiply, Factor, or Solve

Directions: The procedures for multiplying polynomials, factoring polynomials, and solving quadratic equations using factoring are most likely stored very near to each other in your memory because the procedures are entwined. Here is some practice to make sure that the procedures are clearly separated in your mind.

1. Factor: $y^2 - 49$

2. Solve: $21x^2 - 25x - 4 = 0$

3. Multiply: $(x-2)^2$

4. Solve: $x^2 - 16 = 0$

5. Factor: $2x^5 - 2y^2$

6. Solve: $x^2 = 8x - 16$

7. Multiply: $x(x-2)(x+3)$

8. Factor: $x^2 - 3xy - 4y^2$

9. Multiply: $(x-5)(x^2 + 5x + 25)$

10. Factor: $2x^4 + 16x^3 - 40x^2$

Describe what each of the directions mean in your own words.

Factor:

Multiply:

Solve:

Student Activity C

Section 6.7: Parentheses, Brackets, and Braces, Oh My!

Some mathematical situations require very specific notation with parentheses, brackets, and braces. In these cases, the notation is **not** interchangeable.

- Ordered pairs are written with parentheses. For example, $(-2,5)$ is an ordered pair with x-coordinate -2 and y-coordinate 5.
- Interval solutions use parentheses and brackets. For example $(-2,5]$ is an interval in which 5 is included in the solution.
- Solution sets use braces. For example $\{-2,5\}$ shows two solution values, -2 and 5.

1. Match the solution with the appropriate graph:

___ **a.** $(-2,5)$ ___ **b.** $\{-2,5\}$ ___ **c.** $(-2,5]$

I

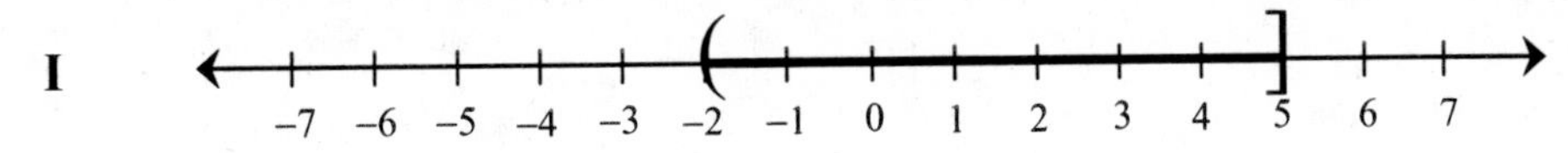

II

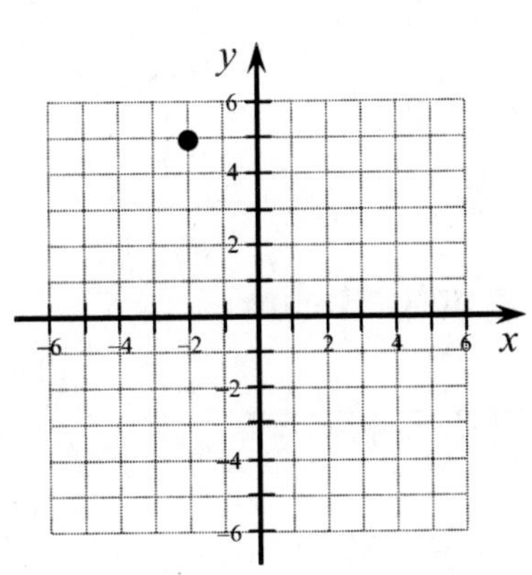

III

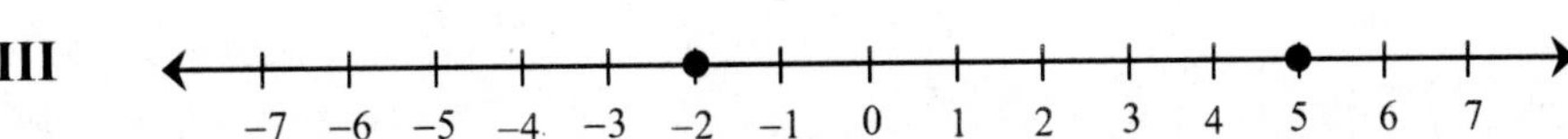

2. Solve $3-2x \leq 9$ and write the solution in interval notation.

3. Find the ordered pair solution for $3x - y = 4$ if $x = 3$.

4. Solve $x^2 - 5x - 36 = 0$ and write the answer as a solution set.

5. Solve $-2 \le x + 4 \le 7$ and write the solution in interval notation.

6. Solve $5(x-2) + 3x = 4x - 8$ and write the solution as a solution set.

7. Is $(-1, 3)$ a solution of $4x + y = 1$?

8. Write the solution that is represented by the graph below.

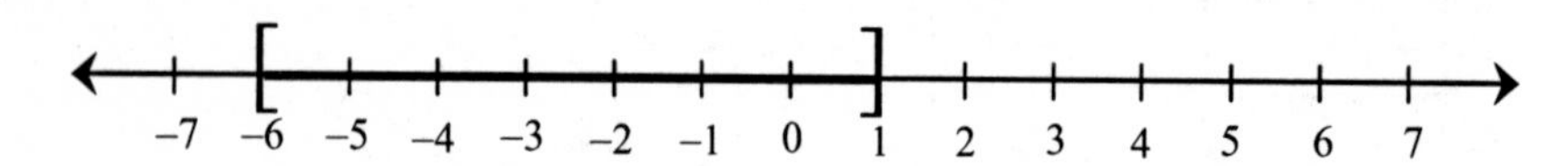

9. Write the ordered pair that is represented by the graph below.

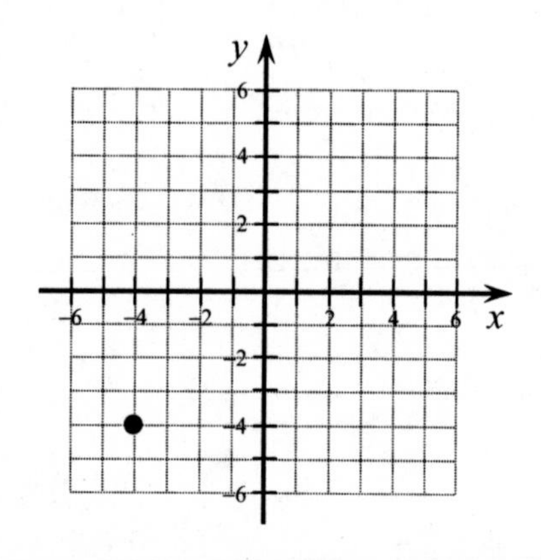

10. Write the solution that is represented by the graph below.

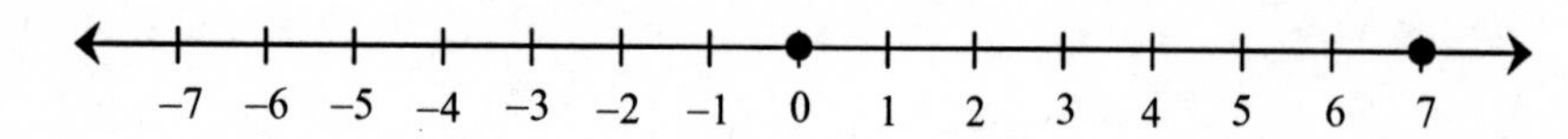

Guided Learning Activity

Section 6.8: Problem Solving Strategy for Quadratic Equations

Problem 1: The area of a vegetable garden is 32 m^2. If the length is 4 meters longer than the width, what are the dimensions (length and width) of the garden?

Analyze the problem: What is given? What are you being asked to find? Can you draw a picture or diagram? Can you construct a table?

Form an equation: Define the variable and write the equation.

Solve the equation:

State the conclusion:

Check the result:

Problem 2: The area of a right triangle is 10 ft^2. If the base of the triangle is 1 foot longer than the height, what are the dimensions of the triangle?

Analyze the problem: What is given? What are you being asked to find? Can you draw a picture or diagram? Can you construct a table?

Form an equation: Define the variable and write the equation.

Solve the equation:

State the conclusion:

Check the result:

Problem 3: The product of two consecutive positive integers is 72. Find the two integers.

Analyze the problem: What is given? What are you being asked to find? Can you draw a picture or diagram? Can you construct a table?

Form an equation: Define the variable and write the equation.

Solve the equation:

State the conclusion:

Check the result:

Problem 4: The base of a right triangle is one less than twice the height. If the length of the hypotenuse is 17 cm, what are the measurements of the other two sides?

Analyze the problem: What is given? What are you being asked to find? Can you draw a picture or diagram? Can you construct a table?

Form an equation: Define the variable and write the equation.

Solve the equation:

State the conclusion:

Check the result:

Student Activity A

Section 6.8: Proof of the Pythagorean Theorem

Materials required: Paper for tracing, scissors, and tape.

This is a geometric proof of the Pythagorean Theorem (by dissection) given by the Arabian Mathematician Thâbit ibn Kurrah, who lived from 836 to 901 A.D. in the Middle East.

Sources: http://mathworld.wolfram.com/PythagoreanTheorem.html, http://www.britannica.com/eb/article-9071897/Thabit-ibn-Qurra

---

Step 1: Trace the figure below onto a **new sheet of paper**.
Cut out the square with side *a* and the square with side *b*.

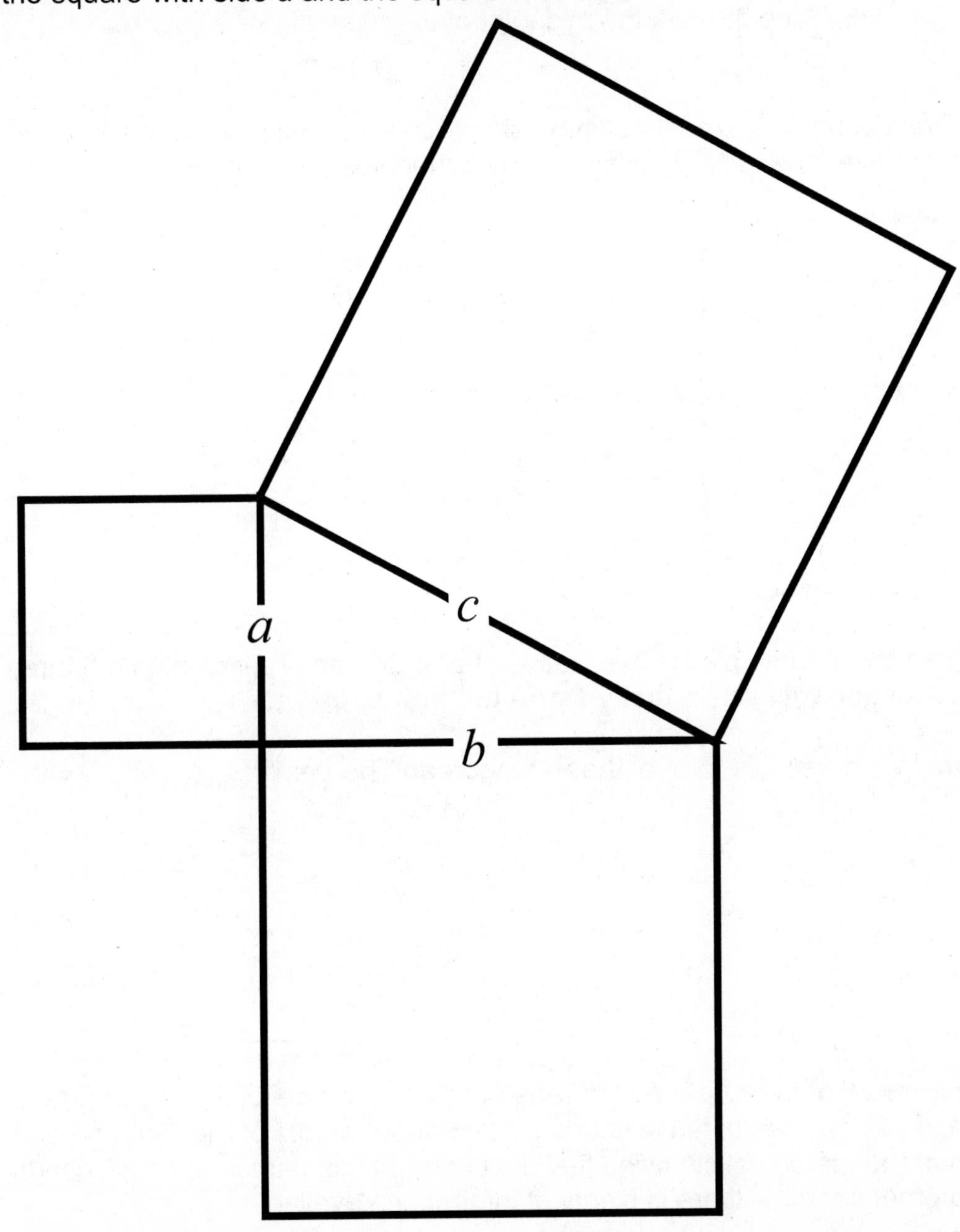

Step 2: Securely tape these two squares together like so.

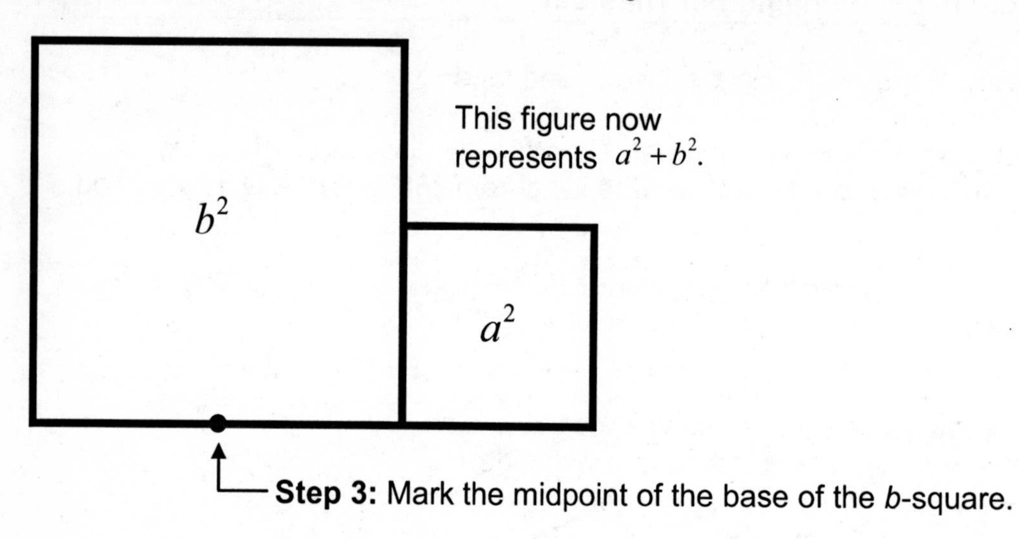

Step 3: Mark the midpoint of the base of the *b*-square.

Step 4: Draw in these two dashed lines (shown below), and **cut** along these lines. You should now have three geometric figures (two triangles and one with an irregular shape

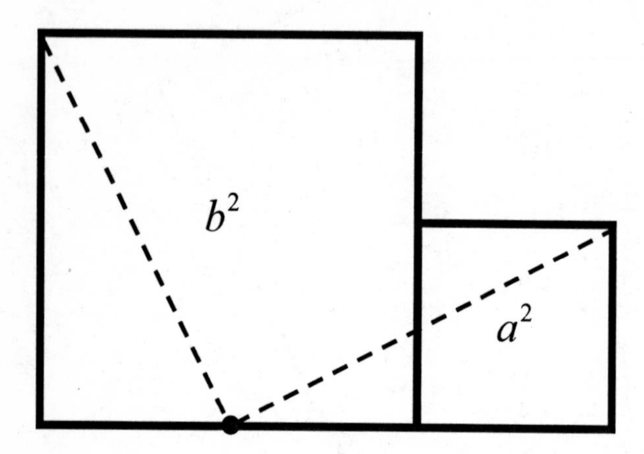

Step 5: Use these three pieces like pieces of a puzzle and reassemble the three pieces to fit in the square with side *c* **that was on the first page**.

Question: Why does this "prove" the Pythagorean Theorem?

More Information: There are hundreds of different proofs of the Pythagorean Theorem. If you are interested in learning more about proofs of the Pythagorean Theorem or the mathematician who first documented this particular proof, conduct a simple Internet search – there is plenty of information available.

Student Activity B

Section 6.8: Pumpkin Launch

Ballistics is the study of the flight characteristics of projectiles. For ballistic flights, Littlewood's Law can be used to *approximate* the peak height of a projectile by using the total flight time.

Littlewood's Law: $h = \frac{g}{8} \cdot t^2$ where h is the peak height in meters, t is the total flight time in seconds, and g is the gravity constant, which is $g = 9.8 \text{ m/s}^2$ near the surface of the earth.

The "Pumpkin Rocket II"
(source: http://www.et.byu.edu/~wheeler/benchtop/pumpkin2.php)

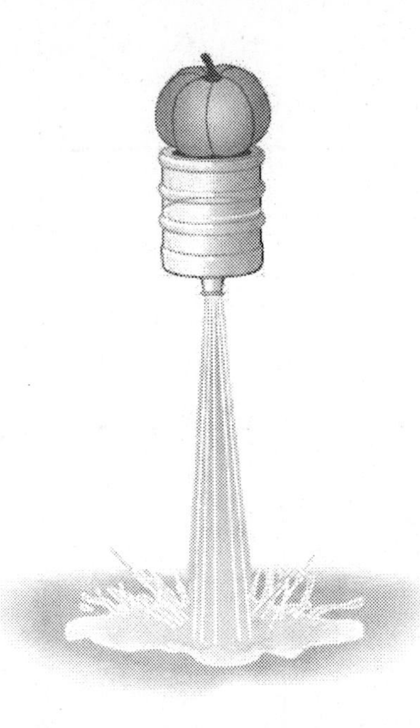

The "Pumpkin Rocket II" was launched using a hydrogen-power method to pressurize the launch vehicle (a large water bottle).

Question 1: The pumpkin and bottle reached a peak height of 18 meters. Try substituting different values of t into Littlewood's Law to find how long the flight must have been to reach this peak height.

t (seconds)	h (meters)
1	
2	
3	
4	
5	
6	

Question 2: Assuming we could build a pumpkin rocket that launch a pumpkin to a height of 78.4 meters, what would the total flight time be for this rocket? **This time, simplify the equation and write it in standard quadratic form, then use factoring to solve the problem.**

Assessment 6C

Chapter 6: End-of-Chapter Assessment for Understanding

For each of the following, describe the type of problem and the strategies and key steps to remember while doing the problem. You do **not** have to complete the problems.

	Type of Problem	Strategies, Rules and Key steps
1. Factor: $4x^2+11x-20$.		
2. Solve: $x^2-8x+12=3(x-4)$.		
3. The product of two positive consecutive even integers is 48. Find the two integers.		
4. Factor: $8x^3+125$.		
5. Factor: $xy^2-9x+4y^2-36$.		
6. Solve: $(x+4)(4x-3)=0$.		
7. The area of a rectangle is $18\,\text{cm}^2$. If the length is twice the width, find the dimensions of the rectangle.		
8. Is -2 a solution of $x^2-5x=14$?		
9. Factor: $5x^2-80$.		
10. Factor: $x^3y^3-64z^6$.		

Assessment 6D

Chapter 6: Metacognitive Skills Assessment

Metacognitive skills refer to the ability to judge how well you have learned something and to effectively direct your own learning and studying. This is a self evaluation tool designed to help you focus your studying and to improve your metacognitive skills with regards to this math class.

Fill the 1st column out before you begin studying.
Fill the 2nd column out after you study and before you take the test.
Go back to this page after your test and circle any of the ratings that you would now change – this identifies the "disconnects" between what you think you know well and what you actually know well.

Use the scale below to assign a number to each topic.

5 *I am confident I can do any problems in this category correctly.*
4 *I am confident I can do most of the problems in this category correctly.*
3 *I understand how to do the problems in this category, but I still make a lot of mistakes.*
2 *I feel unsure about how to do these problems.*
1 *I know I don't understand how to do these problems.*

Topic or Skill	Before Studying	After Studying
Finding the GCF for several terms.		
Factoring out the GCF.		
Factoring out a -1. Why is it sometimes necessary to factor out a -1?		
Factoring by grouping and knowing when to employ this method.		
Distinguishing between trinomials of the form x^2+bx+c and the form ax^2+bx+c.		
Factoring a trinomial of the form x^2+bx+c.		
Factoring a trinomial of the form ax^2+bx+c.		
Knowing the numbers that are perfect squares and perfect cubes.		
Recognizing expressions that are perfect squares and perfect cubes.		
Recognizing binomials that are either a sum or difference of squares.		
Factoring a difference of squares.		
Knowing about the factorability of a sum of squares.		
Recognizing binomials that are either a sum or difference of cubes.		
Factoring a sum or difference of cubes.		
Recognizing a perfect-square trinomial and properly factoring it.		
Using the general factoring strategy. What should you always do first?		
How can the number of terms in a polynomial help you to make decisions about how to factor it?		
Understanding when an expression is not yet completely factored, and how to complete it. Ex: Is $x(2x+4)(x^2-9)$ completely factored?		

Continued on next page.

Use the scale below to assign a number to each topic.

5 *I am confident I can do any problems in this category correctly.*
4 *I am confident I can do most of the problems in this category correctly.*
3 *I understand how to do the problems in this category, but I still make a lot of mistakes.*
2 *I feel unsure about how to do these problems.*
1 *I know I don't understand how to do these problems.*

Topic or Skill	**Before Studying**	**After Studying**
Being able to check the answer to a factoring problem by multiplying out the answer.		
Understanding why you must first get "=0" on one side of a quadratic equation before solving by factoring.		
Rewriting a quadratic equation in standard quadratic form.		
Solving a quadratic equation by factoring.		
Knowing how to properly write a solution set.		
Checking the solution to a quadratic equation.		
Distinguishing between the instructions: multiply, factor, or solve.		
Knowing basic geometry formulas for perimeter, area, and volume.		
Solving application problems involving geometry.		
Knowing the Pythagorean Theorem. What kind of triangles does the Pythagorean Theorem apply to?		
Solving application problems involving the Pythagorean Theorem.		
Knowing how to write consecutive integers algebraically.		
Solving application problems involving consecutive integers.		
Given a quadratic equation model, solve an application problem.		

Student Workbook: Chapter 7

Table of Contents: *Rational Expressions and Equations*

Assessment 7A

Pretest and Diagnostic Tool: Rational Expressions and Equations

Directions: Complete this assessment without looking back at your notes or your book. **Do not use a calculator for this assessment.**

1. Evaluate x^2+3x-4 for $x=2$.

2. Evaluate x^2-9 for $x=-3$.

3. Evaluate: $\dfrac{-3+4}{3+(-3)}$.

4. Solve: $2x-3=0$.

5. Factor: $12x^2+30x$.

6. Factor: $x^2-10x+16$.

7. Factor: x^2-16.

8. Factor: $2x^2+3x-20$.

9. Evaluate: $\dfrac{-10}{25}$.

10. Evaluate: $\dfrac{2+4}{2+6}$.

11. Evaluate: $\dfrac{-5+10}{5+10}$.

12. $-(x+3)=x-3$

    True or false? _______

13. $x^2+2x+1=1+2x+x^2$

    True or false? _______

14. Factor -1 out of $-x^2-3x+4$.

15. Multiply: $\dfrac{24}{27}\cdot\dfrac{45}{16}$.

16. Divide: $\dfrac{2}{3}\div\dfrac{4}{5}$.

17. Subtract: $\dfrac{7}{8}-\dfrac{1}{8}$.

18. Add: $\dfrac{1}{5}x+\dfrac{2}{5}x$.

19. Find the LCD for $\dfrac{5}{12}$ and $\dfrac{7}{18}$.

    ______

20. Find the LCD for $\dfrac{3}{5}$ and $-\dfrac{12}{25}$.

    ______

21. Add: $\frac{3}{4}+\left(-\frac{5}{6}\right)$.

22. Subtract: $2-\frac{5}{3}$.

23. $-5=\frac{-5}{-1}$

True or false? ________

24. $\frac{\frac{1}{2}}{\frac{1}{5}}=\frac{1}{2}\div\frac{1}{5}$

True or false? ________

25. Simplify: $4\left(\frac{x}{2}\right)$.

26. Simplify: $20x^2\left(\frac{3}{5x}\right)$.

27. Simplify: $3(x+2)-2(x-4)$.

28. Solve: $x+10=x^2+4x$.

29. Solve: $x^2+4x+6=x^2-x+16$.

30. Sally swims $\frac{1}{2}$-mile in 40 minutes. What is her swimming rate in miles per hour?

Guided Learning Activity

Section 7.1: Undefined Expressions

A **rational expression** is an expression of the form $\frac{A}{B}$ where A and B are polynomials and B does not equal 0. To evaluate a rational expression, you may find it helpful to first create the parentheses skeleton.

Example 1: Evaluate $\frac{x^2+4x-5}{x^2-9}$ for $x=2$ and $x=-3$.

Parentheses skeleton: $\frac{(\quad)^2+4(\quad)-5}{(\quad)^2-9}$

Evaluate for $x=2$: $\frac{(2)^2+4(2)-5}{(2)^2-9}=\frac{4+8-5}{4-9}=\frac{7}{-5}$

Evaluate for $x=-3$: $\frac{(-3)^2+4(-3)-5}{(-3)^2-9}=\frac{9-12-5}{9-9}=\frac{-8}{0}=$ undefined

Expression	**Parentheses Skeleton**	**Evaluate the expression for ...**				
		$x=1$	$x=4$	$x=0$	$x=-2$	$x=\frac{1}{3}$
$\frac{3x-12}{x^2+2x}$						
$\frac{3x^2+8x-3}{x^2-5x+4}$						
$\frac{x^2-16}{x^2+2x}$						

It is often easier to perform the evaluation if the rational expression is factored first. This way, you can quickly see the values that create a factor of zero in the numerator or denominator.

Example 2: Evaluate $\frac{x^2+4x-5}{x^2-9}$ for $x=0$, $x=1$, and $x=3$.

Factored form: $\frac{(x+5)(x-1)}{(x+3)(x-3)}$.

Evaluate for $x=0$: $\frac{(0+5)(0-1)}{(0+3)(0-3)}=\frac{(5)(-1)}{(3)(-3)}=\frac{-5}{-9}=\frac{5}{9}$

Evaluate for $x=1$: $\frac{(1+5)(1-1)}{(1+3)(1-3)}=\frac{(6)(0)}{(4)(-2)}=\frac{0}{-8}=0$

Evaluate for $x=3$: $\frac{(3+5)(3-1)}{(3+3)(3-3)}=\frac{(8)(2)}{(6)(0)}=\frac{16}{0}=$ undefined

As soon as you see a **factor** of zero in the numerator or denominator, you can quickly find the value of the answer.

Expression	Factored Form of the Expression	Evaluate the expression for ...				
		$x=1$	$x=4$	$x=0$	$x=-2$	$x=\frac{1}{3}$
$\frac{3x-12}{x^2+2x}$						
$\frac{3x^2+8x-3}{x^2-5x+4}$						
$\frac{x^2-16}{x^2+2x}$						

Recall that we can solve an equation like $(x+3)(x-5)=0$ using the zero property. Because of the Zero Factor Property, either $x+3=0$ or $x-5=0$. This leads us to solutions of -3 and 5.

Example 3: Where is the expression $\frac{x^2+4x-5}{x^2-9}$ undefined?

This expression is undefined when the denominator is equal to zero.
We can answer the question by solving the equation $x^2-9=0$.
Factor first: $(x+3)(x-3)=0$
Set each factor equal to zero and solve:

$x+3=0$ OR $x-3=0$

$x=-3$ $\quad$ $x=3$

The expression $\frac{x^2+4x-5}{x^2-9}$ is undefined for -3 and 3.

Expression	Factored Form of the Expression	Set the denominator = 0. Solve the resulting equation.	Where is the expression undefined?
$\frac{3x-12}{x^2+2x}$			
$\frac{3x^2+8x-3}{x^2-5x+4}$			
$\frac{x^2-16}{x^2+2x}$			

Student Activity A

Section 7.1: Match Up on Simplifying Rational Expressions

Match-up: Match each of the expressions in the squares in the table below with an equivalent simplified expression from the top. If an equivalent expression is not found among the choices A through E, then choose F (none of these).

A 1 **B** -1 **C** $x+5$ **D** $\frac{x}{3}$ **E** $3x$ **F** None of these

$\frac{x-1}{1-x}$	$\frac{x^2+x}{3x+3}$	$\frac{9x^3+15x}{3x^2+5}$	$\frac{x^2-25}{x-5}$
$\frac{(3x+2)(x+1)}{3x^2+5x+2}$	$\frac{x-1}{1+x}$	$\frac{3x^3-6x^2}{x-2}$	$\frac{3x+1}{1+3x}$
$\frac{3x^3-27x}{(x+3)(x-3)}$	$\frac{x+5}{x^2+10x+25}$	$\frac{x^2+10x+25}{x+5}$	$\frac{3x-1}{1-3x}$
$\frac{2x^2+x+5}{-2x^2-x-5}$	$\frac{x^2+25}{x^2-25}$	$\frac{x^3+2x^2+x}{3x^2+6x+3}$	$\frac{18x^2-3x}{-1+6x}$
$\frac{x^2+6x+5}{x+1}$	$\frac{2x-3}{3-2x}$	$\frac{2x^2+3x}{3x+2x^2}$	$\frac{x-8}{-x+8}$

Student Activity B

Section 7.1: The Ones Recycling Center

Directions: In each pair of expressions, make a decision about whether the expression is equivalent to 1 or -1. Then sort each expression into the correct recycling bin below. If an expression does not belong in either recycling bin, just leave it out! The first one has been done for you.

The 1 Bin: The numerator and denominator are equivalent.
The -1 Bin: The numerator and denominator are opposites.

$\frac{x-1}{1-x}$	$\frac{x+2}{2+x}$	$\frac{x-3}{x+3}$	$\frac{x^2-1}{1-x^2}$
$\frac{(x+3)^2}{x^2+6x+9}$	$\frac{3(x+4)}{(x+4)(3)}$	$\frac{v^2-1}{v^2+1}$	$\frac{-x-3}{x+3}$
$\frac{-x+3}{3-x}$	$\frac{y^2+1}{1+y^2}$	$\frac{x^2+2x-1}{x^2+2x+1}$	$\frac{x^2-2x-3}{3+2x-x^2}$
$\frac{a-b}{-a+b}$	$\frac{x^2+3x+1}{3x+1+x^2}$	$\frac{5-x}{5-x}$	$\frac{(x+3)^2}{x^2+9}$

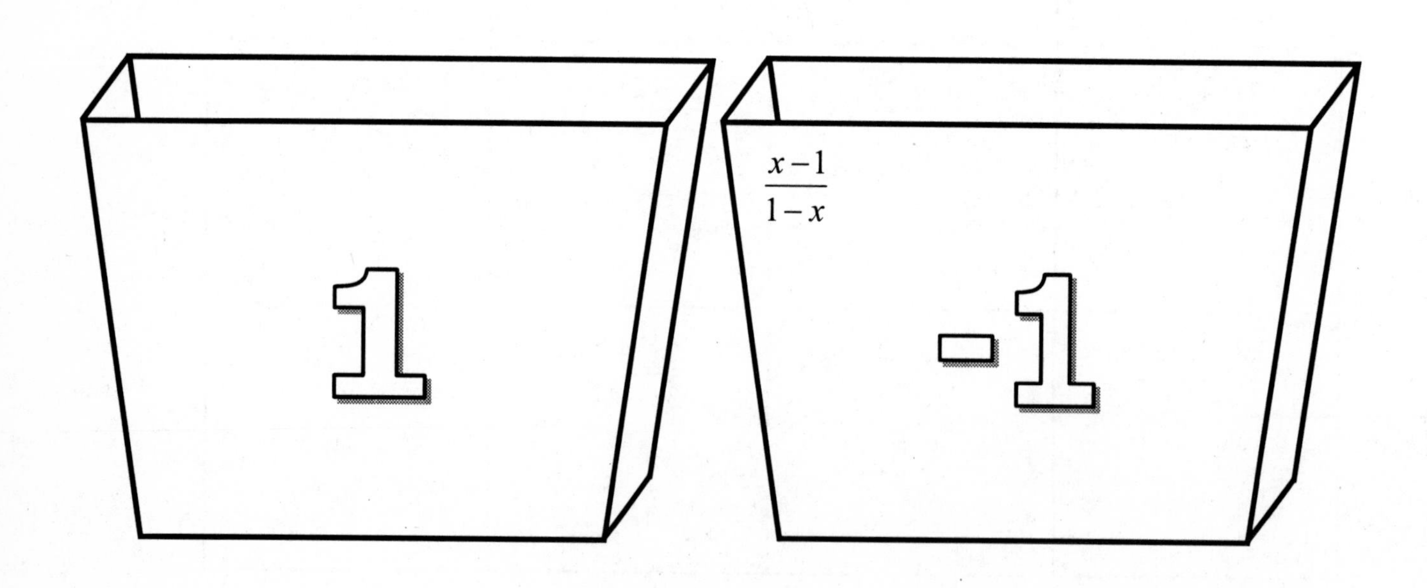

Student Activity C

Section 7.1: Which of These is Not Like the Others?

Directions: When you look at your answer to a problem and compare it to the answer in the back of the book or a friend's answer, you might find that they are not quite the same. This does not mean that one of them is *really* different though. In each row of the table, all the expressions are equivalent except for one of them. **Circle the expressions that are the same and place an X over the "oddball" expression in each row.** The first one has been done for you.

Here's two hints if you're really stuck:

- You could evaluate all the expressions for the same given value and see which expressions have the same result.
- You could try factoring out a -1 if the numerator and denominator look suspiciously like they might be opposites.

1.	$\frac{-4}{-x}$ (X)	$\frac{-4}{x}$ (circled)	$\frac{4}{-x}$ (circled)	$-\frac{4}{x}$ (circled)
2.	$-\frac{x+3}{x-4}$	$\frac{-x-3}{x-4}$	$\frac{-x-3}{-x+4}$	$\frac{x+3}{4-x}$
3.	$\frac{x-5}{5-x}$	$\frac{x-5}{x-5}$	$-\frac{x-5}{x-5}$	$\frac{x-5}{-x+5}$
4.	$-\frac{-x-4}{-x-4}$	$\frac{x+4}{x+4}$	$\frac{x+4}{4+x}$	$-\frac{-x-4}{x+4}$
5.	$-\frac{x+1}{1-x}$	$\frac{1+x}{x-1}$	$\frac{x+1}{x-1}$	$\frac{1+x}{1-x}$
6.	$\frac{x-4}{(x+1)(x+3)}$	$\frac{x-4}{x^2+4x+3}$	$\frac{4-x}{x^2+4x+3}$	$-\frac{4-x}{x^2+4x+3}$

Student Activity A

Section 7.2: Does Position Matter?

Directions: Simplify each expression below. Circle the multiplication example that matches the result of the shaded division problem.

Expression	Simplified Fraction	Decimal Equivalent
$\frac{1}{5}\cdot\frac{3}{4}$	$\frac{3}{20}$	0.15
$\frac{1}{5}\cdot\frac{4}{3}$		
$\frac{5}{1}\cdot\frac{4}{3}$		
$\frac{5}{1}\cdot\frac{3}{4}$		
$\frac{1}{5}\div\frac{3}{4}$		

Expression	Simplify the Expression
$\frac{1}{x+3}\cdot\frac{x+3}{2}$	
$\frac{1}{x+3}\cdot\frac{2}{x+3}$	
$\frac{x+3}{1}\cdot\frac{x+3}{2}$	
$\frac{x+3}{1}\cdot\frac{2}{x+3}$	
$\frac{1}{x+3}\div\frac{2}{x+3}$	

Expression	Simplify the Expression
$\frac{2x+3}{x}\cdot\frac{1}{x}$	
$\frac{x}{2x+3}\cdot\frac{1}{x}$	
$\frac{2x+3}{x}\cdot\frac{x}{1}$	
$\frac{x}{2x+3}\cdot\frac{x}{1}$	
$\frac{2x+3}{x}\div\frac{1}{x}$	

Question: When dividing fractions, does it matter whether you use the reciprocal of the first or second fraction? In other words, do you get the same result?

Student Activity B

Section 7.2: Paint by Factors of 1

Directions: For each expression, perform the multiplication or division and simplify by factoring the numerator and denominator **completely** and removing factors equal to 1.

As you remove factors of 1, like $\frac{x+2}{x+2}$ or $\frac{x^2}{x^2}$, shade in the corresponding square in the grid below. The first one has been done for you. There's a surprise when you're finished!

1. Multiply: $\frac{2x-4}{x+3}\cdot\frac{x+3}{2x+8}=\frac{2(x-2)(x+3)}{2(x+3)(x+4)}=\frac{\cancel{2}^{1}(x-2)\cancel{(x+3)}^{1}}{\cancel{2}_{1}\cancel{(x+3)}_{1}(x+4)}=\frac{x-2}{x+4}$ (shade $\frac{x+3}{x+3}$ & $\frac{2}{2}$)

2. Divide: $\frac{3x^2-6x}{x^3+2x^2}\div\frac{-x^2-2x+8}{x^4+6x^3+8x^2}$

3. Multiply: $\frac{5x^4+2x^3}{2x^4+3x^3}\cdot\frac{2x^3+x^2-3x}{5x^2-3x-2}$

4. Multiply: $\frac{5x^2-125}{x^3+5x^2}\cdot\frac{6x^2+2x}{5x-25}$

5. Divide: $\frac{21x^2-49x-42}{4x^2-28x+48}\div\frac{21x+14}{8x^2-32x}$

$\frac{12}{12}$	$\frac{4x-3}{4x-3}$	$\frac{x+3}{x+3}$	$\frac{x^2}{x^2}$	$\frac{2-5x}{2-5x}$	$\frac{x^2+16}{x^2+16}$
$\frac{7x+3}{7x+3}$	$\frac{x+2}{x+2}$	$\frac{x^3}{x^3}$	$\frac{x-2}{x-2}$	$\frac{-9}{-9}$	$\frac{10x+1}{10x+1}$
$\frac{3}{3}$	$\frac{x-7}{x-7}$	$\frac{5}{5}$	$\frac{2x+3}{2x+3}$	$\frac{4x+3}{4x+3}$	$\frac{8x+3}{8x+3}$
$\frac{4-5x}{4-5x}$	$\frac{5x-7}{5x-7}$	$\frac{x+5}{x+5}$	$\frac{7}{7}$	$\frac{13}{13}$	$\frac{25}{25}$
$\frac{7x+2}{7x+2}$	$\frac{x+25}{x+25}$	$\frac{3x+2}{3x+2}$	$\frac{x+4}{x+4}$	$\frac{3x-4}{3x-4}$	$\frac{x+8}{x+8}$
$\frac{2-9x}{2-9x}$	$\frac{8x-3}{8x-3}$	$\frac{4}{4}$	$\frac{2}{2}$	$\frac{x^2+36}{x^2+36}$	$\frac{4x-17}{4x-17}$
$\frac{4x-5}{4x-5}$	$\frac{x+10}{x+10}$	$\frac{x-3}{x-3}$	$\frac{x-4}{x-4}$	$\frac{x^2+4}{x^2+4}$	$\frac{7x-3}{7x-3}$
$\frac{3x}{3x}$	$\frac{5x+2}{5x+2}$	$\frac{x-5}{x-5}$	$\frac{x}{x}$	$\frac{x-1}{x-1}$	$\frac{2x-1}{2x-1}$

Student Activity C

Section 7.2: Testing Equality

Directions: In each table there is a rational expression and a simplified expression. Use a test value to check the equality (or equivalency) of the two expressions. The first one has been started for you!

Note: Because of some unique properties of 0, 1 and 2 ($0+0=0\cdot 0$, $1\cdot 1=1$, $2+2=2\cdot 2$), it is usually best **not** to use these for test values.

1.

Given expression:		**Possible simplification:**
$\frac{x^2+3}{x}\cdot\frac{1}{x+2}$	$\stackrel{?}{=}$	$\frac{x^2+3}{x^2+2}$

What values make the denominator of the **given** expression zero? $0, -2$

Choose a simple test value for x that does **not** result in a zero denominator: Let $x=3$.

Evaluate the rational expression:		**Evaluate the possible simplification:**
$\frac{(3)^2+3}{(3)}\cdot\frac{1}{(3)+2}=\frac{12}{3}\cdot\frac{1}{5}=\frac{12}{15}=\frac{4}{5}$	$\stackrel{?}{=}$	

Are the two expressions equal? _____

2.

Given expression:		**Possible simplification:**
$\frac{x^3+3x^2}{x}\div\frac{x}{x+7}$	$\stackrel{?}{=}$	$\frac{x^2}{(x^3+3x^2)(x+7)}$

What values make the denominator of the **given** expression zero? ________

Choose a simple test value for x that does **not** result in a zero denominator: _______

Evaluate the rational expression:		**Evaluate the possible simplification:**
	$\stackrel{?}{=}$	

Are the two expressions equal? _____

3. **Given expression:**

$$\frac{x^2-x-12}{x-2} \div \frac{x+3}{x-2}$$

$\overset{?}{=}$

Possible simplification:

$x-4$

What values make the denominator of the **given** expression zero? ________

Choose a simple test value for x that does **not** result in a zero denominator: ______

Evaluate the rational expression:

$\overset{?}{=}$

Evaluate the possible simplification:

Are the two expressions equal? _____

4. **Given expression:**

$$\frac{2x^2-5x-12}{x-4} \cdot \frac{2x+3}{x-4}$$

$\overset{?}{=}$

Possible simplification:

$$\frac{4x^2+12x+9}{x-4}$$

What values make the denominator of the **given** expression zero? ________

Choose a simple test value for x that does **not** result in a zero denominator: ______

Evaluate the rational expression:

$\overset{?}{=}$

Evaluate the possible simplification:

Are the two expressions equal? _____

Student Activity A

Section 7.3: Thread of Like Terms

Directions: Simplify each of the expressions that follow.

1.

7 miles + 5 miles	$7x+5x$	$7x^2+5x^2$	$7x^2y+5x^2y$
$\frac{7}{13}+\frac{5}{13}$	$\frac{7}{x}+\frac{5}{x}$	$\frac{7}{xy}+\frac{5}{xy}$	$\frac{7}{x+2}+\frac{5}{x+2}$

2.

$\frac{9}{11}-\frac{2}{11}$	$9w^2-2w^2$	$\frac{9}{ab^2}-\frac{2}{ab^2}$	$\frac{9}{y}-\frac{2}{y}$
$\frac{9}{x-4}-\frac{2}{x-4}$	$9ab-2ab$	$9y-2y$	$9\text{ ft}^2-2\text{ ft}^2$

3.

$\frac{8}{a^2b}-\frac{1}{a^2b}$	$8xy-xy$	$\frac{8}{z}-\frac{1}{z}$	$8a^2-a^2$
$\frac{8}{x+8}-\frac{1}{x+8}$	$8w-w$	8 cm − 1 cm	$\frac{8}{9}-\frac{1}{9}$

4.

$\frac{4}{a}-\frac{3}{a}$	$4\text{ in}^3-3\text{ in}^3$	$\frac{4}{5}-\frac{3}{5}$	$\frac{4}{5x^2}-\frac{3}{5x^2}$
$\frac{4}{x+1}-\frac{3}{x+1}$	$4x^2y^2-3x^2y^2$	$4a-3a$	$4b^3-3b^3$

Question: What is the lesson to be learned here about addition and subtraction in algebra and mathematics in general?

Student Activity B

Section 7.3: Match Up with Like Denominators

Match-up: Complete each addition or subtraction problem and simplify the result. Then choose the letter that corresponds to the **numerator** of this result. If the numerator is not among the choices, then choose E (None of these). The first one has been done for you.

A $x+1$ **B** -1 **C** 1 **D** $x-1$ **E** None of these

$\frac{x}{3x+3}+\frac{1}{3x+3}$ $=\frac{x+1}{3x+3}=\frac{(x+1)}{3(x+1)}$ $=\frac{1}{3}$ C	$\frac{1}{x^3}-\frac{x^2+1}{x^3}$	$\frac{x^2}{x^2+x}-\frac{1}{x^2+x}$	$\frac{-3x^2+1}{(2x+1)^2}+\frac{3x^2+2x}{(2x+1)^2}$
$\frac{2x-1}{x}-\frac{3x-2}{x}$	$\frac{1-2x}{x}-\frac{2-3x}{x}$	$\frac{x^2+x}{x^2+1}+\frac{1-x^2}{x^2+1}$	$\frac{x^2+x}{x^2-1}+\frac{1-x^2}{x^2-1}$
$\frac{x^2}{x^2-1}+\frac{2x+1}{x^2-1}$	$\frac{x}{2x+1}-\frac{x+1}{2x+1}$	$\frac{2x^2-2x}{(x-1)^2}-\frac{2x-2}{(x-1)^2}$	$\frac{2x^2-2x}{2(x-1)^2}-\frac{2x-2}{2(x-1)^2}$

Student Activity C

Section 7.3: Determining the LCD

Directions: For each of the expressions below, start by factoring both denominators. If a denominator **doesn't** factor, write that denominator **in parentheses**. Then find the LCD for the expression.

For example, the denominators in $\frac{x}{x+1}+\frac{3}{4x+4}$ would be factored like this:

$\frac{x}{(x+1)}+\frac{3}{4(x+1)}$ and the LCD would be $4(x+1)$.

	Expression	Expression with Factored Denominators	LCD
1.	$\frac{4}{x+3}+\frac{2}{x^2+3x}$	$\frac{4}{\_\_\_\_}+\frac{2}{\_\_\_\_}$	
2.	$\frac{5x}{x^2-25}-\frac{5}{x+5}$	$\frac{5x}{\_\_\_\_}-\frac{5}{\_\_\_\_}$	
3.	$\frac{3}{x^2+x-12}+\frac{2x}{x^2-x-6}$	$\frac{3}{\_\_\_\_}+\frac{2x}{\_\_\_\_}$	
4.	$\frac{x+3}{5x^3+25x^2}+\frac{2x}{3x+15}$	$\frac{x+3}{\_\_\_\_}+\frac{2x}{\_\_\_\_}$	
5.	$\frac{x}{x-2}-\frac{3}{2-x}$	$\frac{x}{\_\_\_\_}-\frac{3}{\_\_\_\_}$	
6.	$\frac{10}{8x^2-14x-15}-\frac{15}{8x^2-20x}$	$\frac{10}{\_\_\_\_}-\frac{15}{\_\_\_\_}$	
7.	$\frac{x^2+9}{3x^4-27x^2}+\frac{3x}{x^3-3x^2}$	$\frac{x^2+9}{\_\_\_\_}+\frac{3}{\_\_\_\_}$	
8.	$\frac{4}{15x^3+12x^2}-\frac{y}{3x^2}$	$\frac{4}{\_\_\_\_}-\frac{y}{\_\_\_\_}$	

Student Activity A

Section 7.4: The Missing Form of 1

Directions: In each equation below, there is a missing multiple with a form of 1 and a missing numerator. Fill in the empty spaces to complete each expression.

Example: $\frac{x-2}{x+3}\cdot\left[\frac{\quad}{\quad}\right]=\frac{\quad}{(x+3)(x-2)}$ becomes $\frac{x-2}{x+3}\cdot\left[\frac{x-2}{x-2}\right]=\frac{x^2-4x+4}{(x+3)(x-2)}$.

Hint for problems 6-10: Try factoring the denominator on the right hand side first.

1. $\frac{x+1}{x+2}\cdot\left[\frac{\quad}{\quad}\right]=\frac{\quad}{3x(x+2)}$

2. $\frac{x-5}{x+5}\cdot\left[\frac{\quad}{\quad}\right]=\frac{\quad}{(x+5)(x-5)}$

3. $\frac{x}{2-x}\cdot\left[\frac{\quad}{\quad}\right]=\frac{\quad}{x-2}$

4. $\frac{x^2+3x}{3x+2}\cdot\left[\frac{\quad}{\quad}\right]=\frac{\quad}{(3x+2)(x+3)}$

5. $\frac{4}{x+3}\cdot\left[\frac{\quad}{\quad}\right]=\frac{\quad}{(x+3)(x+4)(x+5)}$

6. $\frac{3x}{x+1}\cdot\left[\frac{\quad}{\quad}\right]=\frac{\quad}{x^2-1}$

   $(x-1)(x+1)$

7. $\frac{2x+3}{x+7}\cdot\left[\frac{\quad}{\quad}\right]=\frac{\quad}{x^2+14x+49}$

8. $\frac{16x}{4-x}\cdot\left[\frac{\quad}{\quad}\right]=\frac{\quad}{16-8x-x^2}$

9. $\frac{x^2}{5x+4}\cdot\left[\frac{\quad}{\quad}\right]=\frac{\quad}{30x^3+24x^2}$

10. $\frac{8+x}{4x+3}\cdot\left[\frac{\quad}{\quad}\right]=\frac{\quad}{20x^2+7x-6}$

Student Activity B

Section 7.4: The Reunion

Directions: Now that we've made it through addition, subtraction, multiplication, and division, it seems only fair to bring the whole gang back together for a reunion. Simplify each expression according to it's operation!

$\frac{x-1}{x+2}+\frac{x+3}{x+2}$	
$\frac{x-1}{x+2}-\frac{x+3}{x+2}$	
$\frac{x-1}{x+2}\cdot\frac{x+3}{x+2}$	
$\frac{x-1}{x+2}\div\frac{x+3}{x+2}$	

$\frac{x}{x-3}+\frac{x-1}{3-x}$	
$\frac{x}{x-3}-\frac{x-1}{3-x}$	
$\frac{x}{x-3}\cdot\frac{x-1}{3-x}$	
$\frac{x}{x-3}\div\frac{x-1}{3-x}$	

$\frac{12}{x+2}+\frac{4x+8}{x+1}$	
$\frac{12}{x+2}-\frac{4x+8}{x+1}$	
$\frac{12}{x+2}\cdot\frac{4x+8}{x+1}$	
$\frac{12}{x+2}\div\frac{4x+8}{x+1}$	

Student Activity C

Section 7.4: Tempting Expressions

Directions: In all the expressions below, you will find yourself tempted to do incorrect mathematics by visually pleasing expressions. Of course, just because it looks good, that doesn't mean it is the right thing to do. ☺ Think carefully about the steps involved in each expression.

1. Add: $\frac{5}{x}+\frac{x}{5}$

2. Subtract: $\frac{2}{x+2}-\frac{1}{x}$

3. Add: $\frac{x}{x+4}+\frac{x+4}{x}$

4. Divide: $\frac{x-1}{x+6}\div\frac{x+6}{x^2-36}$

5. Simplify: $\frac{2x+8}{2x+4}$

6. Multiply: $\frac{x^2+4x-12}{x^2+4x-32}\cdot\frac{x^2+10x+16}{x^2+10x+24}$

7. Subtract: $\frac{3x}{x+2}-\frac{x+2}{3x+3}$

8. Simplify: $\frac{5x^2+25x}{5x^2-125}$

9. Subtract: $\frac{x^2}{x^2-9}-9$

10. Add: $\frac{2}{x+1}+\frac{2}{1-x}$

Assessment 7B

Chapter 7: Mid-Chapter Assessment for Understanding

For each of the following, describe the type of problem and the strategies and key steps to remember while doing the problem. You do **not** have to complete the problems.

	Type of Problem	Strategies and Key Steps
1. Subtract: $\frac{4x}{x+1}-\frac{2x+3}{x+1}$.		
2. Find the LCD for $\frac{5}{2x-6}$ and $\frac{3}{x^2-9}$.		
3. Simplify: $\frac{x^2-2x-35}{7x+35}$.		
4. Add: $\frac{3}{x+2}+\frac{4}{x+3}$.		
5. Multiply: $\frac{x^2+8x}{6x+18}\cdot\frac{12x-24}{x^2+6x-16}$.		
6. Where is $\frac{x+3}{x^2-16}$ undefined?		
7. Add: $\frac{4x+5}{3x+5}+\frac{2x+5}{3x+5}$.		
8. Divide: $\frac{x^2+4x}{x+8}\div\frac{x^3+4x^2}{x^2+6x-16}$.		
9. Simplify: $\frac{10-x}{x-10}$.		
10. Subtract: $\frac{x^2+3}{x^2-4}-\frac{x+5}{x-2}$.		

Student Activity A

Section 7.5: Tic-Tac-Toe on Complex Fraction Pieces

Directions: In every box in the tic-tac-toe grid, there is a rational expression and a simplified form. If the two are equivalent, then circle the simplified form (thus placing an **O** on the square). If the expression has **not** been simplified correctly, then put an **X** over the incorrect simplification.

Tic-tac-toe Game#1:

$x^2\left(\frac{3}{x}\right)$ $3x$	$(2-x)\left(\frac{1}{x-2}\right)$ -1	$\left(\frac{2}{x}\right)\left(\frac{x^2}{4}\right)$ $2x$
$\left(24x^3\right)\left(\frac{3}{8x^2}\right)$ $9x$	$(15x)\left(\frac{4}{25x^2}\right)$ $\frac{4}{5x}$	$\left(\frac{x+3}{2(x-3)}\right)(x-3)$ $\frac{x+3}{2}$
$\left(\frac{3-2x}{x^3}\right)\left(x^3\right)$ $3x^3-2x^4$	$\left(81x^4\right)\left(\frac{9}{9x^4}\right)$ 81	$\left(\frac{1}{x-1}\right)(1-x)$ 1

Tic-tac-toe Game#2:

$\left(8x^2\right)\left(\frac{3}{4x}-2\right)$ $6x-16x^2$	$x^2\left(\frac{5}{x^2}-\frac{2}{x}\right)$ $5-\frac{2}{x}$	$x^3\left(\frac{5}{x^2}+4x\right)$ $5x+4x^2$
$x\left(4+\frac{x+1}{x}\right)$ $5x+1$	$\left(\frac{x}{5}\right)\left(\frac{25}{x}+5\right)$ $5+5x$	$\left(\frac{x^2}{3}\right)\left(6-\frac{9}{10x^2}\right)$ $2x^2-\frac{3}{10}$
$\left(32x^4\right)\left(\frac{1}{16x}-\frac{1}{32x^4}\right)$ $2x^3-1$	$\left(\frac{24x}{y}\right)\left(\frac{y}{8x}-y\right)$ $3-24x$	$\left(42x^2\right)\left(\frac{1}{6x}+\frac{1}{7x}\right)$ 5

Student Activity B

Section 7.5: Double the Fun on Complex Fractions

Directions: For each of the complex fractions below, simplify the expression using both methods. The result should be the same for both methods. If they are not, go back and look for a mistake. The first one has been **started** for you.

Method I: Using division. **Method II:** Multiplying by the LCD.

1.	**Method I**	$\dfrac{\frac{1}{x}+4}{\frac{5}{x}-1} = \dfrac{\frac{1}{x}+4\left(\frac{x}{x}\right)}{\frac{5}{x}-1\left(\frac{x}{x}\right)} = \dfrac{\frac{1+4x}{x}}{\frac{5-x}{x}} =$
	Method II	$\dfrac{\frac{1}{x}+4}{\frac{5}{x}-1} = \dfrac{\left(\frac{1}{x}+4\right)}{\left(\frac{5}{x}-1\right)} \cdot \dfrac{\frac{x}{1}}{\frac{x}{1}} =$
2.	**Method I**	$\dfrac{5-\frac{1}{x^2}}{\frac{2}{x}+x}$
	Method II	$\dfrac{5-\frac{1}{x^2}}{\frac{2}{x}+x}$
3.	**Method I**	$\dfrac{\frac{1}{x+1}+1}{1+\frac{2}{x}}$
	Method II	$\dfrac{\frac{1}{x+1}+1}{1+\frac{2}{x}}$
4.	**Method I**	$\dfrac{\frac{1}{x}}{\frac{-1}{x+2}+\frac{1}{x}}$
	Method II	$\dfrac{\frac{1}{x}}{\frac{-1}{x+2}+\frac{1}{x}}$

Student Activity C

Section 7.5: One of Us is Wrong!

Directions: Lou and Stu have both simplified the given complex fraction. Unfortunately, they have different answers. 1) Choose a test value for x and use it to evaluate the complex fraction and the students' answers. Give the student that is correct a ☺ for their work. 2) Then determine where the unfortunate student made the error.

1. Given expression	Lou's Work	Stu's Work
$\dfrac{\frac{2x+5}{x+2}}{\frac{x+2}{2x+5}}$ Test value: $x =$ ____	$\dfrac{\frac{2x+5}{x+2}}{\frac{x+2}{2x+5}} = \dfrac{2x+5}{x+2} \div \dfrac{x+2}{2x+5}$ $= \dfrac{2x+5}{x+2} \cdot \dfrac{2x+5}{x+2} = \dfrac{(2x+5)^2}{(x+2)^2}$	$\dfrac{\frac{2x+5}{x+2}}{\frac{x+2}{2x+5}} = \dfrac{\frac{2x+5}{\cancel{x+2}^1}}{\frac{\cancel{x+2}_1}{2x+5}}$ $= \dfrac{\cancel{2x+5}^1}{\cancel{2x+5}_1} = 1$
Expression value:	Expression value:	Expression value:

2. Given expression	Lou's Work	Stu's Work
$\dfrac{2x - \frac{1}{2x}}{\frac{1}{2x} + 2x}$ Test value: $x =$ ____	$\dfrac{2x - \frac{1}{2x}}{\frac{1}{2x} + 2x} = \dfrac{\cancel{2x}^1 - \frac{1}{\cancel{2x}^1}}{\frac{1}{\cancel{2x}^1} + \cancel{2x}^1}$ $= \dfrac{1-1}{1+1}$ $= \dfrac{0}{2} = 0$	$\dfrac{2x - \frac{1}{2x}}{\frac{1}{2x} + 2x} = \dfrac{\left(2x - \frac{1}{2x}\right)}{\left(\frac{1}{2x} + 2x\right)} \cdot \dfrac{\frac{2x}{1}}{\frac{2x}{1}}$ $= \dfrac{2x\left(\frac{2x}{1}\right) - \frac{1}{\cancel{2x}_1}\left(\frac{\cancel{2x}^1}{1}\right)}{\frac{1}{\cancel{2x}_1}\left(\frac{\cancel{2x}^1}{1}\right) + 2x\left(\frac{2x}{1}\right)} = \dfrac{4x^2 - 1}{1 + 4x^2}$
Expression value:	Expression value:	Expression value:

3. Given expression	Lou's Work	Stu's Work
$\dfrac{\frac{x^2}{2}}{\frac{2}{x} + \frac{4}{x}}$ Test value: $x =$ ____	$\dfrac{\frac{x^2}{2}}{\frac{2}{x} + \frac{4}{x}} = \dfrac{x^2}{2} \div \left(\dfrac{2}{x} + \dfrac{4}{x}\right)$ $= \dfrac{x^2}{2} \cdot \left(\dfrac{x}{2} + \dfrac{x}{4}\right) = \dfrac{x^2}{2} \cdot \left(\dfrac{2x}{4} + \dfrac{x}{4}\right)$ $= \dfrac{x^2}{2} \cdot \left(\dfrac{3x}{4}\right) = \dfrac{3x^3}{8}$	$\dfrac{\frac{x^2}{2}}{\frac{2}{x} + \frac{4}{x}} = \dfrac{\frac{x^2}{2}}{\frac{6}{x}} = \dfrac{x^2}{2} \div \dfrac{6}{x}$ $= \dfrac{x^2}{2} \cdot \dfrac{x}{6} = \dfrac{x^3}{12}$
Expression value:	Expression value:	Expression value:

Student Activity A

Section 7.6: Checking Solutions to Rational Equations with a Calculator

You can use a calculator to check if the solution that you find to a rational equation is correct. For example, let's see if -10 is a solution of the equation $\frac{4}{x+2}=\frac{5}{x}$.

a) Construct the parentheses skeleton for the equation: $\frac{4}{(\ )+2}=\frac{5}{(\ )}$

b) Use extra parentheses or brackets to group the numerator or denominator if they contain more than one term: $\frac{4}{[(\ )+2]}=\frac{5}{(\ )}$

c) Insert the possible solution: $\frac{4}{[(-10)+2]}=\frac{5}{(-10)}$

d) Enter each side into your calculator, using extra parentheses to group the numerator or denominator if there is more than one term:

Left side: **4/((-10)+2)** Right side: **5/(-10)**

In this case, both sides result in -0.5, so -10 **is** a solution to the equation $\frac{4}{x+2}=\frac{5}{x}$.

Directions: Fill in the table below. If the one side of the equation is *undefined* for a value, say so (your calculator will likely give you an "error" message if this is the case). The first row has been done for you.

Equation	Parentheses skeleton and extra parentheses needed for calculator	Evaluate for…	Left side of equation	Right side of equation	Is this value a solution?
$\frac{5}{3+2x}=\frac{2}{x+1}$	$\frac{5}{[3+2(\ )]}=\frac{2}{[(\ )+1]}$	$x=1$	1	1	yes
		$x=0$			
		$x=-3$			
$\frac{9}{x}=\frac{12}{x-1}$		$x=1$			
		$x=0$			
		$x=-3$			
$\frac{1-x}{x}=\frac{x-1}{3}$		$x=1$			
		$x=0$			
		$x=-3$			
$\frac{x}{x+3}=3x+\frac{x}{x+1}$		$x=1$			
		$x=0$			
		$x=-3$			

Student Activity B

Section 7.6: Can't Use That!

Directions: Examine each of the equations below and decide what values cannot be allowed because of a division by zero problem. For the "disallowed" values, shade in the corresponding value in the grid below. The first one has been done for you.

1. $\frac{x+5}{x-2}=\frac{3}{x}$; $x \neq 0,2$
2. $\frac{x-4}{x+3}+\frac{2}{x-2}=5$
3. $\frac{x}{3}+\frac{2}{x-6}=1$
4. $\frac{3}{2x-1}=\frac{4}{x+5}$
5. $\frac{3}{4x}+\frac{1}{2}=\frac{2-x}{4-x}$
6. $\frac{5}{x+7}-\frac{3x}{x-10}=6$
7. $\frac{x+3}{x-3}=\frac{4}{3x+2}$
8. $1+\frac{x}{5}=\frac{5x}{x-8}$
9. $\frac{1}{4x-1}=\frac{4}{x}$
10. $\frac{9}{9+x}-2=\frac{1-x}{2x}$
11. $\frac{4}{x}=\frac{2x+1}{10}$
12. $\frac{x+6}{x-7}-\frac{9}{3x}=4$
13. $\frac{x+1}{x+5}+\frac{2}{5x}=3$
14. $\frac{x-3}{3x-4}+\frac{2x}{x+3}=8$
15. $\frac{x}{12}+5=\frac{x+2}{x-2}$

$x \neq 0,1,4$	$x \neq -3,0,2$	$x \neq 0,2,4$	$x \neq -5,0,2$	$x \neq 0,2$
$x \neq 0,\frac{1}{4},1$	$x \neq -3,2,4$	$x \neq 0,8$	$x \neq 0,4$	$x \neq 2,4$
$x \neq 0,\frac{1}{4}$	$x \neq -3,2$	$x \neq 8$	$x \neq 6$	$x \neq -5,0$
$x \neq -2,3$	$x \neq -3,-\frac{2}{3},3$	$x \neq -7,10$	$x \neq 0,6$	$x \neq 3,6$
$x \neq 0$	$x \neq -\frac{2}{3},3$	$x \neq 0,7$	$x \neq \frac{4}{3},3$	$x \neq 2$
$x \neq 0,1,9$	$x \neq -5,\frac{1}{2}$	$x \neq -6,0$	$x \neq -7,0,10$	$x \neq -2,0,2$
$x \neq 0,9$	$x \neq -5,1$	$x \neq 0,-1$	$x \neq -3,0,3,4$	$x \neq -1,0,5$

Student Activity C

Section 7.6: One Step at a Time for Rational Equations

When you are solving rational equations, the first steps can be the hardest ones. In this activity we practice taking those first few steps. Find the LCD, multiply by the LCD on both sides of the equation, distribute (where necessary), simplify each side of the equation, and then you finally have an equation that can be solved with prior methods.

Example:

	Equation: $\frac{3}{x}+\frac{4}{x+2}=3-\frac{1}{x}$ LCD: $x(x+2)$ $x(x+2)\left(\frac{3}{x}+\frac{4}{x+2}\right)=\left(3-\frac{1}{x}\right)\cdot x(x+2)$	
	Left side	**Right side**
Distribute	$x(x+2)\left(\frac{3}{x}\right)+x(x+2)\left(\frac{4}{x+2}\right)$	$(3)(x)(x+2)-\frac{1}{x}(x)(x+2)$
Simplify	$3(x+2)+4x$ $3x+6+4x$ $7x+6$	$3x(x+2)-(x+2)$ $3x^2+6x-x-2$ $3x^2+5x-2$
And reunite!	$7x+6=3x^2+5x-2$	

1.

	Equation: $\frac{3x+1}{x-3}=\frac{6x+2}{x^2-4x+3}$ LCD: __________	
	Left side	**Right side**
Distribute		
Simplify		
And reunite!		

2.

	Equation: $1-\frac{27}{x^2+x-12}=-\frac{1}{x-3}$ LCD: ______________	
	Left side	**Right side**
Distribute		
Simplify		
And reunite!		

3.

	Equation: $\frac{2x+5}{4x+5}=\frac{-2}{2x+1}-\frac{5}{4x+5}$ LCD: ______________	
	Left side	**Right side**
Distribute		
Simplify		
And reunite!		

Student Activity D

Section 7.6: Match Up on Simple Rational Equations

Match-up: Solve each rational equation and check the result. Then choose the letter that corresponds to this result. Some problems may have more than one answer. If there is no solution, then choose E (No Solution). If the result is not among the choices, then choose F (None of these). The first one has been done for you.

A 2 **B** -1 **C** 1 **D** $\frac{1}{2}$ **E** No Solution **F** None of These

$\frac{5x+5}{x+4}=2$ $5x+5=2(x+4)$ $5x+5=2x+8$ $3x=3$ $x=1$ C	$3=\frac{8-x}{x+2}$	$\frac{4x-2}{x+2}=4$
$\frac{x-3}{2x+3}=5$	$\frac{3}{x}+\frac{4}{x+2}=-\frac{1}{x}$	$\frac{5}{2}=\frac{11+x}{x+5}$
$\frac{x+2}{x-4}=\frac{2}{x-3}$	$\frac{x-4}{x+1}=-\frac{2x+7}{(x+2)(x+1)}$	$\frac{3x^2+5}{7x+5}=2$

Student Activity A

Section 7.7: Working with the Smaller Pieces

Practice with Uniform Motion Problems

We use the formula $d = rt$ to solve motion problems, where *d* is the distance, *r* is the rate, and *t* is the time.

1. A husband and wife go jogging together. The husband can run 2 miles an hour faster than the wife.

a. If the wife runs at a rate of x miles per hour, what is the rate of the husband?

____

b. If they run 10 miles, write an expression that describes the *time* it takes for the wife to run this distance. __________

c. If they run 10 miles, write an expression that describes the *time* it takes for the husband to run this distance. __________

2. Two trains (the Midnight Express and the Night Special) leave a station traveling in opposite directions. The Midnight Express is 10 mph faster than the Night Special, which travels at 50 mph. How far apart are the trains after two hours? __________ Every hour the trains are __________ more miles apart.

3. The same two trains as the previous problem leave a station traveling in the *same* direction. How far apart are the trains after two hours? __________ Every hour the trains are __________ more miles apart.

Practice with Shared-work Problems

Work completed, work rate, and time can be related using the formula: $w = rt$
Often the "work completed" is represented by 1 (for 1 job completed).

4. Suppose Joann can clean her house in 2 hours and Marty can clean the house in 5 hours. If JoAnn and Marty work together to clean the house, will it take

a. less than 2 hours **b.** between 2 and 5 hours **c.** more than 5 hours

5. If it takes you 2 hours to clean one pool, how long will it take you to clean 3 pools of the same size? __________

6. It normally takes you 2 hours to clean the pool. If you have help to clean it, will it take you

a. more than 2 hours **b.** less than 2 hours

7. Saundra can type 5 pages in 30 minutes. How long should it take her to type 8 pages? __________

8. Shell can paint 500 square feet of wall in 40 minutes. What is his work rate in square feet per minute? __________

Guided Learning Activity

Section 7.7: Understanding Shared-Work Problems

Work completed = rate of work · time completed or $W = rt$ where

W = fraction of one job that is completed

r = Rate of work = $\dfrac{\text{amount of work}}{\text{time}}$

t = time (in the same units as the rate of work)

Start each problem below by first designating what a "job" is. Then fill in the blank boxes in the table, write an equation that can be used to solve the problem, and find the solution.

---

1. If the community pool is filled from the reserve water tower, it will take 6 hours to fill the pool. If the pool is filled from the city water pipes, it will take 4 hours to fill the pool. How long will it take to fill the pool if water from both the reserve water tower and the city water pipes are used?

Job:	Rate	Time	Work completed
Reserve water tower			
City water pipes			

Equation: **Solution:**

2. It takes Mr. Dupree 5 hours to audit 12 bank files. His associate, Ms. Carmichael can audit 16 files in 4 hours. If they work together, how long will it take them to audit 30 files?

Job:	Rate	Time	Work completed
Mr. Dupree			
Ms. Carmichael			

Equation: **Solution:**

3. Macon Construction can erect 100 square feet of rock facing on an exterior wall in 8 hours. Thomson Masonry can erect 200 square feet of rock facing in 12 hours. If you hire the two crews to work together, but Thomson Masonry starts working 2 hours after Macon Construction, how long will it take them to erect 250 square feet of rock facing?

Job:	Rate	Time	Work completed
Macon Construction			
Thomson Masonry			

Equation: **Solution:**

Student Activity A

Section 7.8: Working with the Language of Proportions

Directions: Fill in the table using the language of proportions. Find the units for each ratio and set up the proportion. You do not need to solve the problem. The first one has been done for you.

Problem	Fill in the Missing Values		Proportion $\frac{A}{B} = \frac{C}{D}$
	Ratio 1	Ratio 2	
1. If 12 doughnuts cost \$2.50, how many doughnuts can you buy for \$5.00?	$\frac{12 \text{ doughnuts}}{\$2.50}$	$\frac{x \text{ doughnuts}}{\$5.00}$	$\frac{12}{2.50} = \frac{x}{5.00}$
2. A chocolate chip cookie recipe uses 3 cups of flour to make 5 dozen cookies, how much flour is required to make 12 dozen cookies?	$\frac{\square \text{ cups flour}}{\square \text{ dozen cookies}}$	$\frac{\square \text{ cups flour}}{\square \text{ dozen cookies}}$	
3. If it takes Joel 4.5 hours to run a 26.2 mile marathon, how many minutes does it take him to run 1 mile?	$\frac{\square \text{ miles}}{\square \text{ minutes}}$	$\frac{\square \text{ miles}}{\square \text{ minutes}}$	
4. If jeans are on sale at three pairs for \$50, how much will seven pairs cost?	$\frac{\square \text{ pairs}}{\$\square}$	$\frac{\square \text{ pairs}}{\$\square}$	
5. If Jamie gets paid \$1000 for a 40-hour work week, how many hours will he have to work to earn \$800?	$\frac{\$\square}{\square \text{ hours}}$	$\frac{\$\square}{\square \text{ hours}}$	
6. There are 1160 calories in a 32 oz. chocolate shake from McDonalds. How many calories are in the 24 oz. shake? Source: www.mcdonalds.com	$\frac{\square \text{ calories}}{\square \text{ oz.}}$	$\frac{\square \text{ calories}}{\square \text{ oz.}}$	
7. According to American Red Cross guidelines, CPR should consist of cycles of 30 chest compressions followed by 2 breaths. If a rescuer has given 270 compressions, how many breaths should she have given? Source: www.redcross.org	$\frac{\square \text{ compressions}}{\square \text{ breaths}}$	$\frac{\square \text{ compressions}}{\square \text{ breaths}}$	

Guided Learning Activity

Section 7.8: Orienting Similar Triangles

Similar triangles are triangles with the same shape, but not necessarily the same size. These pairs of triangles can be easy to spot when the orientation of the triangles is the same.

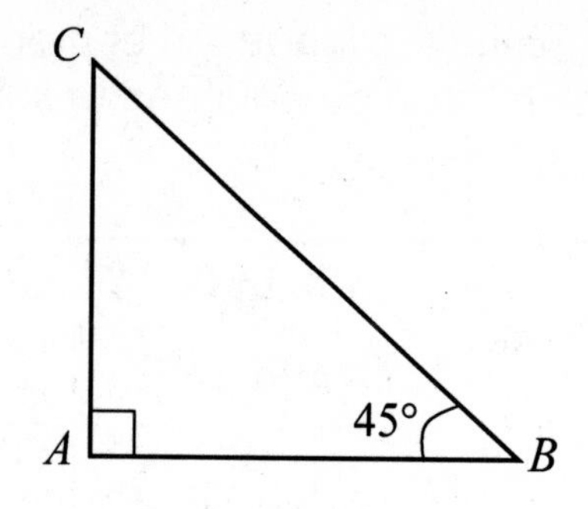

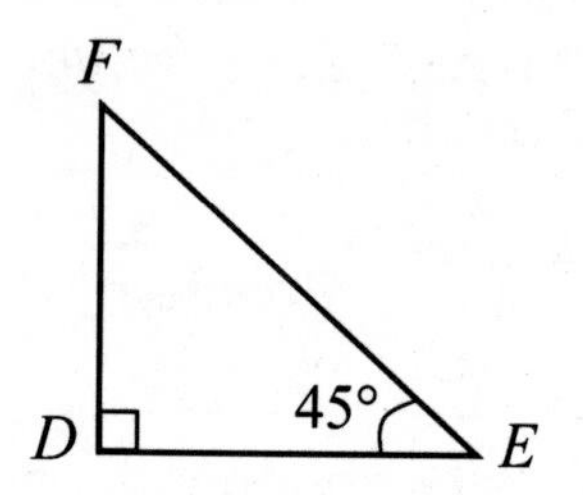

To the right, the two triangles are similar, and we would write that relationship like this: $\Delta ABC \sim \Delta DEF$.

For two triangles to be similar, two of the angle measures in the triangles must be the same (this forces all three angles to be the same).

Example:

The sum of the angles in a triangle is ______.

Find the missing angles in triangles ΔJKL and ΔMNP to the right.

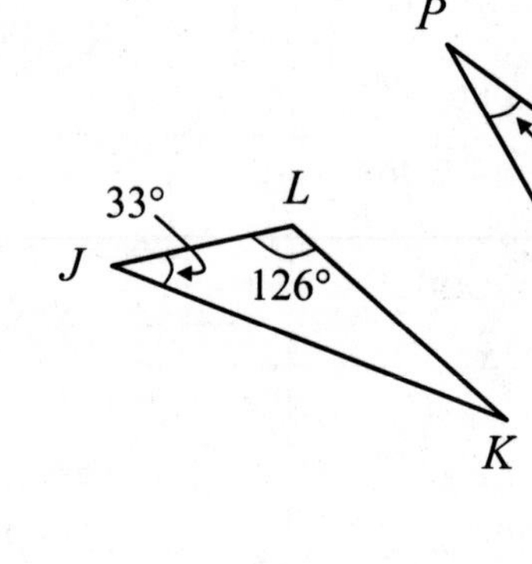

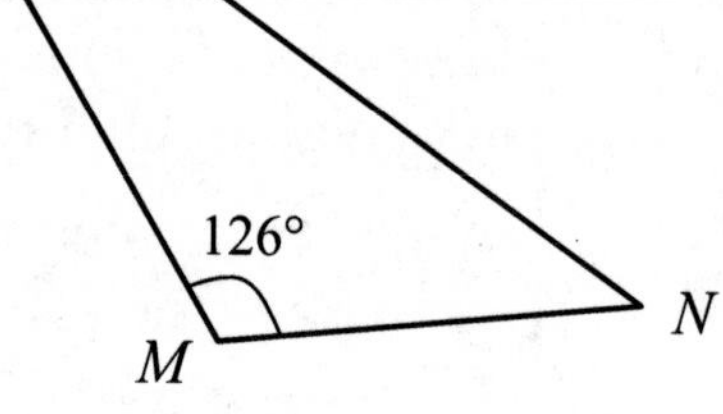

Are the two triangles similar? _____

Which angles are equal? $\angle J =$ _____ $\angle K =$ _____ $\angle L =$ _____

Use a different color to highlight each pair of similar angles.

Use a different color to highlight each pair of similar sides.

Which sides are similar? $JK \sim$ ____ $KL \sim$ ____ $LJ \sim$ ____

Because of these side similarities, we can set up equal ratios of similar sides (fill in the missing denominators): $\frac{JK}{\quad} = \frac{KL}{\quad} = \frac{LJ}{\quad}$

This can be written as three different proportions. What are they?

$\frac{\quad}{\quad} = \frac{\quad}{\quad}$ $\frac{\quad}{\quad} = \frac{\quad}{\quad}$ $\frac{\quad}{\quad} = \frac{\quad}{\quad}$

Triangles	Are the triangles similar? If so, state the side similarities.	Find the lengths of the side(s) indicated by variables.
1. 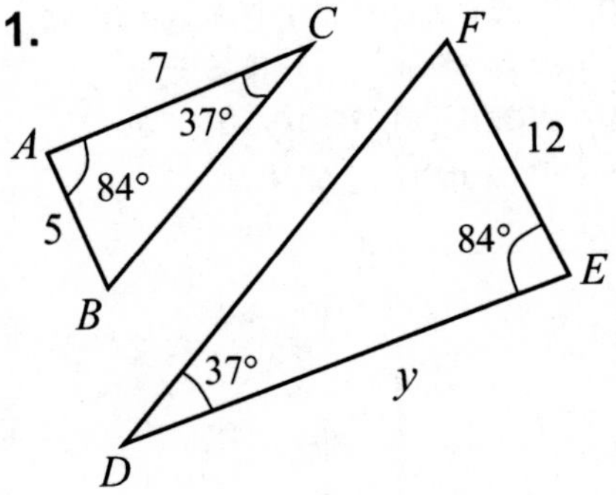 A, B, C, D, E, F; 7; 5; 12; y; 37°; 84°; 84°; 37°		
2. 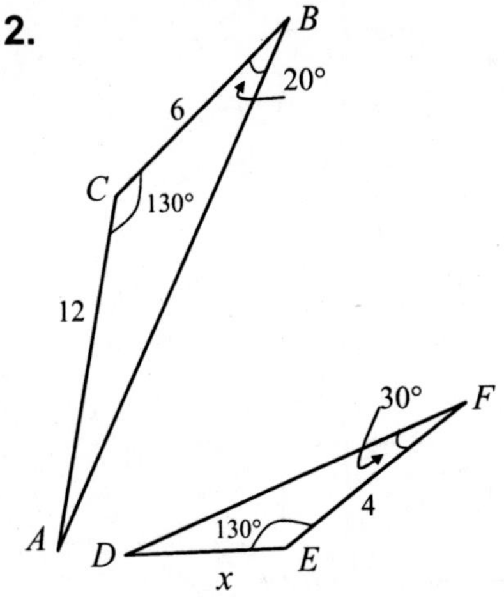 A, B, C, D, E, F; 6; 12; 4; x; 20°; 130°; 30°; 130°		
3. A, B, C, D, E, F; 20; 21; 15; 8; 46°; 62°		
4. A, B, C, D, E, F; 6; y; z; x; 2.5; 1.5; 55°; 55°		

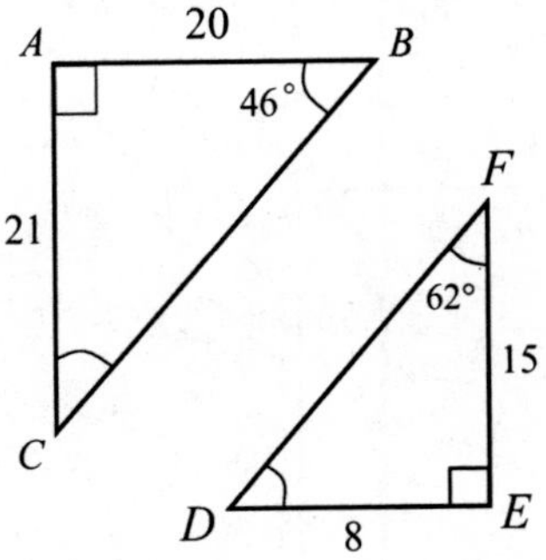

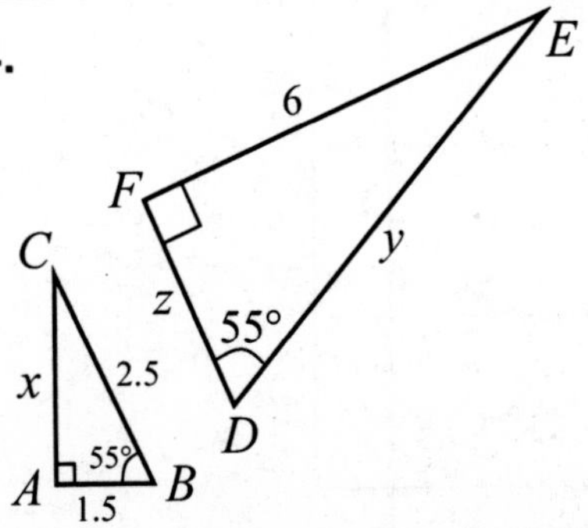

Student Activity B

Section 7.8: Proportion Heteronyms

Directions: In writing, there are words that are spelled the same but have different pronunciations and different definitions; these are called heteronyms. Many mathematical expressions and equations look similar but are really very different (almost like mathematical heteronyms). In each set of "proportion heteronyms" below, first **identify** whether you are being given an equation or an expression. If it is an expression, simplify it. If it is an equation, solve it. **Do your work (simplifying or solving) on another sheet of paper.** The first line has been done for you.

Heteronym Variation I			
	$\frac{5}{x}=\frac{2}{3}$	Equation.	$x=\frac{15}{2}$
	$\frac{5}{x}-\frac{2}{3}$		
	$\frac{5}{x}+\frac{2}{3}$		
	$\frac{5+2}{x+3}$		
	$\frac{5}{x}-\frac{2}{3}=0$		
	$\frac{5}{x}\div\frac{2}{3}$		

Heteronym Variation II			
	$\frac{x+1}{8}+\frac{x}{2}$		
	$\frac{x+1}{8}\div\frac{x}{2}$		
	$\frac{x+1}{8}=\frac{x}{2}$		
	$x+\frac{1}{8}=\frac{x}{2}$		
	$\frac{x}{8}+1=\frac{x}{2}$		
	$\frac{x}{8}+1+\frac{x}{2}$		

Student Activity C

Section 7.8: Equation Lineup

Directions: We have solved many different types of equations at this point in our learning. For each equation below, categorize the type of equation, write the strategies and key steps for remembering how to solve it, then solve the equation. The first one has been started for you.

Equation Types: Linear, quadratic, rational

1. $3x^2 - x = 4$
Type of Equation? Quadratic
Strategies and Key Steps Get =0 on one side of the equation, then factor to solve.
Solve it!

2. $-7(x+2) = 1 + 5(x-1)$
Type of Equation?
Strategies and Key Steps
Solve it!

3. $\frac{x+4}{x-4} = \frac{x-1}{x+3}$
Type of Equation?
Strategies and Key Steps
Solve it!

4. $-\frac{3}{8} = \frac{1}{4} - \frac{5x}{16}$
Type of Equation?
Strategies and Key Steps
Solve it!

5. $2x^2 = 3(x+3)$
Type of Equation?
Strategies and Key Steps
Solve it!

6. $\frac{x}{x+4} - \frac{1}{x+1} = 1$
Type of Equation?
Strategies and Key Steps
Solve it!

7. $5(2x+1) = 3(x+4) + 7x$
Type of Equation?
Strategies and Key Steps
Solve it!

8. $x^2 = 10x$
Type of Equation?
Strategies and Key Steps
Solve it!

Assessment 7C

Chapter 7: End-of-Chapter Assessment for Understanding

For each of the following, describe the type of problem and the strategies and key steps to remember while doing the problem. You do **not** have to complete the problems.

	Type of Problem	Strategies, Rules and Key steps
1. Multiply: $\frac{3x^2+12x}{x-6}\cdot\frac{x^2-4x-12}{x^2+2x}$		
2. If you add the same number to the numerator and denominator of $\frac{7}{9}$, the result is $\frac{5}{6}$. What is the number?		
3. Add: $\frac{4}{x+3}+\frac{1}{3x}$		
4. Solve: $\frac{2}{x+6}+\frac{5}{x^2+6x}=1$		
5. Subtract: $\frac{3x+6}{x+1}-\frac{x+4}{x+1}$		
6. Simplify: $\dfrac{\frac{2}{3}+\frac{5}{x}}{\frac{1}{2x}-4}$		
7. Simplify: $\frac{x^2+9x+14}{4-x^2}$		
8. On a map, a distance of 1.5 inches represents 1.125 miles. How many miles does 2.2 inches on the map represent?		

Assessment 7D

Chapter 7: Metacognitive Skills Assessment

Metacognitive skills refer to the ability to judge how well you have learned something and to effectively direct your own learning and studying. This is a self evaluation tool designed to help you focus your studying and to improve your metacognitive skills with regards to this math class.

Fill the 1st column out before you begin studying.
Fill the 2nd column out after you study and before you take the test.
Go back to this page after your test and circle any of the ratings that you would now change – this identifies the "disconnects" between what you think you know well and what you actually know well.

Use the scale below to assign a number to each topic.

5 I am confident I can do any problems in this category correctly.
4 I am confident I can do most of the problems in this category correctly.
3 I understand how to do the problems in this category, but I still make a lot of mistakes.
2 I feel unsure about how to do these problems.
1 I know I don't understand how to do these problems.

Topic or Skill	Before Studying	After Studying
Factoring polynomials.		
Correctly identifying opposite factors.		
Simplifying a rational expression.		
Finding the values where a rational expression is undefined.		
Multiplying rational expressions.		
Dividing rational expressions.		
Simplifying expressions that involve units of measurement.		
Adding or subtracting rational expressions with like denominators.		
Finding the LCD of two (or more) rational expressions.		
Building an equivalent rational expression with a desired denominator.		
Understanding why you need a common denominator when you add or subtract rational expressions.		
Identifying denominators that are opposites.		
Adding or subtracting rational expressions with unlike denominators.		
Identifying a complex fraction.		
Simplifying a complex fraction using Method I (simplifying the numerator & denominator and then dividing).		
Simplifying a complex fraction using Method II (finding the LCD for all terms in the complex fraction and then multiplying by LCD/LCD).		
Finding the values where a rational *equation* is undefined and knowing why it is important to find these values.		
Clearing the fractions from a rational equation.		

Continued on the next page.

Use the scale below to assign a number to each topic.

5 I am confident I can do any problems in this category correctly.
4 I am confident I can do most of the problems in this category correctly.
3 I understand how to do the problems in this category, but I still make a lot of mistakes.
2 I feel unsure about how to do these problems.
1 I know I don't understand how to do these problems.

Topic or Skill	**Before Studying**	**After Studying**
Understanding why you cannot "clear the fractions" from a rational expression.		
Solving a linear equation that may result from a rational equation.		
Solving a quadratic equation that may result from a rational equation.		
Checking the solutions of a rational equation.		
Solving a formula containing rational expressions for one of the variables.		
Setting up a table of information for a uniform motion application problem.		
Setting up a table of information for a shared-work application problem.		
Setting up a table of information for an investment application problem.		
Converting the information in a table to an equation that can be solved.		
Writing ratios from sentences.		
Creating a proportion from an application problem.		
Solving a proportion.		
Understanding the specific properties that allow a proportion equation to be solved differently from other rational equations.		
Knowing the types of triangles that you can apply the Pythagorean Theorem to, and being able to correctly set up the Pythagorean Theorem for those triangles.		
Knowing what makes two triangles similar.		
Setting up a proportion of sides for two similar triangles.		

Student Workbook: Chapter 8

Table of Contents: *Transition to Intermediate Algebra*

Assessment 8A

Pretest and Diagnostic Tool: Transition to Intermediate Algebra

Directions: Complete this assessment without looking back at your notes or your book. **Do not use a calculator on this assessment.**

1. Solve: $-3x+2=14$

2. Solve: $2(x+4)=5+2x$

3. In which problems do you multiply or divide by a negative number to solve? ___________

a. Solve: $-2x=6$

b. Solve: $-2+x=6$

c. Solve: $\frac{1}{2}x=6$

d. Solve: $-x=6$

4. Write $\{x \mid x \leq 4\}$ in interval notation. __________

5. $A=\{3,5,7\}$ $\quad B=\{3,4,5\}$
Which numbers are in both A and B?

____

6. $A=\{3,5,7\}$ $\quad B=\{3,4,5\}$
List all the numbers that appear in either A or B: _____________

7. Solve: $2 \leq x-5 \leq 8$ and write the answer in interval notation.

8. $|2-5|=$____

9. $|-3 \cdot 2+1|-5=$____

10. Is 2 a solution of $|x-5|>2$? _____

11. Is -3 a solution of $|x|+2=-1$?

_____

12. Factor: $x^2-8x+15$

13. Factor: $15x^3-45x^2$

14. Factor: $ax+2a+bx+2b$

15. Factor: $3x^2-11x-20$

16. Factor: x^2-25

17. Factor: x^3-125

18. Factor: x^2+16

19. Solve: $\frac{1}{3}+\frac{2}{15x}=\frac{4}{5x}$

20. Simplify: $\dfrac{x^2-4}{x^2-2x-8}$

21. Simplify: $\dfrac{\frac{1}{x}+2}{4+\frac{3}{x}}$

22. Subtract: $\dfrac{5x+6}{x-3}-\dfrac{4x-1}{x-3}$

23. Find the slope between $(5,225)$ and $(8,300)$.

24. Identify the slope and y-intercept in $2x+3y=12$.

Slope _______ y-int _______

25. Write the slope-intercept equation of the line with slope -2 that passes through $(1,5)$.

26. Let $f(x)=x^2-4x$, find $f(-3)$.

27. Let $g(x)=\dfrac{x}{1-x}$, find $g(2)$.

28. Simplify: $(x+5)^2-2(x+5)$

29. Solve for k: $2.5=\dfrac{k}{3}$

30. Solve for k in $y=kx^2$ if $x=2$ and $y=20$.

Guided Learning Activity

Section 8.1: Remembering Inequalities

Recall that solving an inequality is the same as solving an equation, with one exception:

When you multiply or divide by a negative number, reverse the direction of the inequality.

Also, when we solve inequalities, we can write answers in two ways: set-builder notation or interval notation. Here are some reminders about interval notation:

- Use $-\infty$ and ∞ to denote the "ends" of the number line.
- Always give intervals from LEFT to RIGHT on the number line.
- Use (or) to denote an endpoint that is approached, but not included.
- Use [or] to denote an endpoint that is included.

Directions: Solve each inequality. Then graph the solution, and write the solution in both set-builder and interval notation.

1. Solve: 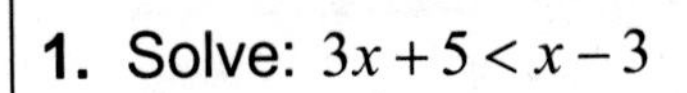 $3x+5<x-3$

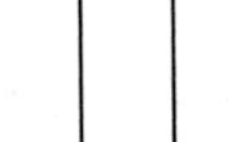

Graph: 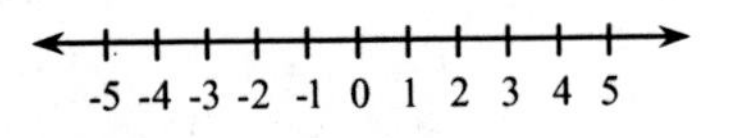

Set-builder notation:

Interval notation:

2. Solve: 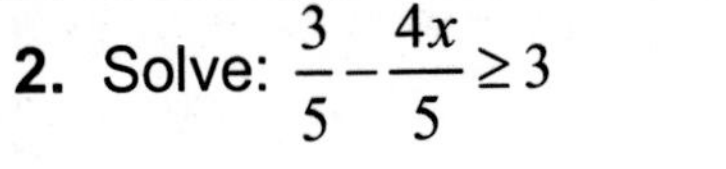 $\frac{3}{5}-\frac{4x}{5}\geq 3$

Graph: 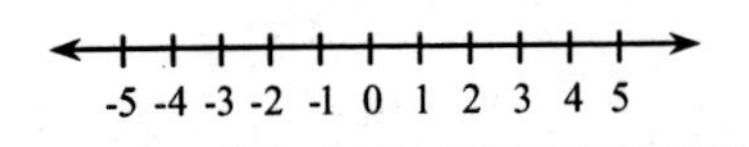

Set-builder notation:

Interval notation:

3. Solve: $3(x+4)<15+3x$

Graph: 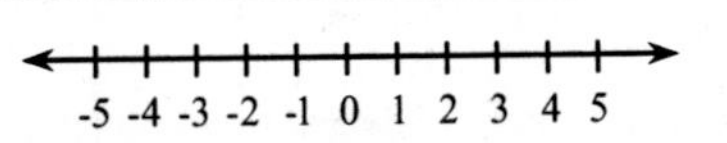

Set-builder notation:

Interval notation:

4. Solve: 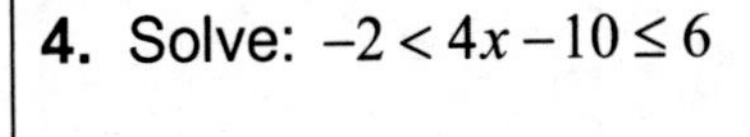 $-2<4x-10\leq 6$

Graph: 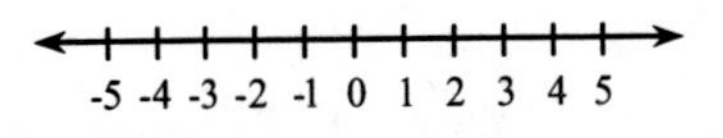

Set-builder notation:

Interval notation:

Student Activity A

Section 8.1: Match up on Equations and Inequalities

Match-up: Match each of the equations and inequalities in the squares of the table below with an equivalent equation or inequality from the top. If one is not found among the choices A through D, then choose E (none of these).

A $x > -1$ **B** $x = -1$ **C** $x < 4$ **D** $x = 4$ **E** None of these

$2x + 4(2x - 1) = -14$	$-8(x - 1) = 4x + 5$	$-4x + 1 > -15$	$\frac{x}{3} - \frac{x}{4} > -\frac{1}{12}$
$3x + 7(2x + 5) = 103$	$\frac{1}{3}x - 2 < \frac{1}{4}x - \frac{5}{3}$	$19x - 21 = -(4x - 71)$	$\frac{x+1}{3} > \frac{-3x-3}{2}$
$\frac{5(x+2)}{2} + 1 = 4x$	$-\frac{5(2-x)}{3} < 2x - 3$	$0.03x + 0.08 > 0.2x - 0.6$	$\frac{3}{2}(x+5) = \frac{17+5x}{2}$
$\frac{11 + 3(2x-1)}{2} = 20 - x$	$6 > \frac{5}{4}x + 1$	$-2(x - 3) < -(x - 2)$	$\frac{4x+1}{3} = \frac{2x-3}{5}$
$-2x - 1 > 5x + 6$	$-\frac{3}{4}x + 1 = -2 + \frac{9}{4}x$	$-2x > 7 - (13x + 18)$	$1.2x + 7.3 = -5.1x + 1$

Student Activity B

Section 8.1: Whose Turn?

Directions: For each formula given below, you will solve for each of the variables.

1. $r = c + m$ Already solved for: $\underline{r}$	
Solve for c:	Solve for m:

2. $I = Prt$ Already solved for: ___		
Solve for P:	Solve for r:	Solve for t:

3. $\omega = \dfrac{\theta}{t}$ Already solved for: _____	
Solve for θ:	Solve for t:

4. $y = mx + b$ Already solved for: _____		
Solve for m:	Solve for x:	Solve for b:

5. $F = ma$ Already solved for: __	
Solve for m:	Solve for a:

6. $E = mgh$ Already solved for: ___		
Solve for m:	Solve for g:	Solve for h:

7. $PV = nRT$				
Solve for P:	Solve for V:	Solve for n:	Solve for R:	Solve for T:

8. $m(x_2 - x_1) = y_2 - y_1$				
Solve for m:	Solve for x_2:	Solve for x_1:	Solve for y_2:	Solve for y_1:

Student Activity C

Section 8.1: Equations vs. Expressions II

Tic-Tac-Toe: For the problems below, decide whether it is an expression or an equation – if it's an expression, simplify the expression and put an **X** in the box. If it's an equation, then solve the equations and circle the solution (thus making an **O**).

$-7x-(-2x+1)$	$-3x+2=-4$	$-1-x=14$
$3(9x-1)+107-14x$	$-3x+2-4$	$\frac{3}{5}x-\frac{1}{2}+\frac{x}{5}=\frac{3}{2}$
$3(9x-1)+107=14x$	$\frac{1}{2}(3x-1)+6=\frac{1}{3}(5x+16)$	$\frac{3}{5}x-\frac{1}{2}+\frac{x}{5}-\frac{3}{2}$

QUESTION: Do the answers to equations look any different than the answers to expressions? Why or why not?

Student Activity A

Section 8.2: Working with Intersection and Union

Section I: Working with Sets of Numbers

	Set A	Set B	**Intersection** $A \cap B$ (A and B)	**Union** $A \cup B$ (A or B)
a.	$\{-2,0,2,4\}$	$\{0,1,2,3\}$		
b.	$\{-3,-2,-1\}$	$\{1,2,3\}$		
c.	$\{-3,1,2,4\}$	$\{-2,-1,2\}$		
d.	$\{1,2,3\}$	$\{1,2,3\}$		
e.	$\{0\}$	$\{-1,0,1\}$		
f.	$\mathbb{Z}$	$\mathbb{N}$		
g.	$\{...,1,2,3\}$	$\{2,3,4,...\}$		

Section II: Working with Number-line Graphs

a. A:

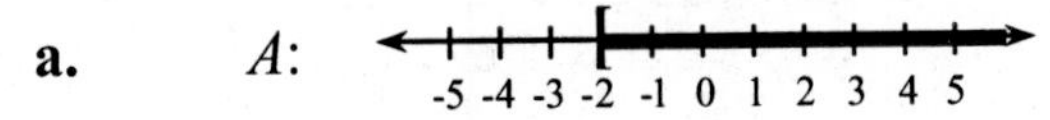

B: -5 -4 -3 -2 -1 0 1 2 3 4 5

$A \cap B$: -5 -4 -3 -2 -1 0 1 2 3 4 5

b. A:

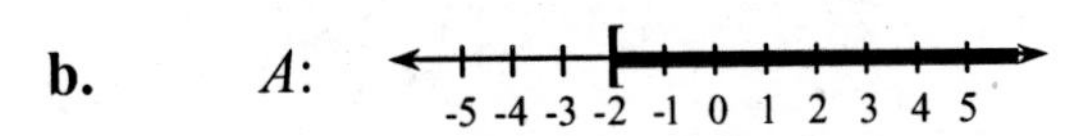

B: -5 -4 -3 -2 -1 0 1 2 3 4 5

A or B: -5 -4 -3 -2 -1 0 1 2 3 4 5

c. A: $x < 0$ -5 -4 -3 -2 -1 0 1 2 3 4 5

B: $x \leq -2$ -5 -4 -3 -2 -1 0 1 2 3 4 5

$A \cap B$: -5 -4 -3 -2 -1 0 1 2 3 4 5

d. A: $x < 0$

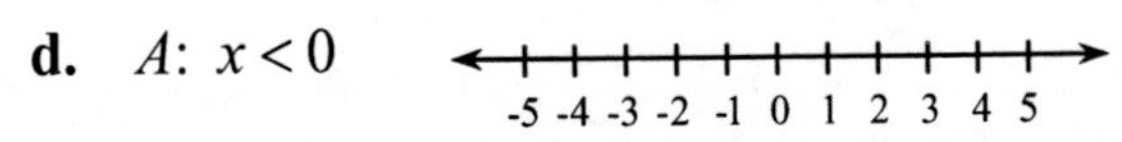

B: $x \leq -2$ -5 -4 -3 -2 -1 0 1 2 3 4 5

$A \cup B$: -5 -4 -3 -2 -1 0 1 2 3 4 5

e. A: -5 -4 -3 -2 -1 0 1 2 3 4 5

B: -5 -4 -3 -2 -1 0 1 2 3 4 5

A and B: -5 -4 -3 -2 -1 0 1 2 3 4 5

f. A: -5 -4 -3 -2 -1 0 1 2 3 4 5

B: -5 -4 -3 -2 -1 0 1 2 3 4 5

$A \cup B$: -5 -4 -3 -2 -1 0 1 2 3 4 5

g. A: $[-3,3]$ -5 -4 -3 -2 -1 0 1 2 3 4 5

B: $(-2,2)$ -5 -4 -3 -2 -1 0 1 2 3 4 5

A or B: -5 -4 -3 -2 -1 0 1 2 3 4 5

h. A: $[-3,3]$ -5 -4 -3 -2 -1 0 1 2 3 4 5

B: $(-2,2)$ -5 -4 -3 -2 -1 0 1 2 3 4 5

A and B -5 -4 -3 -2 -1 0 1 2 3 4 5

Guided Learning Activity

Section 8.2: Solving Compound and Double Inequalities

Solving a Compound Inequality (AND or OR)

1. Solve each inequality separately.
2. Graph each inequality together on a number line. (*use different colors*)
3. Graph the desired intersection (and) or union (or) on a number line. (*use a 3rd color*)
4. Write the interval solution that represents the graphed solution. (*the 3rd color graph*)

Solving a Double Inequality

Apply the properties of inequality to all three of its parts to isolate x in the middle.

---

1. Solve: $3x \le 15$ and $2x > -6$

Graph the separate pieces:

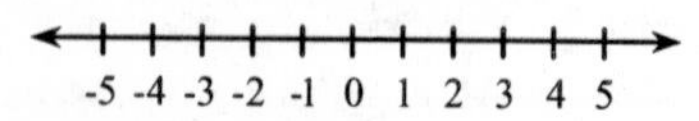

Graph the solution:

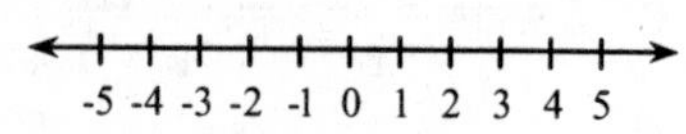

Write the solution: ____________

2. Solve: $3x < 3$ or $-2x < -10$

Graph the separate pieces:

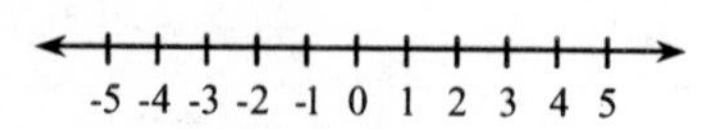

Graph the solution:

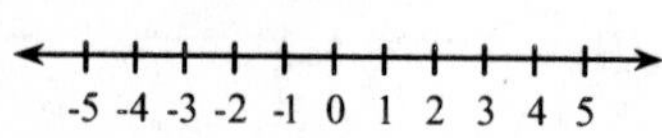

Write the solution: ____________

3. Solve: $5 < 2x + 1 \le 9$

Graph the solution:

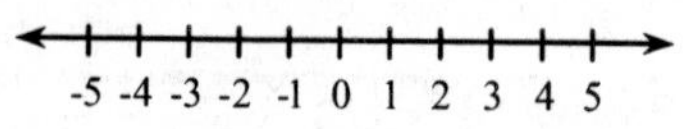

Write the solution: ____________

4. Solve: $2x - 5 < -11$ or $5x + 1 \le 6$

Graph the separate pieces:

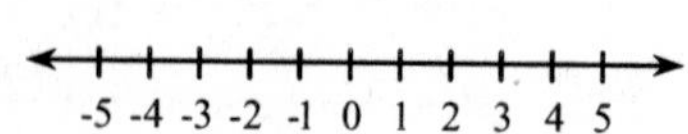

Graph the solution:

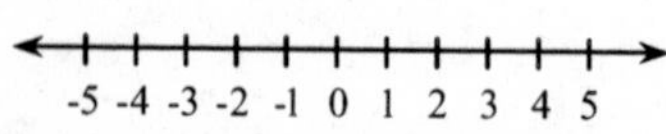

Write the solution: ____________

Student Activity B

Section 8.2: The Match is Out There

Directions: Solve each compound or double inequality and then match the problem up with the graph of its solution.

___ **1.** Solve: $x+2>4$ and $x-1<4$

___ **2.** Solve: $0 \le x+3 \le 5$

___ **3.** Solve: $2+x \le 2$ or $2-3x \ge -4$

___ **4.** Solve: $4x \le -12$ and $\dfrac{x}{2} \le 1$

___ **5.** Solve: $\dfrac{x}{2}+3 \le 1$ or $\dfrac{5x}{4}-1>\dfrac{3}{2}$

___ **6.** Solve: $2x-5<5$ or $3x+1>-2$

a. 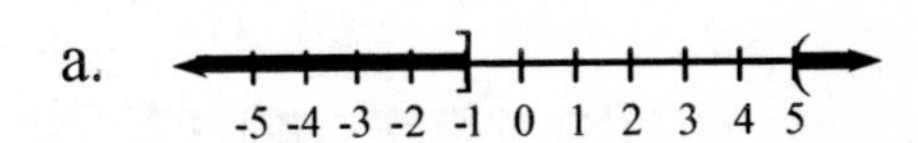

b.
-5 -4 -3 -2 -1 0 1 2 3 4 5

c. Empty set
-5 -4 -3 -2 -1 0 1 2 3 4 5

d.
-5 -4 -3 -2 -1 0 1 2 3 4 5

e.
-5 -4 -3 -2 -1 0 1 2 3 4 5

f.
-5 -4 -3 -2 -1 0 1 2 3 4 5

g. All reals
-5 -4 -3 -2 -1 0 1 2 3 4 5

h.
-5 -4 -3 -2 -1 0 1 2 3 4 5

i.
-5 -4 -3 -2 -1 0 1 2 3 4 5

j.
-5 -4 -3 -2 -1 0 1 2 3 4 5

Guided Learning Activity

Section 8.3: Solving Absolute Value Equations and Inequalities

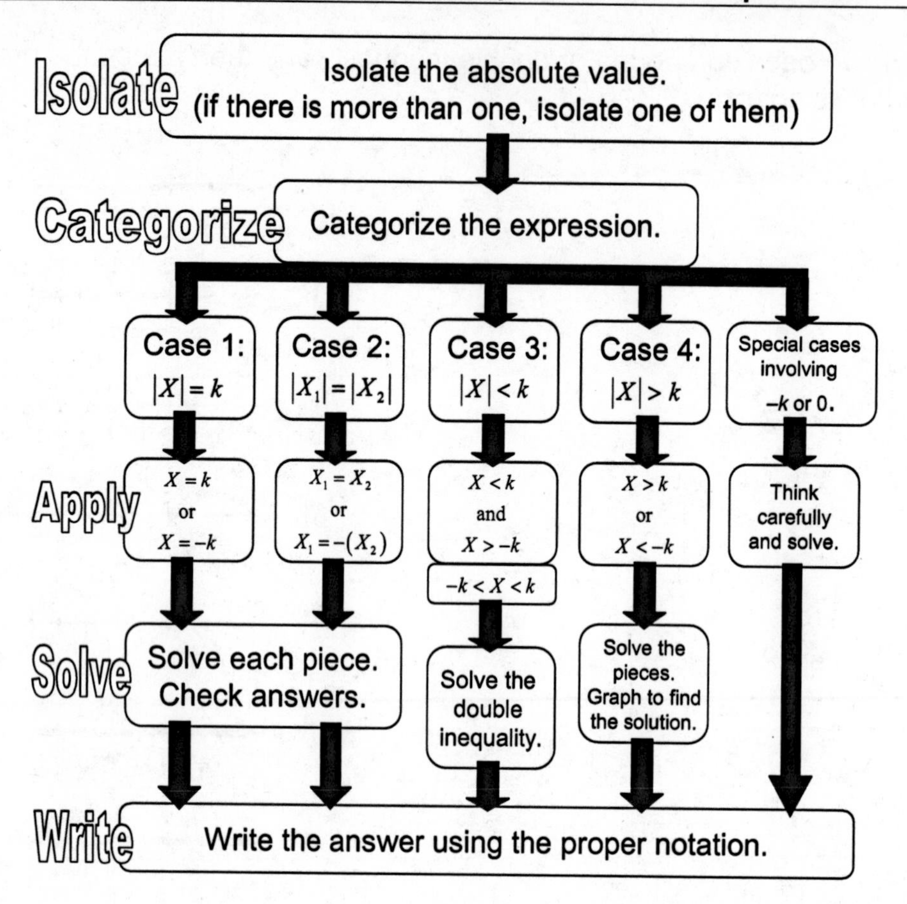

Directions: With the guidance of your instructor, work through the strategy for solving an absolute value equation or inequality with each of the problems below.

1. Solve $|2x| - 3 = 11$

4. Solve $-4|x+1| < 20$

2. Solve $3 + |2x+1| \le 7$

5. Solve $|x+4| = |3x-8|$

3. Solve $\dfrac{|3x+4|}{5} > 1$

6. Let $f(x) = |3x+5|$, For what values of x is $f(x) = 32$?

Student Activity A

Section 8.3: Connecting Absolute Value Problems to Graphs

The graph of $f(x)=|x|$ gives a V-shaped curve. We can interpret why the absolute value equations and inequalities come out the way they do by looking at graphs of functions that correspond to the equations. For example:

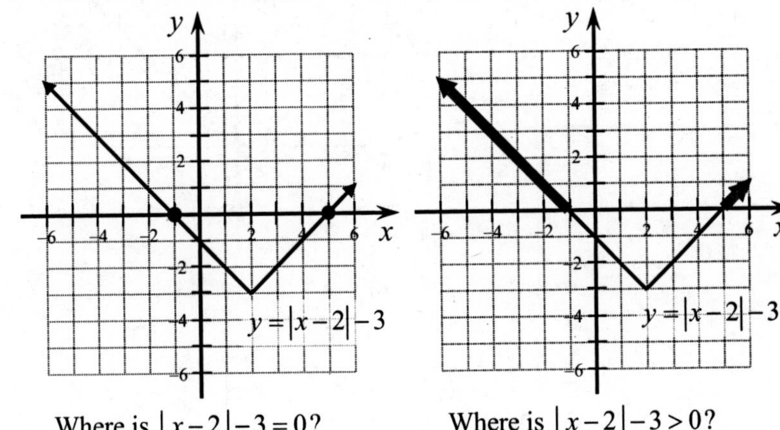

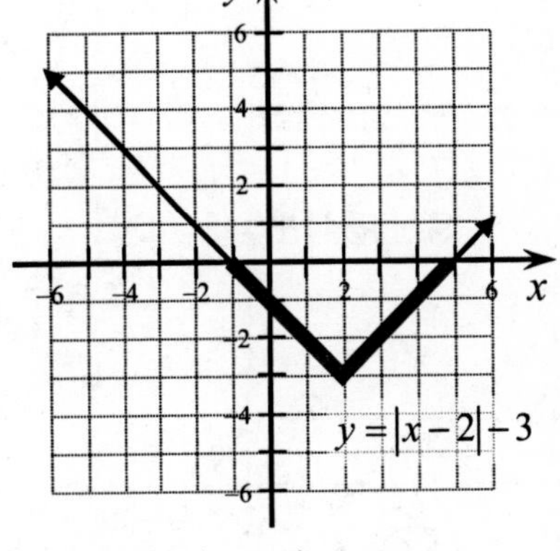

Where is $|x-2|-3=0$?
Where $x=-1$ or $x=5$.

Where is $|x-2|-3>0$?
Where $x<-1$ or $x>5$.
Intervals: $(-\infty,-1)\cup(5,\infty)$

Where is $|x-2|-3<0$?
Where $-1<x<5$.
Interval: $(-1,5)$

Now solve the corresponding equations and inequalities using the techniques we have learned in this section. Remember to isolate the absolute values first!

1. a. Solve: $|x-2|-3=0$ **b.** Solve: $|x-2|-3>0$ **c.** Solve: $|x-2|-3<0$

As you answer the following problems, you might find it helpful to trace each answer on the graph in a different color.

2. a. Where is $|x|-4=0$? ________________

b. Where is $|x|-4<0$? ________________

c. Where is $|x|-4\geq 0$? ________________

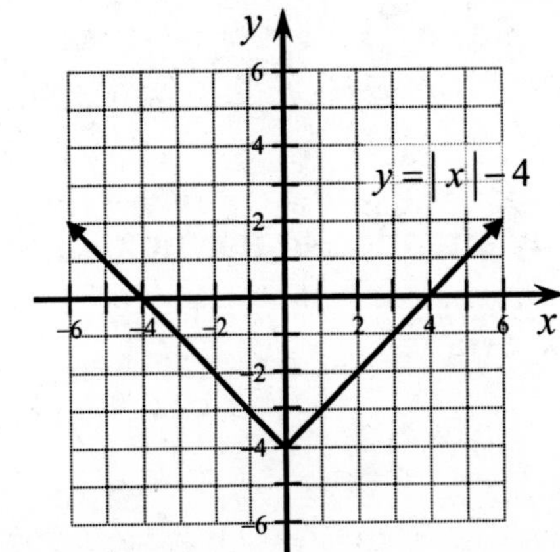

3. a. Where is $|x+1|-4=0$? ______________

Solve: $|x+1|-4=0$

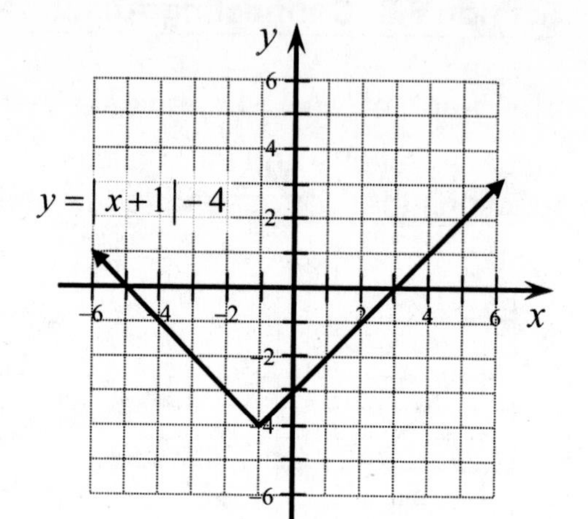

b. Where is $|x+1|-4 \le 0$? ______________

Solve: $|x+1|-4 \le 0$

---

4. a. Where is $|x+3|+1<0$? ______________

Solve: $|x+3|+1<0$

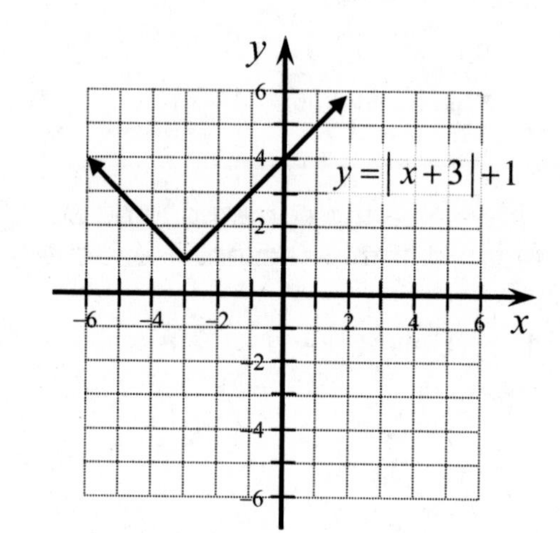

b. Where is $|x+3|+1 \ge 0$? ______________

Solve: $|x+3|+1 \ge 0$

---

5. a. Where is $|x-3| \le 0$? ______________

Solve: $|x-3| \le 0$

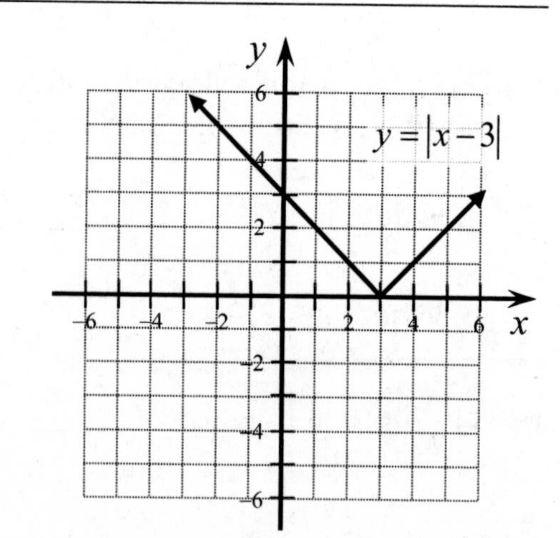

Pay attention to this next one!

b. Where is $|x-3| \ge 2$? ______________

Solve: $|x-3| \ge 2$

Student Activity B

Section 8.3: Matching Up the Different Cases

Directions: Begin by isolating the absolute value in each of the equations and inequalities below. Then categorize the problem as one of the four cases or a special case. The first one has been done for you.

Let k be a **positive** constant and let X, X_1, and X_2 be mathematical expressions.

Case 1:	Case 2:	Case 3:	Case 4:	Special Case:
$\lvert X \rvert = k$	$\lvert X_1 \rvert = \lvert X_2 \rvert$	$\lvert X \rvert < k$ $\lvert X \rvert \le k$	$\lvert X \rvert > k$ $\lvert X \rvert \ge k$	If the constant is negative or zero.

$2\lvert x-3 \rvert < 4$ $\lvert x-3 \rvert < 2$ **CASE 3**	$\lvert x-3 \rvert - 4 \ge 2$	$\lvert 2a-3 \rvert + 5 > 3$	$\frac{\lvert x \rvert}{3} - 2 = 4$
$\lvert x+4 \rvert = \lvert 2x-2 \rvert$	$1 = \lvert 2t-4 \rvert + 4$	$5 - \lvert 2x \rvert < 2$	$4\lvert x+3 \rvert - 3 \le 2$
$-2\lvert 4y \rvert < 10$	$\frac{\lvert 3x-4 \rvert}{-6} \le -2$	$1 > \lvert 3-b \rvert - 6$	$\lvert 2x \rvert = \lvert x+3 \rvert$
$5 - \lvert x+4 \rvert = 2$	$6 - 2\lvert x+4 \rvert \le 0$	$\lvert x-4 \rvert + 5 = 5$	$3 < \frac{\lvert d+12 \rvert}{5}$

Student Activity A

Section 8.4: Finding Factors

Directions: Factor each polynomial and mark which factors it contained. The first one has been done for you.

Factor these polynomials.	Mark the factors that appear in the polynomials.							
	x	$x+2$	$x-3$	$x+5$	$2x+1$	$3x-2$	x^2+4	x^2-2
1. $x^2+7x+10$ $(x+2)(x+5)$		X		X				
2. x^3-2x								
3. $x^2+2x-15$								
4. $2x^2-5x-3$								
5. x^3-3x^2-2x+6								
6. $6x^2-x-2$								
7. $3x^3-2x^2+12x-8$								
8. $2x^2+5x+2$								
9. x^4+2x^2-8								
10. $3x^3-11x^2+6x$								

Student Activity B

Section 8.4: Factors in Hiding

Directions: Factor each polynomial and then find each factor in the grid at the bottom of the page. You should find every factor in the grid.

1. Factor: $(a+b)^2+5(a+b)+6$ HINT: Treat it like a trinomial.

2. Factor: $(a-b)^2+9(a-b)-22$

3. Factor: $(a+b)^2-4(a+b)+4$

4. Factor: $4(a-b)+11(a-b)-3$

5. Factor: $(a-b)(x+y)+2(x+y)$ HINT: Is there a GCF?

6. Factor: $(a+b)^2+x(a+b)$

7. Factor: $a^2b^2+5a^2b+3a^2+10b^2+50b+30$ HINT: Factor by grouping.

8. Factor: $a^2b+4ab+2b-3a^2-12a-6$

$a+b-2$	$a+b-2$	$a-b-2$	$a+b$
$a+b+2$	b^2+5b+3	$b-3$	$x+y$
a^2+10	$a-b+11$	$a-b+2$	$a-b+3$
$a+b+3$	$4a-4b-1$	a^2+4a+2	$a+b+x$

Guided Learning Activity

Section 8.5: Learning the Cubes Procedure

The best way to learn the sum and difference of cubes formulas is to learn a procedure for constructing the formulas rather than memorizing the formulas. Here are two examples, with the description of each step:

Example 1: Factor: $x^3 - 125$

Step 1: Construct a skeleton of the difference of cubes: $(\quad)^3 - (\quad)^3$

Step 2: Fill in the skeleton with the appropriate bases: $(x)^3 - (5)^3$

Step 3: Construct the skeleton of the cubes formula: $(\quad - \quad)(\quad + \quad + \quad)$

- $-$: same sign as original expression
- first $+$: opposite of first sign
- second $+$: always positive

Step 4: Insert the bases: $(x-5)(\quad + \quad + \quad)$

Step 5: Insert the squares of those bases: $(x-5)(x^2 + \quad + 25)$

- x^2 : $(x)^2$
- 25 : $(5)^2$

Step 6: Insert the product of the bases: $(x-5)(x^2 + 5x + 25)$

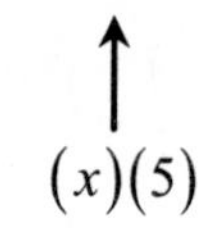

- $5x$: $(x)(5)$

Example 2: Factor: $8w^3 + 27x^6$

Step 1: Construct a skeleton of the sum of cubes: $(\quad)^3 + (\quad)^3$

Step 2: Fill in the skeleton with the appropriate bases: $(2w)^3 + (3x^2)^3$

Step 3: Construct the skeleton of the cubes formula: $(\quad + \quad)(\quad - \quad + \quad)$

- $+$: same sign as original expression
- $-$: opposite of first sign
- $+$: always positive

Step 4: Insert the bases: $(2w+3x^2)(\quad - \quad + \quad)$

Step 5: Insert the squares of those bases: $(2w-3x^2)(4w^2 + \quad + 9x^4)$

- $4w^2$: $(2w)^2$
- $9x^4$: $(3x^2)^2$

Step 6: Insert the product of the bases: $(2w-3x^2)(4w^2 + 6x^2w + 9x^4)$

- $6x^2w$: $(2w)(3x^2)$

Now you try these!

1. Factor: $x^3 - 8$

Step 1: Construct a skeleton of the sum or difference of cubes: $(\quad)^3 - (\quad)^3$

Step 2: Fill in the skeleton above with the appropriate bases.

Step 3: Construct the skeleton of the cubes formula below (put in the signs):

$$(\qquad)(\qquad\qquad\qquad)$$

$$\uparrow \qquad \uparrow \qquad \uparrow$$

$$(\quad)^2 \quad (\quad)(\quad) \quad (\quad)^2$$

Step 4: Insert the bases in the skeleton above.

Step 5: Insert the squares of those bases into the skeleton above.

Step 6: Insert the product of the bases into the skeleton above.

2. Factor: $64y^3 + 1$

$$64y^3 + 1 = (\quad)^3 + (\quad)^3$$

$$= (\qquad)(\qquad\qquad\qquad)$$

$$\uparrow \qquad \uparrow \qquad \uparrow$$

$$(\quad)^2 \quad (\quad)(\quad) \quad (\quad)^2$$

3. Factor: $a^9 + 8b^6$

$$a^9 + 8b^6 = (\quad)^3 + (\quad)^3$$

$$= (\qquad)(\qquad\qquad\qquad)$$

$$\uparrow \qquad \uparrow \qquad \uparrow$$

$$(\quad)^2 \quad (\quad)(\quad) \quad (\quad)^2$$

4. Factor: $512R^3 - 343r^3$

$$512R^3 - 343r^3 = (\quad)^3 - (\quad)^3$$

$$= (\qquad)(\qquad\qquad\qquad)$$

$$\uparrow \qquad \uparrow \qquad \uparrow$$

$$(\quad)^2 \quad (\quad)(\quad) \quad (\quad)^2$$

Student Activity A

Section 8.5: Leftovers

Directions: Factor each of the binomials below and then mark the factors that they contain. Make sure to look for GCFs first! The first one has been done for you.

Factor these binomials.	Mark the factors that appear in the binomials.						
	$x+2$	$x-3$	$x+5$	$x-4$	$2x-3$	x^2+4	Leftovers
1. x^3+125 $(x+5)(x^2-5x+25)$			X				$(x^2-5x+25)$
2. $3x^2-27$							
3. $4x^3-9x$							
4. x^3-64							
5. $2x^4-32$							
6. $4x^3+32$							
7. $2x^3-54$							
8. $8x^3-27$							
9. x^6+64							
10. $128x-8x^5$							

Student Activity B

Section 8.5: Skeletons of Tricky Binomials

Directions: Categorize each of the expressions using the choices below, then write the appropriate skeleton for the problem, and factor the expression. If the expression cannot be factored, say so. The first one has been done for you.

A Sum of Squares: $(\ )^2+(\ )^2$

B Difference of Squares: $(\ )^2-(\ )^2$

C Sum of Cubes: $(\ )^3+(\ )^3$

D Difference of Cubes: $(\ )^3-(\ )^3$

E None of these

Polynomial	Category	Skeleton	Factored Form
1. $x^4-(a+b)^2$	B	$(x^2)^2-(a+b)^2$	$(x^2+(a+b))(x^2-(a+b))$ or $(x^2+a+b)(x^2-a-b)$
1. $(a-b)^2-x^4$			
2. $1+(x+y)^3$			
3. $(a-b)^2+x^2$			
4. $\frac{8}{27}-x^6$			
5. x^3-36			
6. $x^2+2xy+y^2-100$			
7. $(x+y)^2+27$			
8. $(a+b)^3-(c-d)^3$			

Assessment 8B

Chapter 8: Mid-chapter Assessment for Understanding

For each of the following, describe the type of problem and the strategies and key steps to remember while doing the problem. You do **not** have to complete the problems.

	Type of Problem	Strategies and Key Steps
1. Solve: $\lvert x+4\rvert-1<6$		
2. Factor: $5ax^2-2x^2+25a-10$		
3. Solve and graph. $\frac{x}{6}>1$ or $3(x+4)<6$.		
4. Factor: $(2x+1)^3-(x^2)^3$		
5. Solve: $-13\le 5x+2\le 17$		
6. Solve: $\lvert x+1\rvert=\lvert 2x-1\rvert$		
7. Factor: $x^2+16x+48$		
8. Let $A=\{1,3,5,7\}$ and $B=\{5,6,7,8\}$. Find the intersection of A and B.		
9. Solve: $-3x+4(x-1)=2x+5$		
10. Factor: $x^2+4x+4-25y^2$		

Student Activity A

Section 8.6: Rational Reunion

Directions: Rational expressions are back and all the operations are reunited below! Simplify each expression according to its operation!

$\frac{x-2}{x+3}+\frac{x+5}{x+3}$	
$\frac{x-2}{x+3}-\frac{x+5}{x+3}$	
$\frac{x-2}{x+3}\cdot\frac{x+5}{x+3}$	
$\frac{x-2}{x+3}\div\frac{x+5}{x+3}$	

$\frac{x}{x-4}+\frac{x-2}{4-x}$	
$\frac{x}{x-4}-\frac{x-2}{4-x}$	
$\frac{x}{x-4}\cdot\frac{x-2}{4-x}$	
$\frac{x}{x-4}\div\frac{x-2}{4-x}$	

$\frac{6}{x+3}+\frac{3x+9}{x+2}$	
$\frac{6}{x+3}-\frac{3x+9}{x+2}$	
$\frac{6}{x+3}\cdot\frac{3x+9}{x+2}$	
$\frac{6}{x+3}\div\frac{3x+9}{x+2}$	

Student Activity B

Section 8.6: How Things Change

Directions: Each of the rational equations below is **slightly** different. But these slight changes result in very different solutions. So ... pay attention to what you're doing and see how a slight change in the rational equation makes a big difference in the solution.

1. Solve: $\frac{x+3}{x-1}=\frac{-3-x}{x^2-1}$	**2.** Solve: $\frac{x}{x-1}+3=\frac{-3-x}{x^2-1}$
3. Solve: $\frac{x+3}{x-1}=-3-\frac{x}{x^2-1}$	**4.** Solve: $\frac{x+3}{x-1}=\frac{3-x}{x^2-1}$

5. Solve: $\frac{x+1}{x+2}=\frac{x-2}{x^2-4}$	**6.** Solve: $\frac{x}{x+2}+1=\frac{x-2}{x^2-4}$
7. Solve: $\frac{x+1}{x+2}=\frac{2-x}{x^2-4}$	**8.** Solve: $\frac{x+1}{x+2}=\frac{-2-x}{x^2-4}$

Student Activity A

Section 8.7: Winter Severity Index

Deer are adapted to survive most Michigan winter weather conditions. However, research suggests that winter weather can affect the deer population. The Michigan DNR monitors winter conditions by calculating a winter severity index (WSI). The WSI reflects the estimated effect of winter weather conditions on the energy reserves of deer; the higher the index value, the larger the potential impact on the herd. The current WSI System takes advantage of standard weather data available from the National Climate Data Center and a weekly index value is calculated from November through April (for more information visit www.michigan.gov/dnr).

The graph below shows WSI data for the Upper Peninsula, 2006-2007 season. The graph shows the weekly regional WSI value for the current season (dashed line) and the average weekly regional WSI value for the 1996 through 2005 seasons (solid line). Answer the questions using information from the graph. [Graph courtesy of the Michigan Department of Natural Resources.]

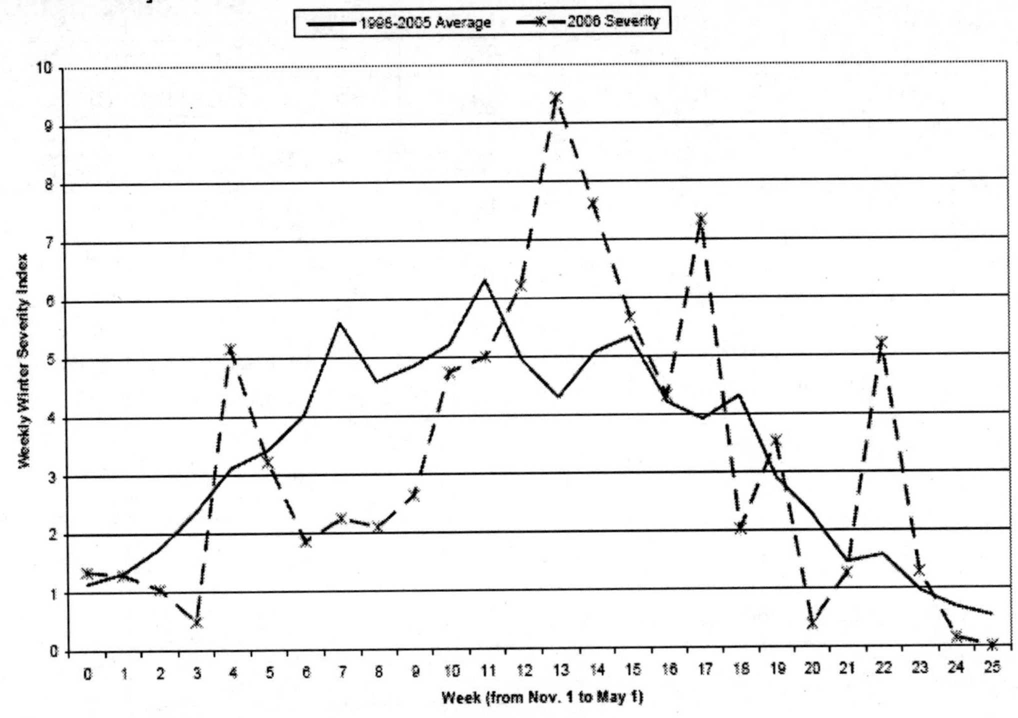

Use the graph to answer the following questions:

1. How many weeks after November 1 was the highest WSI recorded for the 2006-2007 season? ____________
   - Write an ordered pair to represent this point: (____,____)
   - In what month would this peak have occurred? ____________
2. When was the first major cold-snap of the season, and what was the associated WSI?
   - Write an ordered pair to represent this point: (____,____)
3. Between what two weekly readings was there the biggest change in the WSI?
   - Write ordered pairs to represent these points: (____,____) & (____,____)
4. Which 6-week period in 2006-07 was harsher than the same period for the 1998-2005 average? _______

Student Activity B

Section 8.7: Entertaining Rates of Change

It is often helpful to summarize raw data with a rate describing the amount of change in one quantity with respect to the amount of change in another. This is called an **average rate of change**, and is often used to describe change that occurs *over time*.

$$\text{Average Rate of Change} = \frac{\text{change in quantity}}{\text{change in time}}$$

Directions: Answer the questions below by calculating an average rate of change and using this information to answer the follow-up questions.

1. This table contains box office data from 2000 – 2006. (Source: www.boxofficemojo.com)

Year	Total Gross (in millions)	Tickets Sold (in millions)	# of Pics	Ticket Price	#1 Picture of the year
2006	$9,209.4	1400.0	606	$6.58	Dead Man's Chest
2005	$8,840.4	1381.3	547	$6.40	Revenge of the Sith
2004	$9,418.3	1516.6	551	$6.21	Shrek 2
2003	$9,185.9	1523.3	508	$6.03	Return of the King
2002	$9,167.0	1578.0	467	$5.81	Spider-Man
2001	$8,412.5	1487.3	482	$5.66	Harry Potter
2000	$7,661.0	1420.8	478	$5.30	The Grinch

a. What is the average annual rate of change for the total box office gross between 2000 and 2006? Show your calculation and round your answer to the nearest tenth.

b. Complete the statement: *On average, total box office gross increased* ___________ *dollars per year over the 6-year period between 2000 and 2006. The total increase in box office gross during this period was* ____________.

c. What is the average annual rate of change for the total box office gross between 2003 and 2006? Show your calculation and round your answer to the nearest tenth.

d. Complete the statement: *On average, total box office gross increased* ___________ *dollars per year over the 3-year period between 2003 and 2006. The total increase in box office gross during this period was* ____________.

e. In general, Is the average annual rate of change for total box office gross increasing or decreasing? ____________

2. The following tables contain U.S. Music purchasing data for 2005 and 2006.

Units sold (in millions):	2005	2006
Overall Music Sales (Albums, singles, music video, digital tracks)	1,003	1,198
Total Album Sales (Includes CD, CS, LP, Digital albums)	618.9	588.2
Digital Track Sales	352.7	581.9
Overall Album Sales (Includes all albums & track equivalent albums)	654.1	646.4
Internet Album Sales (Physical album purchases via e-commerce sites)	24.7	29.4
Digital Album Sales	16.2	32.6

(Source: www.businesswire.com)

Top Ten Selling Albums of 2006	Units Sold
Soundtrack/High School Musical	3,719,071
Me and My Gang/Rascal Flatts	3,479,994
Some Hearts/Carrie Underwood	3,015,950
All the Right Reasons/Nickelback	2,688,166
Futuresex/Lov…/Justin Timberlake	2,377,127
Back to Bedlam/James Blunt	2,137,142
B'day/Beyonce	2,010,311
Soundtrack/Hannah Montana	1,987,681
Taking the Long Way/Dixie Chicks	1,856,284
Extreme Behavior/Hinder	1,817,350

Show your calculations and round your answer to the nearest hundredth.

a. What is the average *monthly* rate at which **overall music sales** increased from 2005 to 2006?

b. What is the average *monthly* rate at which **total album sales** decreased from 2005 to 2006?

c. What is the average *monthly* rate at which **digital album sales** increased from 2005 to 2006?

3. Based on your results from parts b and c, does it appear that all the people who stopped buying physical albums are now purchasing digital albums? Explain.

4. What is the average *daily* sales rate for the top selling album of 2006? What is the average daily sales rate for the *tenth* top selling album of 2006? (Round your answer to nearest whole number.)

Student Activity C

Section 8.7: Midpoints as Averages

1. Plot the two given points on each graph provided. Connect the two points with a line **segment** using a straight-edge. Then estimate the midpoint of the line segment.

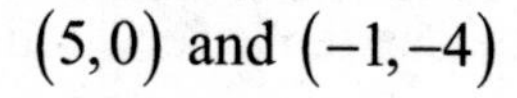

$(5,0)$ and $(-1,-4)$

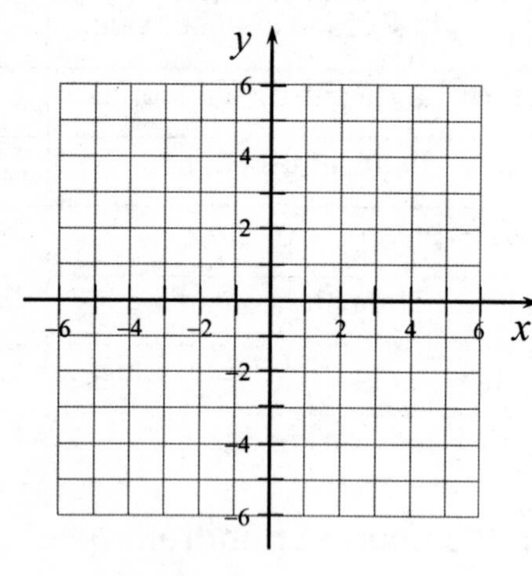

Midpoint: (____,____)

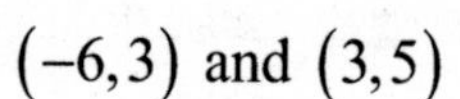

$(-6,3)$ and $(3,5)$

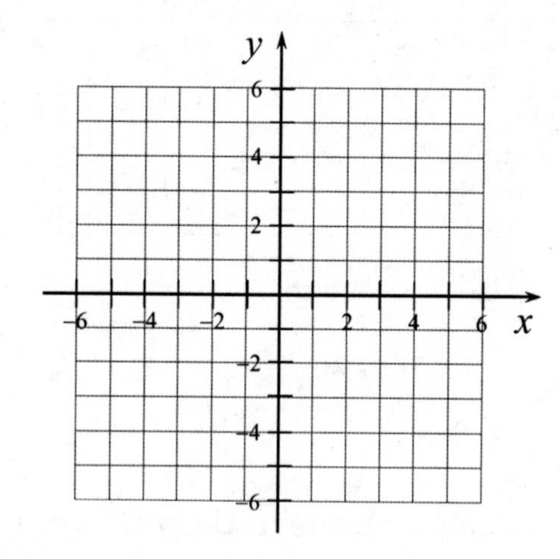

Midpoint: (____,____)

2. We find an average by finding the sum of the values, and then dividing by the number of values. For practice, find the average of 5, 12, 3, and 4.

Average: $\dfrac{\square+\square+\square+\square}{\square} =$

3. We find the midpoint of two points by averaging the *x*-coordinates and averaging the *y*-coordinates. Find the midpoint for $(5,0)$ and $(-1,-4)$ using this technique.

x-coordinate: $\dfrac{\square+\square}{\square} =$

y-coordinate: $\dfrac{\square+\square}{\square} =$

Midpoint: (____,____) Is this consistent with the midpoint you found in #1?

4. Repeat this process for $(-6,3)$ and $(3,5)$

Midpoint: . $\left(\dfrac{\square+\square}{\square}, \dfrac{\square+\square}{\square}\right) =$ (____,____)

5. See if you can find "third-points" for the points $(-2,-6)$ and $(1,6)$. These would be two points that split the line segment into thirds. There are many ways to do it.

Student Activity A

Section 8.8: Everyday Functions

Case #1: During an eight hour shift, hospital patients in the ICU are assigned to a nurse at a ratio of 2:1. If the ICU has eight patients, the nursing assignments would be as follows:

Patient	Nurse
Patient 1	Nurse A
Patient 2	
Patient 3	Nurse B
Patient 4	
Patient 5	Nurse C
Patient 6	
Patient 7	Nurse D
Patient 8	

1. Consider the set of patients to be the domain and the set of nurses to be the range. The situation described here **is** a function. Why?

2. Describe the practical implications if this relation were **not** a function.

3. If the domain was the set of nurses and the range was the set of patients, would the relation still be a function? Why or why not?

Case #2: Suppose the exchange rate for the US Dollar to Mexican Peso is $1=10.880 pesos. The table below shows the conversion of several dollar amounts to pesos.

US Dollars	Mexican Pesos
5	54.4
24	261.12
78	848.64
110	1,196.8
500	5,440
1250	1,360

4. Let the domain be US Dollars and the range be Mexican Pesos. This relation **is** a function. Why?

5. Describe the practical implications if this relation were **not** a function.

6. If the domain and range were reversed, would it still be a function? Why or why not?

Case #3: These ordered pairs represent the identification numbers of commercial fishing boats and the profit (or loss) per boat in thousands of dollars.

$(1,37)$ $(2,43)$ $(5,37)$ $(7,13)$ $(8,-5)$ $(10,37)$ $(12,-117)$

7. Let the domain be ______________________________ and the range be ______________________.

8. Is this a function? Why or why not?

9. Describe the practical implications if this relation were **not** a function.

10. If the domain and range were reversed, would it still be a function? Why or why not?

Student Activity B

Section 8.8: Skeleton Crew

Skeleton Method: To evaluate a function that uses function notation, it is helpful to "see" the function without the variables. To make the "skeleton" of the problem, replace the variables with an empty parentheses "skeleton".

For example, $f(x)=x^2-5x+1$ becomes $f(\ )=(\ )^2-5(\ )+1$.

Now if you have to evaluate the expression for $x=-2$, write the -2 inside each parentheses skeleton and evaluate.

So $f(\ )=(\ )^2-5(\ )+1$ becomes $f(-2)=(-2)^2-5(-2)+1$.

Directions: For each row of the table, a function is described in words, in function notation, or as a parentheses skeleton. Complete the table and then evaluate the function for the given value. The first one has been done for you.

Word Description	$f(x)$	Skeleton	$f(-2)$
Multiply by 2 and then subtract 3	$f(x)=2x-3$	$f(\ )=2(\ )-3$	$f(-2)=2(-2)-3$ $=-4-3=-7$
Subtract 3 and then multiply by 2			
	$f(x)=x^3-5$		
Add 3, then square the result			
		$f(\ )=\lvert(\ )-6\rvert$	
			$f(-2)=\frac{(-2)+3}{4}=\frac{1}{4}$
	$f(x)=9-x$		
Take the reciprocal of it (assume it is not zero)			

Guided Learning Activity

Section 8.8: Graphing Linear Functions

Function 1: Fill in the missing values in the table of solutions for the linear function, $f(x) = -2x + 1$, plot the points, and then draw the line through the points.

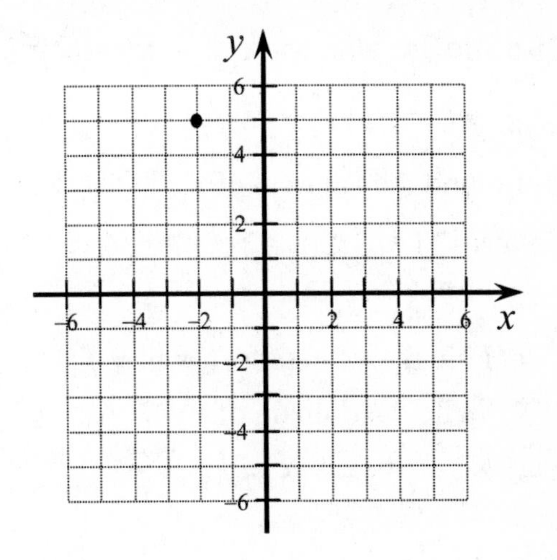

x	$f(x)$
-2	5
-1	
0	
1	
2	

$f(-2) = -2(2) + 1 = 5$

$f(-1) =$

$f(0) =$

$f(1) =$

$f(2) =$

Function 2: Fill in the missing values in the table of solutions for the linear function, $f(x) = x$, plot the points, and then draw the line through the points. This is called the **identity function** because it assigns each real number to itself.

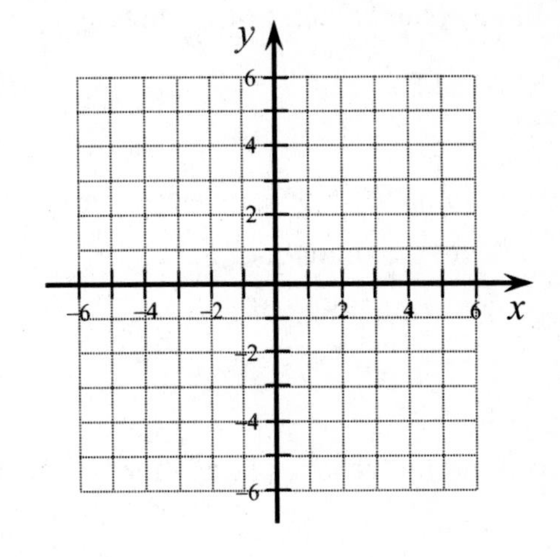

x	$f(x)$
-5	
-3	
0	
4	
6	

$f(-5) =$

$f(-3) =$

$f(0) =$

$f(4) =$

$f(6) =$

Function 3: Fill in the missing values in the table of solutions for the linear function, $f(x) = -4$, plot the points, and then draw the line through the points. This is called a **constant function** because for any input x, the output is a constant number.

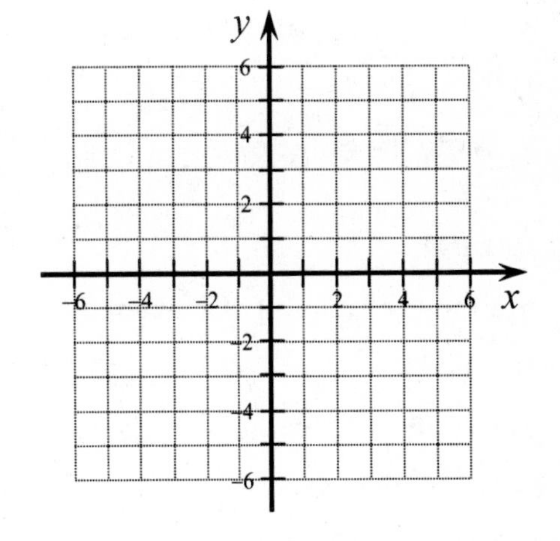

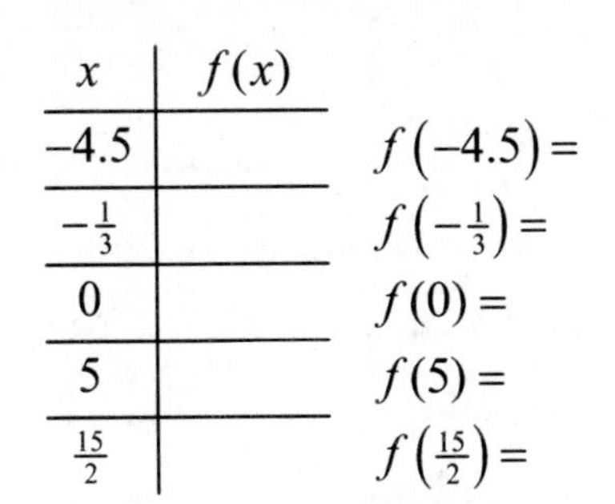

x	$f(x)$
-4.5	
$-\frac{1}{3}$	
0	
5	
$\frac{15}{2}$	

$f(-4.5) =$

$f\left(-\frac{1}{3}\right) =$

$f(0) =$

$f(5) =$

$f\left(\frac{15}{2}\right) =$

4. The graph of the constant function is a _______________ line.

Student Activity C

Section 8.8: Gleaning Information from Graphs

Directions: Use the graph for each problem to answer the questions.

1. a. $f(-2) =$ ____

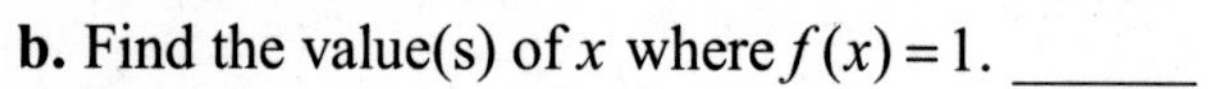

b. Find the value(s) of x where $f(x) = 1$. ______

c. $f(5) =$ ____

d. $f(\_\_\_) = -3$

e. Find the value(s) of x where $f(x) = 2$. ______

f. Write the domain of f: ________________

g. Write the range of f: ________________

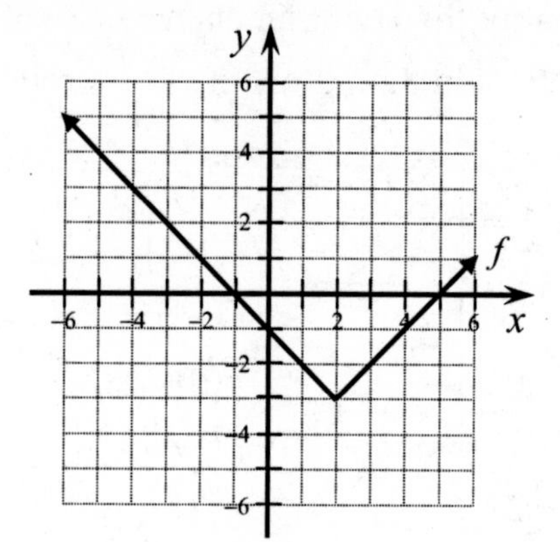

2. a. $g(2) =$ ____

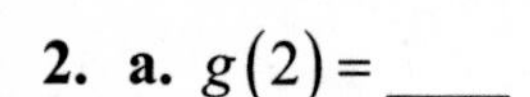

b. Find the value(s) of x where $g(x) = 4$. ______

c. $g(0) =$ ____

d. $g(\_\_\_) = 2$

e. What is the slope, m, of this line? _____

f. Write the domain of g: ________________

g. Write the range of g: ________________

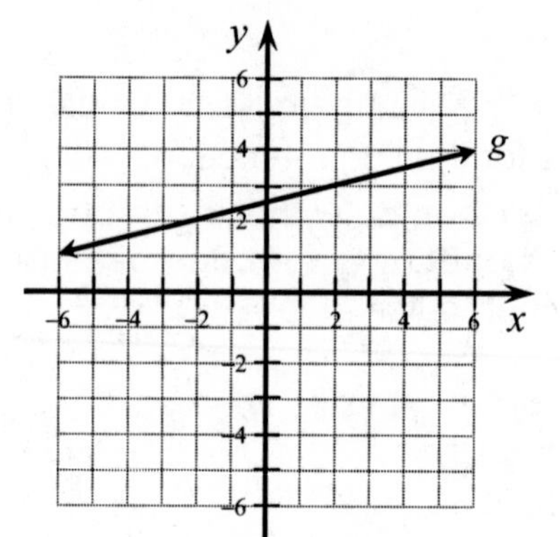

3. a. $h(-1) =$ ____

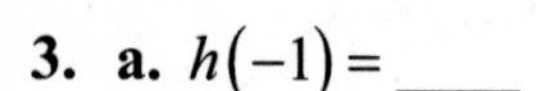

b. Find the value(s) of x where $h(x) = 0$. ______

c. $h(-4) =$ ____

d. $h(\_\_\_) = 2$

e. Find the value(s) of x where $h(x) = -2$. ______

f. Write the domain of h: ________________

g. Write the range of h: ________________

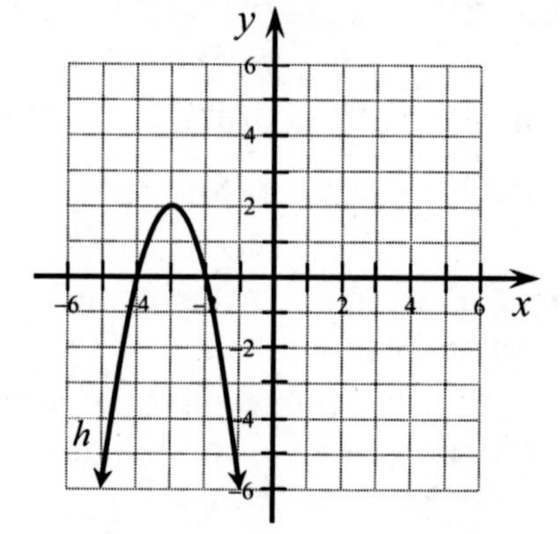

4. a. $p(4) =$ ____

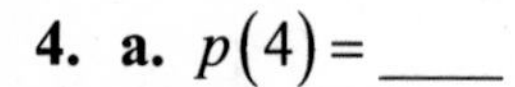

b. Find the value(s) of x where $p(x) = 1$. ________

c. $p(-2) =$ ____

d. Find the value(s) of x where $p(x) = 4$. ______

e. What is the slope of p where $x > 2$? _____

f. Write the domain of p: ________________

g. Write the range of p: ________________

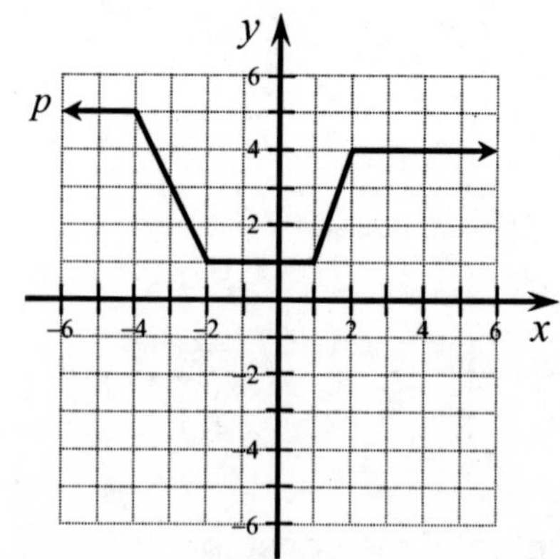

Student Activity D

Section 8.8: Shapeshifters

Directions: For problems 1-4, a function is drawn using a dashed line. Follow the directions for each problem to draw the graph that is described using a **solid line.**

1. Shift the graph of f down 4 units.

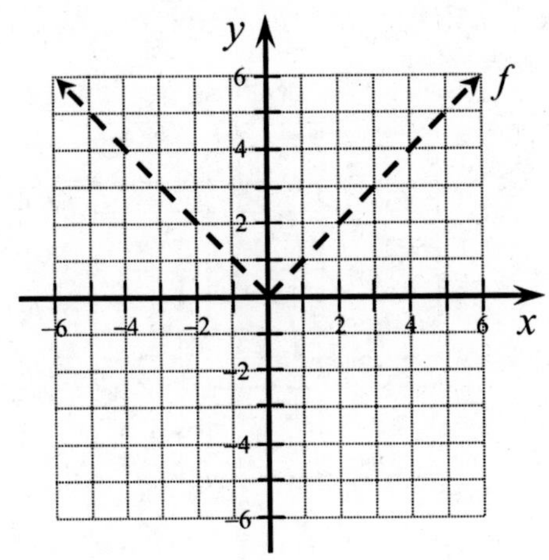

2. Reflect the graph of g over the x-axis.

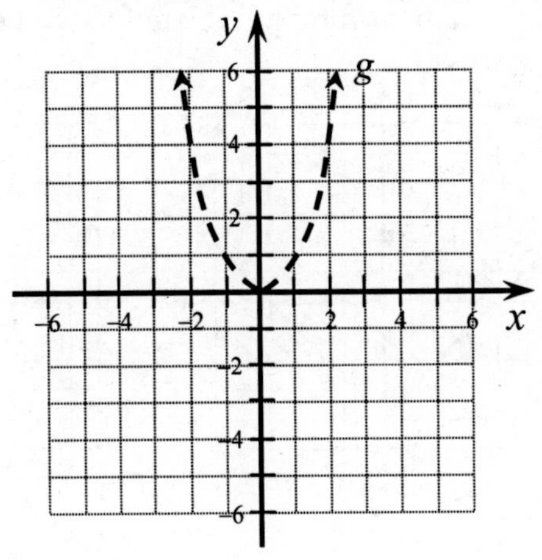

3. Shift the graph of h to the right 2 units.

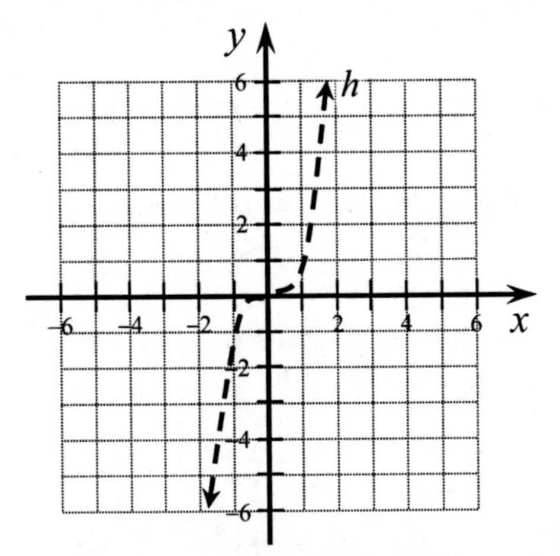

4. Shift the graph of p up 3 units and to the left 5 units.

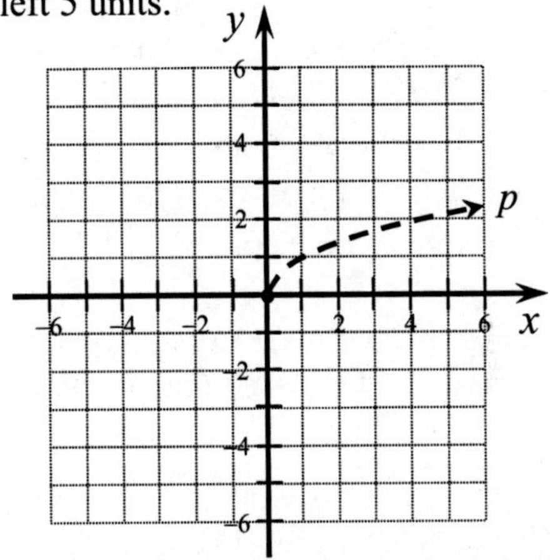

5. Describe the translations and/or reflections of f that must occur to get the graph of g.

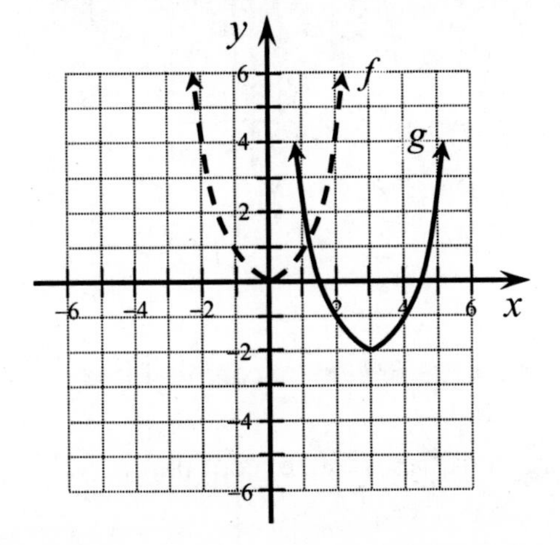

6. Describe the translations and/or reflections of f that must occur to get the graph of g.

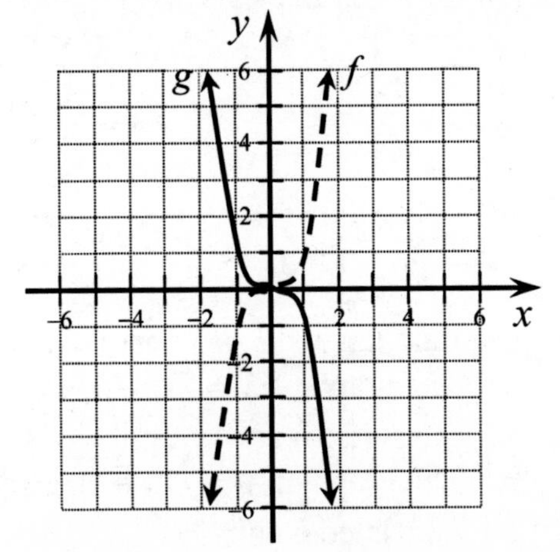

Student Activity E

Section 8.8: Reflecting on Graph Translations

In this activity, we learn about the different types of translations and reflections of function graphs.

1. In the graphs below, examine the functions *f* and *g* in each picture. How is the change in the mathematical function related to the change between the graph of *f* and the graph of *g*?

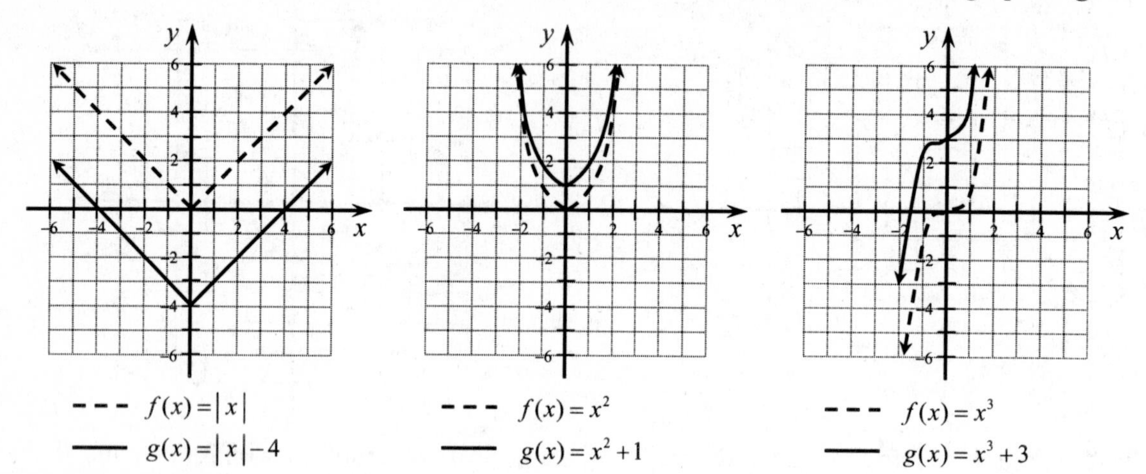

If *f* is a function and *k* is a positive number then

a. The graph of $y = f(x) + k$ is identical to the graph of $y = f(x)$ except that it is translated *k* units ____________.

b. The graph of $y = f(x) - k$ is identical to the graph of $y = f(x)$ except that it is translated *k* units ____________.

2. In the graphs below, examine the functions *f* and *g* in each picture. How is the change in the mathematical function related to the change between the graph of *f* and the graph of *g*?

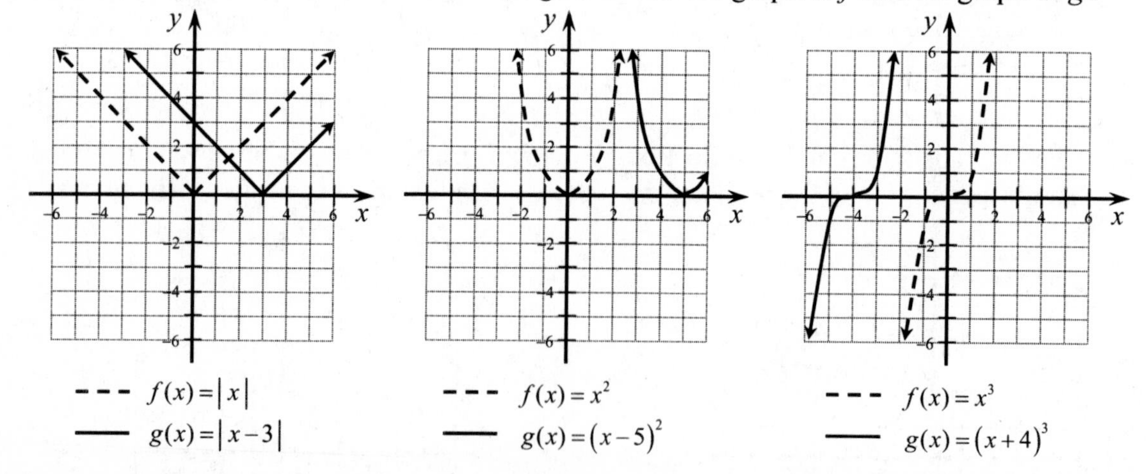

If *f* is a function and *h* is a positive number then

a. The graph of $y = f(x - h)$ is identical to the graph of $y = f(x)$ except that it is translated *h* units ____________.

b. The graph of $y = f(x + h)$ is identical to the graph of $y = f(x)$ except that it is translated *h* units ____________.

3. In the graphs below, examine the functions f and g in each picture.

A vertical reflection is a reflection about the x-axis. These graphs show vertical reflections:

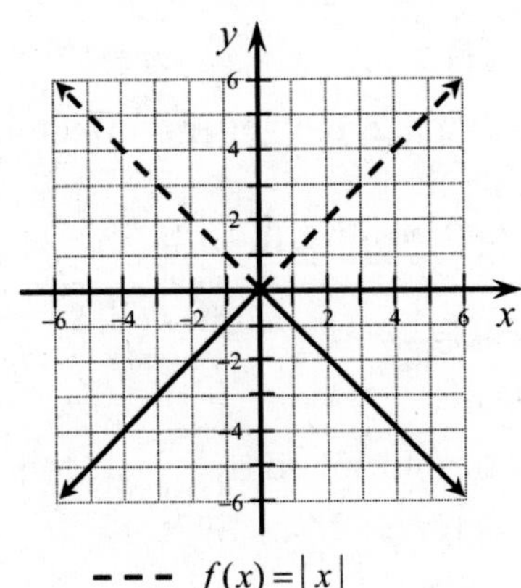

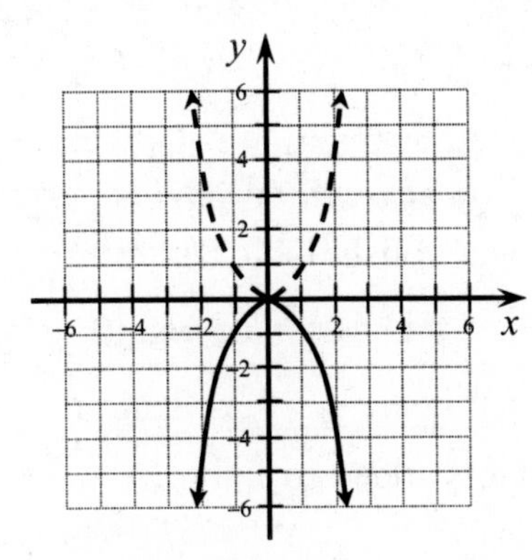

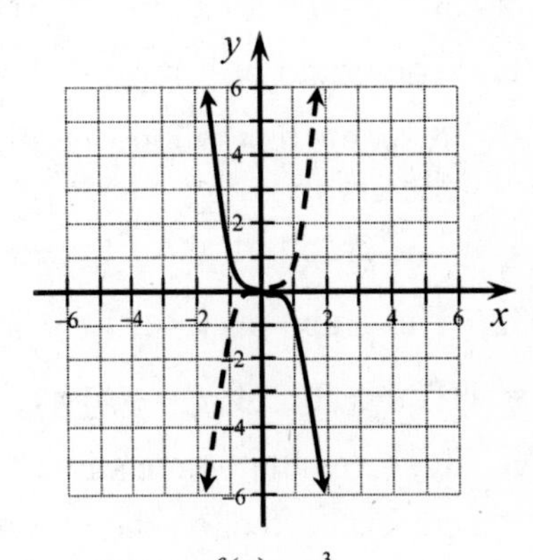

- - - $f(x)=|x|$
— $g(x)=-|x|$

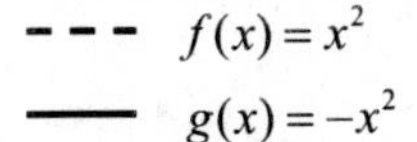

- - - $f(x)=x^3$
— $g(x)=-x^3$

A horizontal reflection is about the y-axis. These graphs show horizontal reflections:

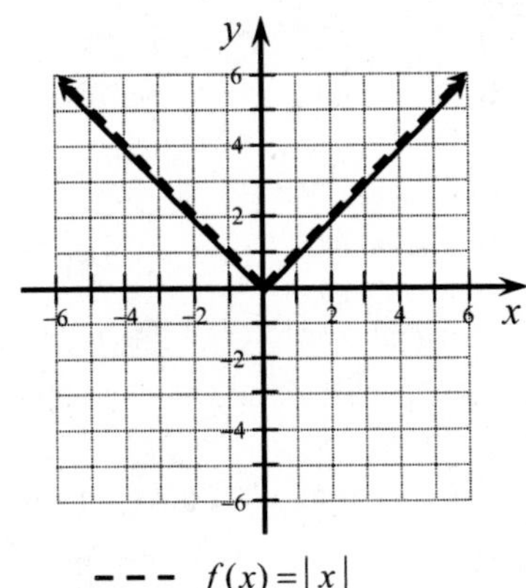

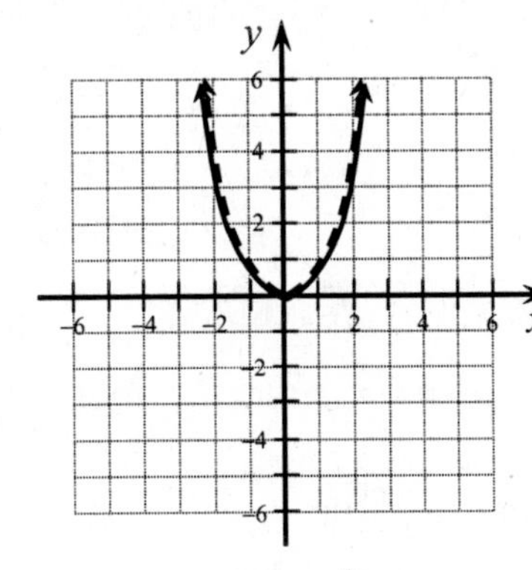

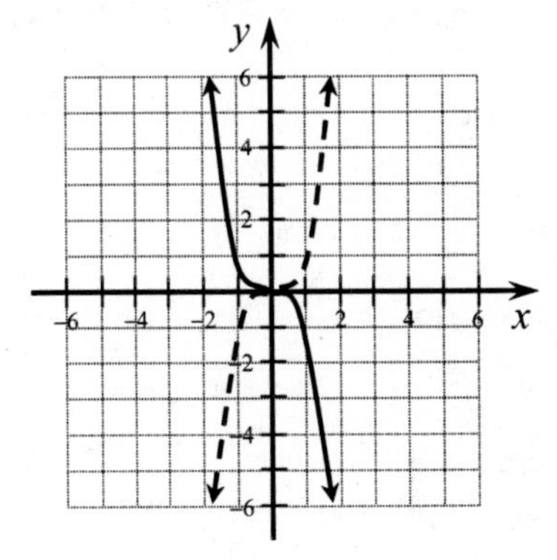

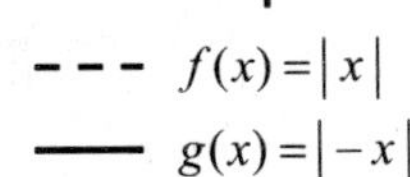

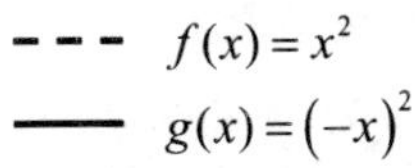

- - - $f(x)=x^3$
— $g(x)=(-x)^3$

a. The graph of $y=-f(x)$ is the graph of $y=f(x)$ reflected __________________.

b. The graph of $y=f(-x)$ is the graph of $y=f(x)$ reflected __________________.

c. Why is the graph of $g(x)=-x^3$ the same as the graph of $g(x)=(-x)^3$?

d. Why is the graph of $f(x)=|x|$ the same as the graph of $g(x)=|-x|$?

4. Let's recap. Let k and h represent positive numbers.

a. The graph of $y = f(x+h)$ is a translation of h units _____________ from $f(x)$.

b. The graph of $y = f(x-h)$ is a translation of h units _____________ from $f(x)$.

c. The graph of $y = f(-x)$ is a __________ reflection of $f(x)$.

d. It seems that when x is replaced by another expression involving x, the general effect on the graph is ____________ (*horizontal or vertical?*).

e. The graph of $y = f(x)+k$ is a translation of h units _____________ from $f(x)$.

f. The graph of $y = f(x)-k$ is a translation of h units _____________ from $f(x)$.

g. The graph of $y = -f(x)$ is a __________ reflection of $f(x)$.

h. It seems that when something is added to, subtracted from, or multiplied by $f(x)$, the general effect on the graph is ____________ (*horizontal or vertical?*).

5. a. $g(x) = (x-1)^2$ is the graph of $f(x) =$ ____ shifted ___ units ___________.

b. $g(x) = |x| + 6$ is the graph of $f(x) =$ ____ shifted ___ units ___________.

c. $g(x) = (x+2)^3$ is the graph of $f(x) =$ ____ shifted ___ units ___________.

d. $g(x) = (x+5)^2 - 2$ is the graph of $f(x) =$ ____ shifted 5 units ____________ and 2 units _____________.

e. $g(x) = -|x|$ is the graph of $f(x) =$ ____ reflected about the ___________.

f. $g(x) = \sqrt{x} + 3$ is the graph of $f(x) = \sqrt{x}$ shifted ____ units ___________.

g. $g(x) = \sqrt{-x}$ is the graph of $f(x) = \sqrt{x}$ reflected about the ___________.

6. Suppose we define a new function called the "box" function, mathematically, it is written like this: $f(x) = \boxed{x}$. Based on the graphing rules we just developed, describe how each of the following functions would be graphed, in comparison with the graph of f. Think carefully about whether x has been replaced with another expression or whether the $f(x)$ portion of the graph has remained intact.

a. $g(x) = \boxed{x+2}$ ______________________________

b. $g(x) = \boxed{x} - 4$ ______________________________

c. $g(x) = 2 + \boxed{x}$ ______________________________

d. $g(x) = -\boxed{x}$ ______________________________

e. $g(x) = \boxed{x-3}$ ______________________________

f. $g(x) = \boxed{-x}$ ______________________________

g. $g(x) = \boxed{x+3} - 1$ ______________________________

Student Activity A

Section 8.9: Charting the Language of Variation

Directions: Fill in the blank spaces in the chart. The first one has been done for you.

In Words	In Equation Form	Type of Variation (direct, inverse, joint, or combined)
1. The amount of sales tax, s, on an item varies directly as the price, p, of the item.	$s = kp$	direct
2. The time, t, it takes to clean the house is ___________ proportional to the number of people, n, who help clean it.	$t = \frac{k}{n}$	
3. The area of a circle, A, varies directly with as square of the radius, r.		
4. The time, t, it takes to arrive at a destination is ___________ proportional to the rate of travel, r.	$t = \frac{k}{r}$	
5. The force of wind, F, on a billboard varies jointly as the area of the billboard, A, and the square of the wind velocity, v.		
6. The voltage, V, in an electrical circuit with a fixed resistance is ___________ proportional to the current, I.	$V = kI$	
7. The mass, m, of an iron solid varies directly as the volume, V.		
8. The number of supports, n, needed to hold up a bridge varies ___________ as its length, ℓ.	$n = k\ell$	
9. The price of gasoline, p, varies inversely as the number of gallons of crude oil, g, that are refined.		
10. The pressure, P, of a gas is directly proportional to the temperature, T, and inversely proportional to the volume, V.		

Guided Learning Activity

Section 8.9: Constant of Variation... an Oxymoron

What does the word *constant* mean? ______________________________

In mathematics, what is a *constant*? ______________________________

What does the word *variation* mean? ______________________________

In mathematics, what is a variable? ______________________________

When we work with variation problems, we use the letter k to represent the constant of variation. But what does it mean when we say *constant of variation*? Let's investigate.

Hourly Pay:

1. Fill in the tables for Monte, Clay, and Lisa to tell them how much gross pay (before taxes and other fees are removed) they will take home for working the given number of hours. Then write an equation for each that relates their gross pay, P, to the hours worked, h.

Monte: $7.50/hour

Hours	Pay
10	
15	
20	
30	
40	

Clay: $22/hour

Hours	Pay
10	
15	
20	
30	
40	

Lisa: $18.25/hour

Hours	Pay
10	
15	
20	
30	
40	

Monte: $P =$ ______ Clay: $P =$ _______ Lisa: $P =$ ______

2. In this example, what does k represent? ______________________________

3. Write an equation for: "gross pay is directly proportional to hours worked" using k as the constant of variation. ___________

a. Lisa makes $365 for working 20 hours. Use this data and the equation above to solve for the constant k.

$k =$ _______ Look familiar?

b. Now write the variation equation with the numeric value of k in place: ___________

c. Use this equation to find Lisa's gross pay if she works 25 hours. ___________

d. Could you use the equation from part b to find gross pay if Monte works for 25 hours? Why or why not?

Density: The density, ρ, of an object is its mass per unit volume, that is $\rho = \frac{m}{V}$. If the object is made of a uniform material, then the density is constant.

1. Calculate the density (in g/cm^3) for the given mass and volume of each material in the tables below.

Balsa Wood:

Mass	12 g	15 g	24 g	39 g	45 g
Volume	100 cm^3	125 cm^3	200 cm^3	325 cm^3	375 cm^3
Density					

Ice:

Mass	229.25 g	366.8 g	458.5 g	495.18 g	687.75 g
Volume	250 cm^3	400 cm^3	500 cm^3	540 cm^3	750 cm^3
Density					

2. What is constant for Balsa wood? ____________ What is constant for ice? ____________

3. Is the density constant for Balsa wood the same as the density constant for ice? _____

Now we'll turn this around.

4. Write an equation for: "mass is directly proportional to volume" using k as the constant of variation. ____________

a. The mass of 100 cm^3 of balsa wood is 12 g. Use this data and the equation you just wrote to solve for *k*.

$k =$ ________ Look familiar?

b. Now write the variation equation with the numeric value of *k* in place: ____________

c. Use this equation to find the volume of a 500 g piece of balsa wood.

d. Could you use the equation from part b to find the volume of 500 g of ice? Why or why not?

5. See if you can now explain what it means when we say *constant of variation.*

Student Activity C

Section 8.9: Equation Lineup

Directions: We have solved many different types of equations at this point in our learning. For each equation below, categorize the type of equation, write the strategies and key steps for remembering how to solve it, then solve the equation. The first one has been started for you.

Equation Types: Linear, quadratic, rational

1. $3x^2 - x = 4$
Type of Equation? Quadratic
Strategies and Key Steps Get =0 on one side of the equation, then factor to solve.
Solve it!

2. $-7(x+2) = 1 + 5(x-1)$
Type of Equation?
Strategies and Key Steps
Solve it!

3. $\dfrac{x+4}{x-4} = \dfrac{x-1}{x+3}$
Type of Equation?
Strategies and Key Steps
Solve it!

4. $-\dfrac{3}{8} = \dfrac{1}{4} - \dfrac{5x}{16}$
Type of Equation?
Strategies and Key Steps
Solve it!

5. $2x^2 = 3(x+3)$

Type of Equation?

Strategies and Key Steps

Solve it!

6. $\dfrac{x}{x+4} - \dfrac{1}{x+1} = 1$

Type of Equation?

Strategies and Key Steps

Solve it!

7. $5(2x+1) = 3(x+4) + 7x$

Type of Equation?

Strategies and Key Steps

Solve it!

8. $x^2 = 10x$

Type of Equation?

Strategies and Key Steps

Solve it!

Assessment 8C

Chapter 8: End-of-chapter Assessment for Understanding

For each of the following, describe the type of problem and the strategies and key steps to remember while doing the problem. You do **not** have to complete the problems.

	Type of Problem	Strategies and Key Steps
1. Graph the line that has slope 1 and passes through $(-2,-3)$.		
2. Multiply: $\frac{3x^2+12x}{x-6}\cdot\frac{x^2-4x-12}{x^2+2x}$		
3. Factor: $x^3y^3-64z^6$		
4. If $f(x)=-3x+1$, find $f(4)$.		
5. Graph $g(x)=(x-1)^2$ and write the domain and range.		
6. Solve: $\frac{2}{x+6}+\frac{5}{x^2+6x}=1$		
7. If y is inversely proportional to x, and $y=8$ when $x=2$, find y when $x=5$.		
8. Find the equation of the line that passes through $(0,3)$ and $(-1,5)$.		
9. Factor: $xy^2-9x+4y^2-36$		
10. Solve: $\lvert x+3\rvert-2>4$		

Assessment 8D

Chapter 8: Metacognitive Skills Assessment

Metacognitive skills refer to the ability to judge how well you have learned something and to effectively direct your own learning and studying. This is a self evaluation tool designed to help you focus your studying and to improve your metacognitive skills with regards to this math class.

Fill the 1st column out before you begin studying.
Fill the 2nd column out after you study and before you take the test.
Go back to this page after your test and circle any of the ratings that you would now change – this identifies the "disconnects" between what you think you know well and what you actually know well.

Use the scale below to assign a number to each topic.

5 *I am confident I can do any problems in this category correctly.*
4 *I am confident I can do most of the problems in this category correctly.*
3 *I understand how to do the problems in this category, but I still make a lot of mistakes.*
2 *I feel unsure about how to do these problems.*
1 *I know I don't understand how to do these problems.*

Topic or Skill	Before Studying	After Studying
Checking whether an ordered pair is a solution to an inequality or a system of inequalities.		
Understanding whether the boundary line for an inequality is dashed or solid and which side of the boundary line to shade.		
Solving for y in a linear inequality (including the special rule that applies to dividing or multiplying by a negative number when solving inequalities).		
Graphing the boundary lines for a system of inequalities.		
Graphing a linear inequality using a check point.		
Determining the proper region to shade as the solution in a system of inequalities.		
Deciding whether an inequality is true or false.		
Solving an inequality.		
Graphing an inequality.		
Writing the answer to an inequality in interval notation.		
Solving an application involving inequalities.		
Understanding intersection and union, including their relationship to the words "and" and "or" and the symbols $\cup$ and $\cap$.		
Solving a compound inequality (one that uses **and** or **or**).		
Solving a double inequality (like $-3 < x + 1 < 5$).		
Isolating the absolute value in an equation or inequality.		
Determining the type of absolute value equation or inequality (Case 1, 2, 3, 4, or special case).		
Remembering the proper procedure after you determine the category of absolute value equation or inequality.		
Reasoning out the answer to a "special case" absolute value equation or inequality.		

(continued on next page)

Use the scale below to assign a number to each topic.

5 I am confident I can do any problems in this category correctly.
4 I am confident I can do most of the problems in this category correctly.
3 I understand how to do the problems in this category, but I still make a lot of mistakes.
2 I feel unsure about how to do these problems.
1 I know I don't understand how to do these problems.

Factoring expressions by taking out a GCF.		
Factoring expressions by grouping terms.		
Factoring trinomials of the form $x^2 + bx + c$.		
Factoring trinomials of the form $ax^2 + bx + c$.		
Factoring a difference of squares.		
Factoring a sum or difference of cubes.		
Factoring expressions that involve a variety of techniques.		
Simplifying a rational expression involving addition or subtraction.		
Simplifying a rational expression involving multiplication or division.		
Solving a rational equation.		
Simplifying a complex fraction.		
Graphing linear equations by plotting points, using the intercept method, or using the slope-intercept form.		
Finding the slope of a line or the slope between two points.		
Writing the equation of a line given some of the information about it.		
Recognizing lines that are parallel or perpendicular using their slopes.		
Writing a linear equation model and solve application problems involving linear equation models.		
Finding the midpoint of a line segment.		
Understanding function notation.		
Finding the domain and range of a relation.		
Evaluating a function written in function notation for a specified value.		
Identifying whether a relation is a function (including the vertical line test).		
Solving application problems that involve function notation.		
Finding information on graphs using function notation.		
Finding the domain and range of a function given its graph.		
Graphing the identity, squaring, cubing, and absolute value functions and recognizing their characteristic shapes.		
Recognizing indicators of translations and reflections in the function equations.		
Understanding what the constant of variation is.		
Converting the language of variation problems to variation equations containing a constant of variation.		
Solving for k in a variation equation.		
Using the variation equation to solve for something else in a problem.		

Student Workbook: Chapter 9

Table of Contents: *Radical Expressions and Equations*

Assessment 9A

Pretest and Diagnostic Tool: Radical Expressions and Equations

Directions: Complete this assessment without looking back at your notes or your book. Assume all variables represent positive real numbers. **Do not use a calculator for this assessment.**

1. $(\_\_\_)^2 = 49$
2. $-5^2 = \_\_\_\_$ and $(-5)^2 = \_\_\_$
3. $0^2 = \_\_\_\_$ and $1^2 = \_\_\_$
4. $(\_\_\_)^2 = -25$
5. $125 = (\_\_\_)^3$
6. $(\_\_\_)^4 = 16$
7. $x^{15} = (\_\_\_)^3$
8. $x^{15} = (\_\_\_)^5$
9. Add: $\frac{1}{4} + \frac{1}{3}$
10. Evaluate: 4^{-1}
11. Evaluate: $\left(\frac{2}{3}\right)^{-2}$
12. Multiply: $\frac{1}{3} \cdot \frac{1}{4}$
13. $45 = (\_)^2 \cdot 5$
14. $8x^6 = (\_\_\_\_\_)^3$
15. $40 = (2)^3 \cdot \_\_\_$
16. $x^7 = (x^2)^3 \cdot \_\_\_$
17. $x + x = \_\_\_\_$, $x \cdot x = \_\_\_\_$
18. Simplify: $3x^2 - x^2$
19. Simplify: $5a + 2b - 7a - 5b$
20. Simplify: $(2x^2 + 3x) - (x^2 - 5x)$
21. Simplify: $3x \cdot 4x^2$
22. Simplify: $\frac{200x^3y^6}{2xy^2}$
23. Multiply: $(x+4)(x-4)$
24. Expand and simplify: $(x-3)^2$
25. Solve: $4x + 1 = 9$
26. Solve: $x^2 - 5x - 14 = 0$
27. Solve: $4x + 6 = 6x - 2$
28. Simplify: $(5-2)^2 + (3-4)^2$
29. Simplify: $[(3-(-1))]^2 + (-2-4)^2$
30. Solve for x^2: $4^2 + x^2 = 5^2$

Student Activity A

Section 9.1: Radicals on Trial

Directions: Each of the radicals in the box is "on trial" for being rational, irrational, or non-real and you are the judge.

If the radicand is real and **is** a perfect square, then sentence the radical to its appropriate rational number box.

If the radicand is real and **is not** a perfect square, then sentence the radical to the appropriate irrational number box by placing it between its two nearest integers.

If the radial expression is non-real, sentence it to life in the non-real numbers box.

Do this activity without a calculator!

$\sqrt{\frac{64}{49}}$	$-\sqrt{-1}$	$\sqrt{\frac{1}{2}}$	$\sqrt{\frac{4}{9}}$	$\sqrt{-9}$	$\sqrt{0}$	$\sqrt{-25}$
$\sqrt{3}$	$\sqrt{16}$	$\sqrt{6}$	$\sqrt{45}$	$\sqrt{-4}$	$\sqrt{\frac{60}{2}}$	$\sqrt{50}$
$\sqrt{\frac{16}{25}}$	$\sqrt{-\frac{1}{2}}$	$\sqrt{\frac{39}{2}}$	$\sqrt{9}$	$\sqrt{14}$	$\sqrt{1}$	$\sqrt{36}$

Rational Numbers

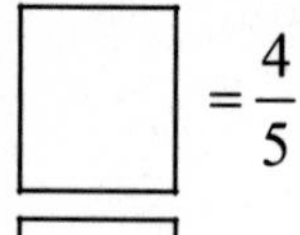

$\square = \frac{4}{5}$

$\square = 4$

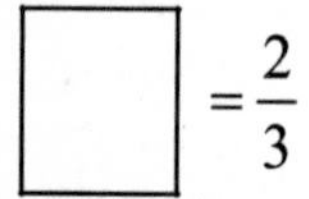

$\square = \frac{2}{3}$

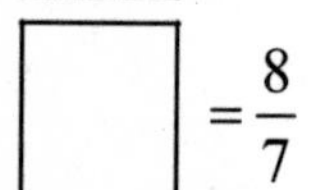

$\square = \frac{8}{7}$

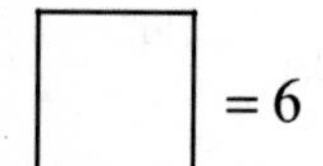

$\square = 6$

$\square = 3$

$\square = 0$

$\square = 1$

Irrational Numbers

$0 < \square < 1$

$1 < \square < 2$

$2 < \square < 3$

$3 < \square < 4$

$4 < \square < 5$

$5 < \square < 6$

$6 < \square < 7$

$7 < \square < 8$

Non-Real (Imaginary) Numbers

Student Activity B

Section 9.1: Square Roots Using a Calculator

Often the $\sqrt{\ }$ function is found ABOVE one of the keys, and students will need to use a [2nd] button to use the $\sqrt{\ }$ function.

Your calculator probably falls into one of two groups:

Group 1: Radicand first

On these calculators, you would find $\sqrt{9}$ by keying in [9] then [$\sqrt{\ }$]. It is likely that this will immediately give you the answer 3 without pressing any other keys.

Group 2: Radical Symbol first

On most graphing calculators and some of the newer scientific calculators, you would do the keystrokes like this: [$\sqrt{\ }$] and then [9]. Then you probably see $\sqrt{\ }(9$ in your display. You have to close the parentheses by pressing [ )] and then [ENTER] to evaluate.

Now fill out the table below, by finding the proper keystrokes to do the evaluation on your calculator. In the last column, square the calculator result to see if you get the radicand back. The first one has been done for you.

	Expression	Calculator Keystrokes and result	Check by squaring the result
a.	$\sqrt{8}$	Keystrokes will vary.	$(2.8284)^2 \approx 7.9998 \approx 8$
b.	$\sqrt{169}$		
c.	$\sqrt{\frac{16}{25}}$		
d.	$\sqrt{\frac{1}{2}}$		
e.	$\sqrt{\frac{2}{3}}$		
f.	$\sqrt{-4}$		
g.	$\sqrt{\frac{4}{9}}$		
h.	$\sqrt{1.5625}$		

Student Activity C

Section 9.1: Match Up on Simplifying Radical Expressions

Match-up: Match each of the expressions in the squares in the table below with its simplified form from the top. Assume all variable represent positive numbers. If the simplified form is not found among the choices A through D, then choose E (none of these).

A $3x^2$ **B** $2x$ **C** $-5x$ **D** $4x^3$ **E** None of these

$\sqrt{9x^4}$	$\sqrt[3]{-125x^3}$	$\sqrt[4]{81x^8}$
$\sqrt[6]{64x^6}$	$\sqrt[3]{64x^9}$	$\sqrt{-25x^2}$
$2x^2 + x^2$	$\sqrt[3]{27x^6}$	$\sqrt{\frac{36x^2}{9}}$
$\sqrt{9x}$	$\sqrt{x^2} + x$	$\sqrt{16x^6}$
$-\sqrt[4]{625x^4}$	$\sqrt[3]{12x^9}$	$2 \cdot \sqrt[3]{8x^9}$

Guided Learning Activity

Section 9.1: Radical Graphing

The graph of $f(x) = \sqrt{x}$, is a curve with a definite endpoint.

1. Draw the graph of $f(x) = \sqrt{x}$ by plotting points.

x	$f(x)$
9	
4	
1	
0	
–1	
–4	
–9	

What is the domain of f? _________

What is the range of f? _________

Why are there no points for $x < 0$?

The graph of $f(x) = \sqrt[3]{x}$, is a curve that has a similar shape to the curve $y = x^3$.

2. Draw the graph of $f(x) = \sqrt[3]{x}$ by plotting points.

x	$f(x)$
8	
1	
0	
–1	
–8	

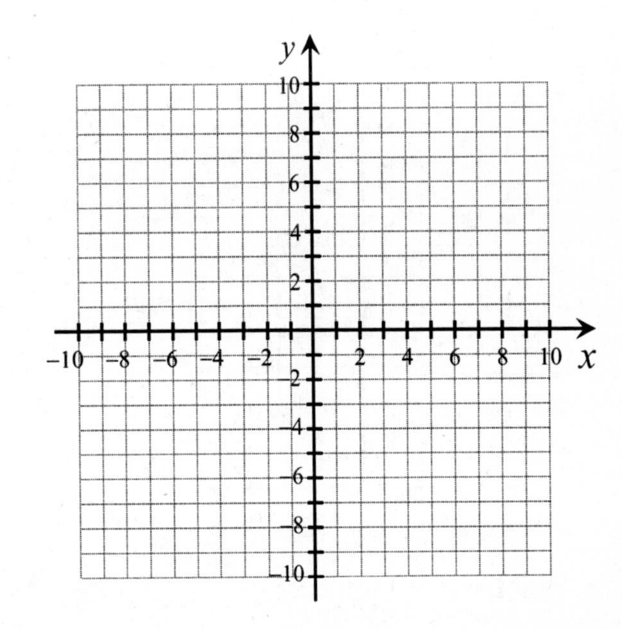

What is the domain of f? _________

What is the range of f? _________

These radical functions can be translated and reflected in the exact same way as $y = |x|$, $y = x^2$, and $y = x^3$.

3. Draw the graph of $g(x) = \sqrt{x} - 3$ by shifting the graph of $f(x) = \sqrt{x}$ ______ 3 units.

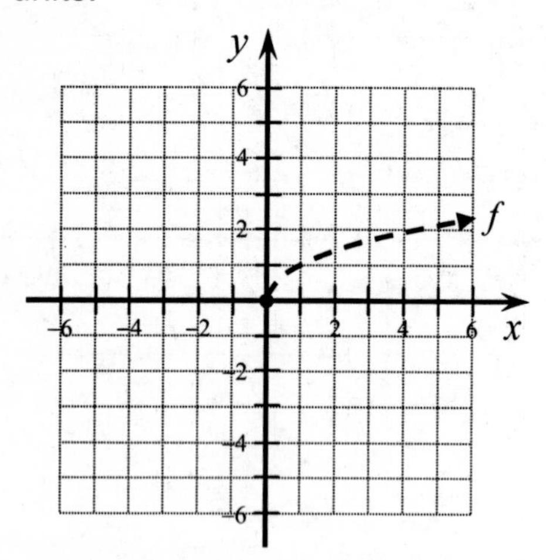

What is the domain of g? ______

What is the range of g? ______

4. Draw the graph of $g(x) = \sqrt[3]{x+2}$ by shifting the graph of $f(x) = \sqrt[3]{x}$ ______ 2 units.

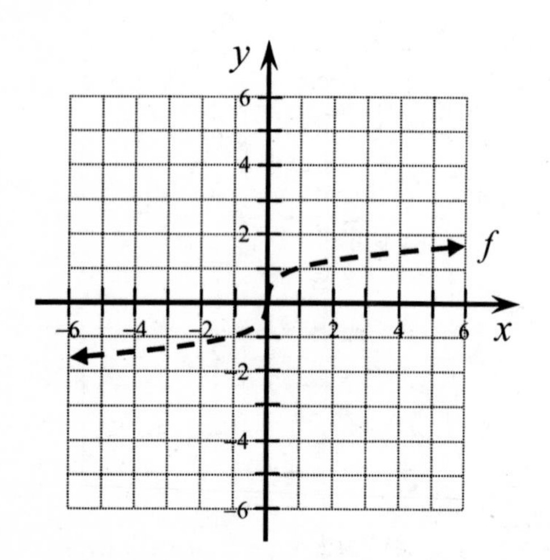

What is the domain of g? ______

What is the range of g? ______

5. Draw the graph of $g(x) = -\sqrt{x}$ by reflecting the graph of $f(x) = \sqrt{x}$ over the __-axis.

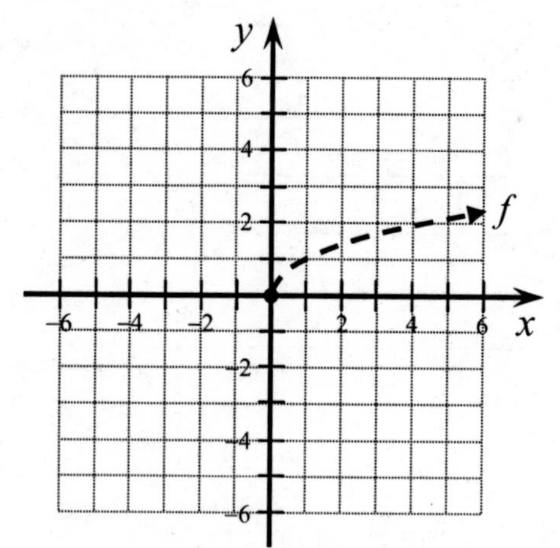

What is the domain of g? ______

What is the range of g? ______

6. Draw the graph of $g(x) = \sqrt[3]{x+1} - 4$ by shifting the graph of $f(x) = \sqrt[3]{x}$ ______ 1 unit and ______ 4 units.

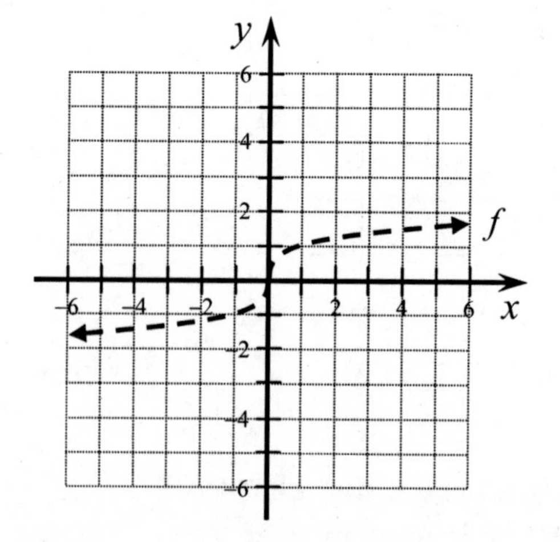

What is the domain of g? ______

What is the range of g? ______

Student Activity D

Section 9.1: Difficult Choices

If the variables can represent any real number and the radical expression represents a real number:

$$\sqrt[n]{x^n} = |x| \text{ if } n \text{ is even.}$$

$$\sqrt[n]{x^n} = x \text{ if } n \text{ is odd.}$$

If the variables only represent positive values, it is not necessary to use absolute values for the even roots.

Directions: Simplify each expression and choose the simplified expression from A or B. If the radical expression represents a non-real number (an imaginary number), then choose C (non-real).

	A	B	C
1. Simplify $\sqrt{36x^2}$ if x can be any real number.	$6x$	$6\|x\|$	Non-real
2. Simplify $\sqrt{-36x^2}$ if x can be any real number.	$6x$	$-6x$	Non-real
3. Simplify $\sqrt{36x^2}$ if x is a positive number.	$6x$	$6\|x\|$	Non-real
4. Simplify $\sqrt[3]{8x^6}$ if x can be any real number.	$2x^2$	$\|2x^2\|$	Non-real
5. Simplify $\sqrt[3]{-8x^6}$ if x can be any real number.	$-2x^2$	$2x^2$	Non-real
6. Simplify $\sqrt{49x^2}$ if x can be any real number.	$7x$	$7\|x\|$	Non-real
7. Simplify $\sqrt{49x^2}$ if x is a positive number.	$7x$	$7\|x\|$	Non-real
8. Simplify $\sqrt{16x^4}$ if x can be any real number.	$4x^2$	$4\|x^2\|$	Non-real
9. Simplify $\sqrt{-16x^4}$ if x can be any real number.	$-4x^2$	$\|-4x^2\|$	Non-real
10. Simplify $\sqrt[3]{64x^3}$ if x can be any real number.	$4x$	$4\|x\|$	Non-real
11. Simplify $\sqrt[4]{16x^4}$ if x is a positive number.	$2x$	$2\|x\|$	Non-real
12. Simplify $\sqrt[4]{16x^4}$ if x can be any real number.	$2x$	$2\|x\|$	Non-real
13. Simplify $\sqrt[4]{-16x^4}$ if x is a positive number.	$2x$	$2\|x\|$	Non-real
14. Simplify $\sqrt[5]{-32x^5}$ if x can be any real number.	$-2x$	$2\|x\|$	Non-real
15. Simplify $\sqrt[5]{32x^5}$ if x can be any real number.	$2x$	$2\|x\|$	Non-real

Student Activity A

Section 9.2: Paint by Rational Exponents

Directions: Convert each expression according to the directions. Then shade in the box that contains your answer. The first one has been done for you. There's a surprise when you're finished!

Convert to radical notation:

1. $(xy)^{1/2} = \sqrt{xy}$
2. $(25+x^2)^{1/2}$
3. $5x^{1/2}$
4. $(-x)^{1/5}$
5. $\dfrac{y}{x^{-1/2}}$

Convert to rational exponent notation:

6. $2\sqrt{xy}$
7. $5\sqrt[4]{x^3}$
8. $\sqrt{3x^5}$
9. $\sqrt[6]{(10x+7)^7}$
10. $\sqrt[5]{-ab^4}$

$3x^{2/5}$	$(3x^5)^{1/2}$	$(3x)^{5/2}$	$\sqrt{25+x^2}$	$5+x$
$\dfrac{1}{\sqrt[5]{x}}$	$\sqrt[5]{-x}$	$-\sqrt{x^5}$	$y\sqrt{x}$	$-\dfrac{y}{\sqrt{x}}$
$(xy)^2$	$10x+7^{6/7}$	$10x^{7/6}+7^{7/6}$	$x\sqrt{y}$	$a^{1/5}b^{-4/5}$
$5x^{4/3}$	$\sqrt{5x}$	$\sqrt{xy}$	$(2xy)^2$	$ab^{-4/5}$
$5x^{3/4}$	$\dfrac{r}{\sqrt{t}}$	$10x+7^{7/6}$	$2(xy)^2$	$(-ab^4)^{1/5}$
$(5x)^{3/4}$	$5\sqrt{x}$	$-t\sqrt{r}$	$2(xy)^{1/2}$	$(2xy)^{1/2}$
$(5x)^{\frac{4}{3}}$	$-r\sqrt{t}$	$(10x+7)^{7/6}$	$(10x+7)^{6/7}$	$2xy^{1/2}$

Student Activity B

Section 9.2: Language of the Other Roots

Directions: For each set, fill in the missing boxes in the first row with numbers that make the two expressions equal. Write out a statement that describes the relationship between the two expressions in the second row. The first two have been started for you.

Expression written with root or power	simplified expression
Write out a statement in words.	

$\sqrt[3]{125}$	5
The cube root of ____ is ____.	

$(\quad)^3$	216
The cube of ____ is ____.	

$(\quad)^{\frac{1}{4}}$	2

$\sqrt[4]{\quad}$	9

	27
The cube of ____ is ____.	

$\sqrt[3]{\quad}$	7

	$\frac{1}{16}$
____ raised to the -2 power is ____.	

$(\quad)^{-2}$	
3 raised to the ____ power is ____.	

$\frac{1}{5^2}$	

$1^{-\frac{3}{4}}$	

$(-1000)^{\frac{1}{3}}$	

$(\quad)^{\frac{3}{2}}$	
____ raised to the $\frac{3}{2}$ is 8.	

Student Activity C

Section 9.2: Rational Exponents Using a Calculator

By now you should be familiar with the keys on your calculator used to evaluate expressions like $\sqrt{196}$, $\sqrt[5]{3125}$. Your calculator can also evaluate expressions containing rational exponents.

Consider the following expression: $\left(\sqrt[4]{625}\right)^3$

To evaluate this expression with a calculator, you would first need to write an equivalent expression with a rational exponent:

$$\left(\sqrt[4]{625}\right)^3 = 625^{3/4}$$

Most calculators have an exponent key that looks like $\boxed{\wedge}$. When entering a rational exponent, you should always use parenthesis around the exponent. Enter each of the following expressions exactly as they are shown:

$625\boxed{\wedge}3/4 =$ ______________

$625\boxed{\wedge}(3/4) =$ ______________

You should see two *very* different answers! Which one is correct? If you calculate $\sqrt[4]{625}$ then cube the result you get 125. This is the correct answer. Your calculator will always follow the order of operations, so without parenthesis around the $3/4$, your calculator will cube 625 *then* divide the result by 4. The result is a very large number!

Now try finding these with your calculator. The first row has been done for you.

Radical Notation	Rational Exponent Notation	Keystrokes to Enter into Calculator	Answer
1. $\left(\sqrt[3]{64}\right)^2$	$64^{2/3}$	$64\boxed{\wedge}(2/3)$	16
2.	$-243^{6/5}$		
3. $\dfrac{1}{\sqrt[3]{8^2}}$			
4.	$1024^{-2/5}$		
5. $\dfrac{1}{\left(\sqrt[4]{1296}\right)^{-3}}$			
6.	$625^{3/4}$		
7. $\left(\sqrt[5]{32}\right)^{-3}$			

Student Activity D

Section 9.2: Escape the Rational Exponent Matrix

Directions: Assume all variables represent positive numbers. Begin at the box marked START. By shading in adjacent pairs of squares that contain equivalent expressions, you will eventually find the path to "escape" this matrix of boxes. The first "step" in the path and a couple steps in the middle have been taken for you.

START $\left(36b^6\right)^{1/2}$	$18b^3$	$9b^3$	$x+\sqrt{x}$	$7^{2/3}$
$6b^3$	$36b^3$	$3y$	$\sqrt[4]{3y}$	$x^{9/5}$
$x^{3/4}\cdot x^2$	$\sqrt[4]{x^{11}}$	$\left(3y^{2/3}\right)^3$	$9y^2$	$\left(\sqrt{2x}\right)^3$
$\sqrt{x^3}$	$x^{4/11}$	$27y^2$	$\sqrt[3]{3y^2}$	$2x^{4/3}$
$x^{4/3}$	$x^{-4/3}$	$\dfrac{x^3}{x^{7/3}}$	$\sqrt[3]{x^2}$	$x^{5/6}\left(x^{1/6}+x^{1/2}\right)$
$x^{1/3}$	$\dfrac{1}{x^7}$	x^7	$x^{1/9}$	$x+\sqrt[3]{x^4}$
7	$\dfrac{x^3}{8}$	$\left(\dfrac{x^5}{32}\right)^{3/5}$	$y^{1/20}$	$\sqrt[4]{\sqrt[5]{y}}$
$\dfrac{1}{y^2}$	$\sqrt[10]{y^5}$	y^2	$y^{5/4}$	y^{20}
$3\sqrt{y}$	$y^{1/2}$	49	ESCAPE $343^{2/3}$	$\sqrt{y^{400}}$

Student Activity E

Section 9.2: Radical Heteronyms

Directions: In writing, there are words that are spelled the same but have different pronunciations and different definitions; these are called heteronyms. Many mathematical expressions look similar but are really very different (almost like mathematical heteronyms). Simplify each of the expressions below, paying close attention to the use of parentheses, the mathematical operations, and notation. Assume all variables represent positive numbers.

1.

9^{-1}	$9^{1/2}$	9^{2}	9^{-2}	-9^{-1}	$9^{-1/2}$

2.

$\sqrt{25-9}$	$(25-9)^{1/2}$	$25^{1/2}-9^{1/2}$	$(25-9)^{-1}$	$25^{-1}-9^{-1}$

3.

$x^{1/2}x^{1/3}$	$\left(x^{1/2}\right)^{1/3}$	$x^{1/2}+x^{1/3}$	$\dfrac{x^{1/2}}{x^{1/3}}$	$\left(\dfrac{1}{2}x\right)\left(\dfrac{1}{3}x\right)$	$\dfrac{1}{3}\left(\dfrac{1}{2}x\right)$

4.

$\left(16a^{4}\right)^{-2}$	$\left(16a^{4}\right)^{1/2}$	$\left(16a^{4}\right)^{1/4}$	$\dfrac{1}{2}\left(16a^{4}\right)$	$16\left(a^{4}\right)^{1/2}$

5.

$\left(\dfrac{9}{4}\right)^{-2}$	$\left(\dfrac{9}{4}\right)^{1/2}$	$\left(\dfrac{9}{4}\right)^{-1/2}$	$\left(\dfrac{9}{4}\right)^{-1}$	$\dfrac{9^{1/2}}{4}$	$\dfrac{9^{-1/2}}{4}$

Student Activity A

Section 9.3: Factor Trees

Directions: Use a factor tree to find the prime factorization for each number. Build branches of the tree using factor pairs. When a prime number is reached in one of the branches, circle it in red to indicate that it is one of the "apples" (prime factors) on the tree. To write the prime-factored form, collect all the "apples" on the tree. The factor tree shown here tells us that the prime factorization of 80 is $(2\cdot 2)(2\cdot 2)5$. To find $\sqrt{80}$, we find pairs of factors in $\sqrt{2\cdot 2\cdot 2\cdot 2\cdot 5}$ and bring them outside the radical as a single factor to get $2\cdot 2\sqrt{5}$ or $4\sqrt{5}$. To make a particularly colorful and "treelike" factor tree below, use brown for all the "branches," green for all the numbers, and red to circle and shade in the prime numbers. In the factor tree below, your "trees" will go in all directions to make the branches. The number 90 has been prime factored for you.

80

8 10

4 2 2 5

2 2

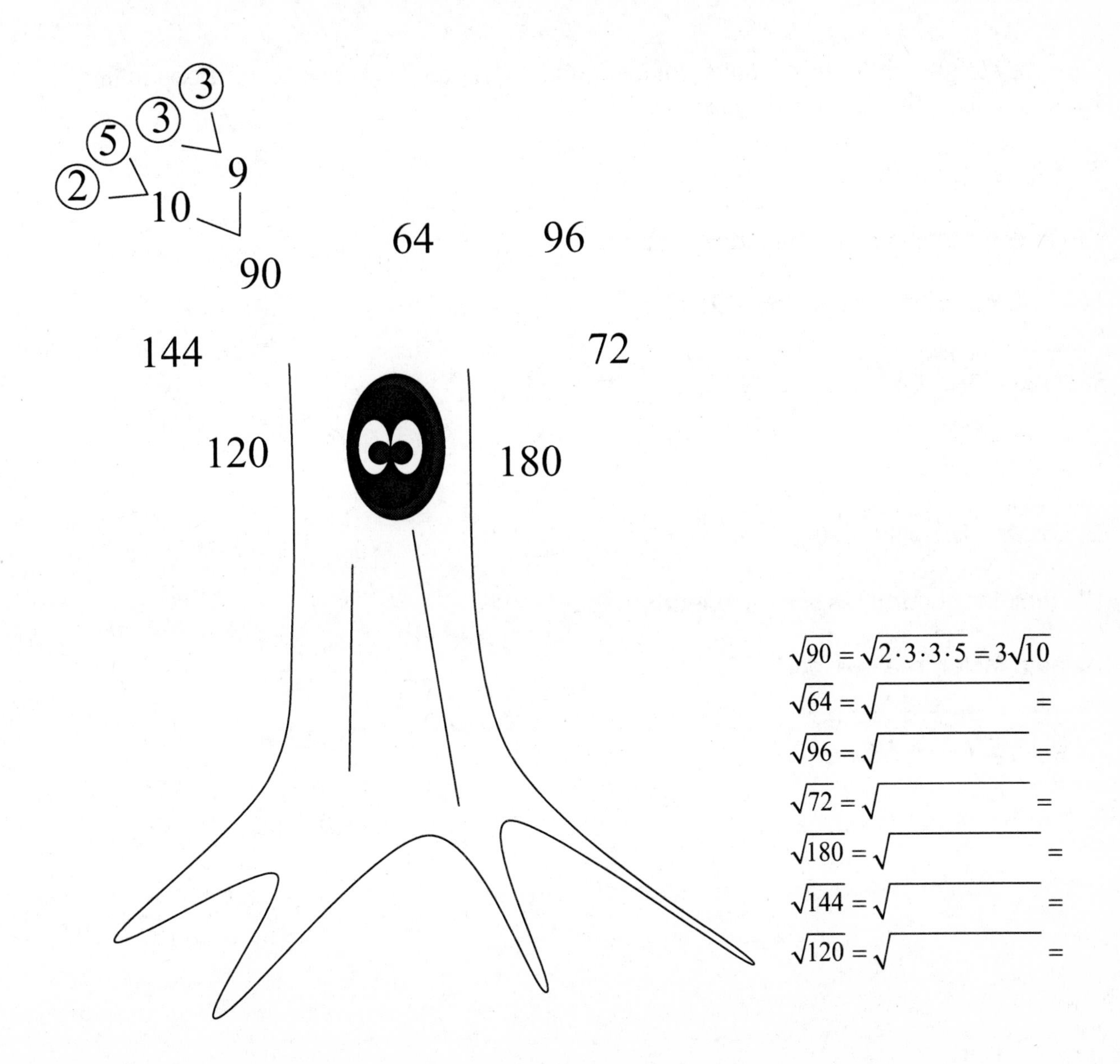

Guided Learning Activity
Section 9.3: Higher on the Factor Tree

Recall that a **square root** is usually written without an index, so $\sqrt{a} = \sqrt[2]{a}$. In a square root we had to find **pairs** of a factor in order to simplify the radicand.

Similarly, in a **cube root**, like $\sqrt[3]{a}$, we would need to find **triples** of a factor in order to simplify the radicand.

And in a **fourth root**, $\sqrt[4]{a}$, we would need to find **quadruples** of a factor in order to simplify the radicand.

Notice that the index of the radical tells us how many matching factors we need to find in order to simplify.

$$\rightarrow\sqrt[2]{a} \quad \rightarrow\sqrt[3]{a} \quad \rightarrow\sqrt[4]{a}$$

One way to search for these pairs, triples, quadruples, etc. is to use factor trees to find the prime factorization for each radicand.

Example: Simplify: $\sqrt[3]{162}$.

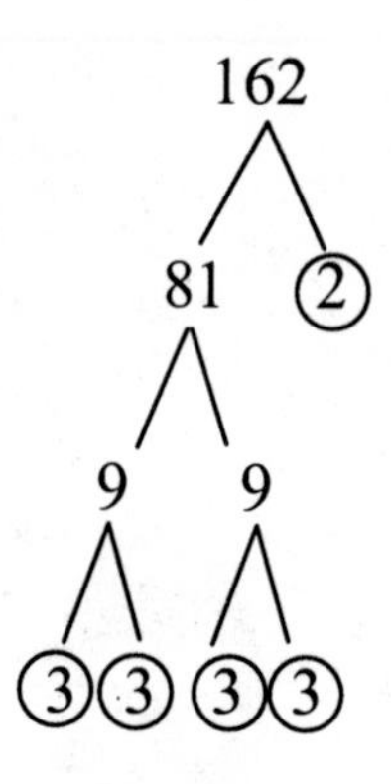

We look for triples since the index is three.

Using a factor tree, we factor 162.

$$\sqrt[3]{162} = \sqrt[3]{3 \cdot 3 \cdot 3 \cdot 3 \cdot 2} = 3\sqrt[3]{3 \cdot 2} = 3\sqrt[3]{6}$$

Example: Simplify: $\sqrt[4]{80}$

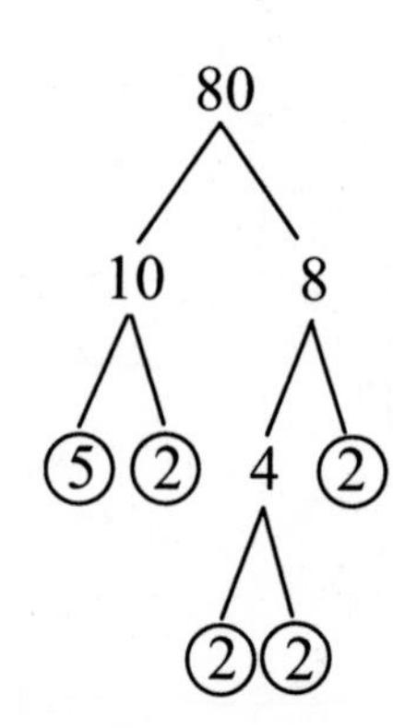

We look for quadruples since the index is four.

Using a factor tree, we factor 80.

$$\sqrt[4]{80} = \sqrt[4]{5 \cdot 2 \cdot 2 \cdot 2 \cdot 2} = 2\sqrt[4]{5}$$

Directions: Use factor trees to help simplify the following higher order roots. The first factor tree has been started for you.

1. $\sqrt[3]{3,000}$

4. $\sqrt[3]{625}$

2. $\sqrt[4]{432}$

5. $\sqrt[4]{512a^8}$

3. $\sqrt[5]{243}$

6. $\sqrt[3]{216x^5}$

Student Activity B

Section 9.3: Match Up on Simplifying Radicals

Match-up: Simplify each radical expression. Then look at the remaining radicand and choose the letter that corresponds to the remaining radicand. For example, for $2\sqrt{3}$, $x\sqrt{3}$, $5\sqrt{3}$, and $x^2\sqrt{3}$, you would choose B. If the radicand is not one of the choices, or there is no radical left then choose E (none of these). Assume that all variables represent nonnegative numbers.

A 2 **B** 3 **C** 5 **D** x **E None of these**

$\sqrt{72}$	$\sqrt{80}$	$\sqrt{81}$	$\sqrt[3]{375}$
$\sqrt{90}$	$\sqrt{x^7}$	$\sqrt{125}$	$\sqrt[3]{x^4}$
$\sqrt{200}$	$\sqrt{45}$	$\sqrt{30}$	$\sqrt[3]{3993x^3}$
$\sqrt{108}$	$\sqrt{99}$	$\sqrt{32}$	$\sqrt[3]{1024}$
$\sqrt{180}$	$\sqrt{x^3}$	$\sqrt{16x}$	$\sqrt[3]{3000x^3}$

Student Activity C

Section 9.3: Thread of Like Terms II

Directions: Simplify each of the expressions that follow.

1.

$3x^2 + 5x^2$	$\frac{3}{11} + \frac{5}{11}$	$3\sqrt{x} + 5\sqrt{x}$	$3\sqrt[3]{3} + 5\sqrt[3]{3}$

2.

$x^3 - 5x^3$	$\frac{1}{3} - \frac{5}{3}$	$\sqrt{5} - 5\sqrt{5}$	$\sqrt[3]{3} - 5\sqrt[3]{3}$

3.

$\frac{8}{x+8} - \frac{1}{x+8}$	$8xy - xy$	$8\sqrt{y} - \sqrt{y}$	$8\sqrt[3]{2} - \sqrt[3]{2}$

4.

$4x^2y^2 - 6x^2y^2$	$\frac{4}{5x^2} - \frac{6}{5x^2}$	$4\sqrt{xy} - 6\sqrt{xy}$	$4\sqrt[3]{15} - 6\sqrt[3]{15}$

5. What is the lesson to be learned here about addition and subtraction in algebra and mathematics in general?

6. Now be a little more careful with this group of expressions. If the expression can't be simplified, just say so!

$8\sqrt{3} + 2\sqrt{3}$	$6\sqrt{2} + 6\sqrt{3}$	$\sqrt{5} + \sqrt{5}$	$\sqrt{5}\sqrt{5}$
$3\sqrt{2} \cdot \sqrt{2}$	$3\sqrt{2} + \sqrt{2}$	$\sqrt[3]{x} - \sqrt[3]{x}$	$\sqrt{3} + \sqrt{3} + \sqrt{3}$
$3\sqrt[3]{ab} - \sqrt[3]{ab}$	$-5\sqrt{2} + 5\sqrt{2}$	$5\sqrt{x} + 5\sqrt{y}$	$\sqrt{4} \cdot \sqrt{4}$

Student Activity D

Section 9.3: Paint by Radicals

Directions: Simplify each expression and shade in the corresponding square in the grid below (that contains the correctly simplified version). Assume all variables represent positive numbers. The first one has been done for you. There's a surprise when you're finished!

1. Simplify $\sqrt{72}-\sqrt{50}=6\sqrt{2}-5\sqrt{2}=\sqrt{2}$
2. Simplify: $\sqrt{72}+\sqrt{50}$
3. Simplify: $\sqrt{16+9}$
4. Simplify: $\sqrt{40}-\sqrt{90}$
5. Simplify: $10\sqrt{20}+\sqrt{5}$
6. Simplify: $\sqrt{300}-10\sqrt{3}$
7. Simplify: $\sqrt{75}+\sqrt{12}$
8. Simplify: $\sqrt{48y^2}-y\sqrt{27}$
9. Simplify: $\sqrt{75x^2+25x^2}$
10. Simplify: $\sqrt{81b^2}+\sqrt{500b}$
11. Simplify: $-13+\sqrt{4x^2}+5\sqrt{2}-5\sqrt{2}$
12. Simplify: $10+10\sqrt[3]{5}+2\sqrt[3]{5}+2$
13. Simplify: $\sqrt{52b^2-3b^2}$
14. Simplify: $\sqrt{900}-\sqrt{80}+\sqrt{363}+\sqrt{80}$
15. Simplify: $\sqrt[3]{54}-\sqrt[3]{64}-3\sqrt[3]{2}$

7	$21\sqrt{5}$	5	0	$11\sqrt{2}$	$10x$
$\sqrt{5}$	$y\sqrt{3}$	$\sqrt{3y}$	$2\sqrt{2xy}$	$\sqrt{122}$	$-13+\sqrt{2}+2x$
$11b\sqrt{5}$	$\sqrt{2}$	$\sqrt{22}$	$30+11\sqrt{3}$	-8	$2x-13$
$-\sqrt{10}$	$7b$	$24\sqrt[3]{5}$	$19b\sqrt{10b}$	-4	$4-6\sqrt[3]{2}$
$\sqrt{-10}$	$12+12\sqrt[3]{5}$	$80+11\sqrt{3}$	$9b+10\sqrt{5b}$	$\sqrt{3}$	$7\sqrt{3}$

Student Activity E

Section 9.3: Radical Addition and Multiplication Tables

Here is a simple addition table with radical inputs. Fill in the missing boxes, use X if the inputs cannot be combined with the given operation.

Addition Table I:

+	$\sqrt{2}$	$\sqrt{3}$	$\sqrt{5}$	$\sqrt{6}$
$\sqrt{2}$	$2\sqrt{2}$	X	X	X
$\sqrt{3}$				
$\sqrt{5}$				
$\sqrt{6}$				

Now try this addition table that also involves variables! Again, use X if you cannot combine the input terms. You may want to simplify some of the radicals before you add. Assume $x > 0$.

Addition Table II:

+	$\sqrt{3x}$	$\sqrt{12x}$	$\sqrt{75x}$	$\sqrt{3x^2}$	$\sqrt{12x^2}$	$\sqrt{75x^2}$
$\sqrt{3x}$						
$\sqrt{12x}$						
$\sqrt{75x}$						
$\sqrt{3x^2}$						
$\sqrt{12x^2}$						
$\sqrt{75x^2}$						

Here is a simple multiplication table with radical inputs. Perform the multiplications and simplify each result.

Multiplication Table I:

$\bullet$	$\sqrt{2}$	$\sqrt{3}$	$\sqrt{5}$	$\sqrt{6}$
$\sqrt{2}$	2	$\sqrt{6}$	$\sqrt{10}$	$2\sqrt{3}$
$\sqrt{3}$				
$\sqrt{5}$				
$\sqrt{6}$				

Now try this multiplication table that also involves variables! Make sure to simplify the result of each multiplication. Assume $x > 0$.

Multiplication Table II:

$\bullet$	$\sqrt{3x}$	$\sqrt{12x}$	$\sqrt{3x^2}$	$\sqrt{12x^2}$	$\sqrt{3x^3}$	$\sqrt{12x^3}$
$\sqrt{3x}$						
$\sqrt{12x}$						
$\sqrt{3x^2}$						
$\sqrt{12x^2}$						
$\sqrt{3x^3}$						
$\sqrt{12x^3}$						

Student Activity F

Section 9.3: Tempting Radical Expressions

Directions: The expressions on this page might try to tempt you away from the strict mathematical rules you have learned about radical expressions. So, work carefully to simplify each expression and think about every move you make! If an expression cannot be simplified, then say so.

1. $\sqrt{25-9}$	**2.** $7\sqrt{3}-\sqrt{3}$	**3.** $5-2\sqrt{2}$
4. $\sqrt{8}+\sqrt{1}$	**5.** $\sqrt{25}\cdot\sqrt{25}$	**6.** $\sqrt{5}+\sqrt{5}$
7. $\sqrt[3]{8+1}$	**8.** $\sqrt[4]{4+12}$	**9.** $\sqrt{169-25}$
10. $5\sqrt{x}-\sqrt{x}$	**11.** $\sqrt{x^2}\cdot\sqrt{x^2}$	**12.** $\sqrt[3]{3}+\sqrt[3]{3}+\sqrt[3]{3}$
13. $\frac{\sqrt{3}}{2}+\frac{\sqrt{3}}{2}$	**14.** $\frac{\sqrt{3}}{2}\cdot\frac{\sqrt{3}}{2}$	**15.** $\sqrt[3]{\frac{8-7}{8}}$

Scrambled Answers:

$\sqrt[3]{9}$	12	$6\sqrt{3}$	x^2	4
$2\sqrt{5}$	$\frac{1}{2}$	$\frac{3}{4}$	Cannot be simplified further	2
$2\sqrt{2}+1$	$4\sqrt{x}$	$\sqrt{3}$	$3\sqrt[3]{3}$	25

Student Activity A

Section 9.4: Multiplication Madness Match Up

Directions: There are so many different ways to combine multiplication and radicals it can be easy to mix up the rules. Carefully work to simplify each of these expressions applying the appropriate rules as you go. Assume all roots are real numbers.

$\sqrt[n]{a} \cdot \sqrt[n]{b} = \sqrt[n]{a \cdot b}$ $\qquad a\sqrt[n]{b} \cdot c\sqrt[n]{d} = ac\sqrt[n]{bd}$ $\qquad \left(\sqrt[n]{a}\right)^n = a$

A $4x$ **B** $6+4\sqrt{2}$ **C** $36-x$ **D** $18\sqrt{10}$ **E** None of these

$\sqrt{2}\left(3\sqrt{2}+4\right)$	$3\sqrt{2}\cdot 6\sqrt{5}$	$\left(\sqrt[3]{4x}\right)^3$	$3\left(2+4\sqrt{2}\right)$	$\left(\sqrt{36}\right)^2-\left(\sqrt{x}\right)^2$
$\left(\sqrt[3]{6}\right)^3-\left(\sqrt[3]{x}\right)^3$	$\sqrt{2x}\cdot\sqrt{8x}$	$2\sqrt{5}\cdot 9\sqrt{2}$	$2\left(3+2\sqrt{2}\right)$	$4\sqrt{x}\cdot\sqrt{x}$
$\sqrt{2x}\cdot 2\sqrt{2x}$	$\left(6-\sqrt{x}\right)\left(6+\sqrt{x}\right)$	$\left(\sqrt{2}+2\right)^2$	$\sqrt{x}\left(36-\sqrt{x}\right)$	$\sqrt{5}\cdot 18\sqrt{2}$
$(6+x)(6-x)$	$\left(6-\sqrt{x}\right)^2$	$\sqrt[3]{64\cdot x^3}$	$\left(9\sqrt{10}\right)^2$	$\left(\sqrt{6}\right)^2+\sqrt{32}$

Student Activity B

Section 9.4: Double the Fun on Radical Expressions

Directions: When you simplify expressions that have multiplication or division of radicals, you sometimes have a choice. You could first simplify each radicand and then apply the operation, or you could first combine the radicals with the operation, and then simplify. For each of the expressions below, simplify the radical expression both ways (let $x > 0$). Your simplified expression should be the same for both methods. If they are not, you'll have to go back and look for a mistake. The first one has been done for you.

	Radical Expression	Simplify the radicands first	Combine the radicals first
1.	$\frac{\sqrt{18x}}{\sqrt{2x^3}}$	$\frac{\sqrt{18x}}{\sqrt{2x^3}} = \frac{3\sqrt{2x}}{x\sqrt{2x}} = \frac{3\cancel{\sqrt{2x}}^{1}}{x\cancel{\sqrt{2x}}_{1}} = \frac{3}{x}$	$\frac{\sqrt{18x}}{\sqrt{2x^3}} = \sqrt{\frac{18x}{2x^3}} = \sqrt{\frac{9}{x^2}} = \frac{3}{x}$
2.	$\sqrt{75x^2} \cdot \sqrt{48x^4}$		
3.	$\frac{\sqrt{108x^6}}{\sqrt{12x^4}}$		
4.	$\frac{\sqrt{625x^4}}{\sqrt{25x^8}}$		
5.	$\sqrt{12x} \cdot \sqrt{24x^2}$		
6.	$\frac{\sqrt[3]{192x^6}}{\sqrt[3]{3x^3}}$		

Student Activity C

Section 9.4: Rationalize Square Roots with the Missing Form of 1

Directions: In each radical expression below, we need to rationalize the denominator. We do this rationalization by multiplying by a **form of one** that changes the denominator from a radical expression to a non-radical expression. Fill in the "ones" to rationalize the denominator of each expression and simplify the result.

Example: $\frac{6}{\sqrt{3}} \cdot \left[\frac{\quad}{\quad}\right]$ becomes $\frac{6}{\sqrt{3}} \cdot \left[\frac{\sqrt{3}}{\sqrt{3}}\right] = \frac{6\sqrt{3}}{3} = 2\sqrt{3}$.

Hint for problems 6-10: Remember to use a conjugate.

1. $\frac{8}{\sqrt{2}} \cdot \left[\frac{\quad}{\quad}\right]$

2. $\frac{5\sqrt{3}}{\sqrt{15}} \cdot \left[\frac{\quad}{\quad}\right]$

3. $\frac{\sqrt{5}}{\sqrt{10}} \cdot \left[\frac{\quad}{\quad}\right]$

4. $\sqrt{\frac{7}{2}} = \frac{\sqrt{\quad}}{\sqrt{\quad}} \cdot \left[\frac{\quad}{\quad}\right]$

5. $\frac{4}{\sqrt{18x}} = \frac{4}{\sqrt{\quad}} \cdot \left[\frac{\quad}{\quad}\right]$

6. $\frac{4}{\sqrt{2}+1} \cdot \left[\frac{\quad}{\quad}\right]$

7. $\frac{12}{3-\sqrt{5}} \cdot \left[\frac{\quad}{\quad}\right]$

8. $\frac{\sqrt{3}}{-2-\sqrt{3}} \cdot \left[\frac{\quad}{\quad}\right]$

9. $\frac{\sqrt{3}-2}{\sqrt{3}+4} \cdot \left[\frac{\quad}{\quad}\right]$

10. $\frac{\sqrt{6}}{x+\sqrt{2}} \cdot \left[\frac{\quad}{\quad}\right]$

Guided Learning Activity

Section 9.4: Rationalizing Higher-order Roots

Believe it or not, rationalizing denominators with square roots is a pretty intuitive procedure. The procedure for rationalizing denominators with higher-order roots is not so obvious. Let's look at a comparison and see what's different.

Example with square roots:

To rationalize $\frac{6}{\sqrt{3}}$ we multiply by $\frac{\sqrt{3}}{\sqrt{3}}$, to get: $\frac{6}{\sqrt{3}} \cdot \frac{\sqrt{3}}{\sqrt{3}} = \frac{6\sqrt{3}}{\sqrt{3^2}} = \frac{6\sqrt{3}}{3} = 2\sqrt{3}$.

Example with cube roots:

If we followed exactly the same procedure as for square roots, here's what happens:

We multiply $\frac{6}{\sqrt[3]{2}}$ by $\frac{\sqrt[3]{2}}{\sqrt[3]{2}}$ to get: $\frac{6}{\sqrt[3]{2}} \cdot \frac{\sqrt[3]{2}}{\sqrt[3]{2}} = \frac{6\sqrt[3]{2}}{\sqrt[3]{2^2}} = ?$

The problem is that $\sqrt[3]{2^2}$ doesn't simplify. However, $\sqrt[3]{2^3}$ would simplify, equaling 2.

Let's try again:

1. Rationalize: $\frac{6}{\sqrt[3]{2}}$

It helps to think first of the radical denominator that you're going to aim for. Then you can figure out what to multiply by to get it. Fill in the blanks in the next few problems.

2. Rationalize $\frac{15}{\sqrt[3]{5}}$ by aiming for $\sqrt[3]{125}$ or $\sqrt[3]{5^3}$ in the denominator.

$$\frac{15}{\sqrt[3]{5}} \cdot \frac{\sqrt[3]{\quad}}{\sqrt[3]{\quad}} = \frac{\quad}{\sqrt[3]{125}} = \frac{\quad}{5} =$$

3. Rationalize $\frac{6}{\sqrt[4]{27}}$ by aiming for $\sqrt[4]{81}$ or $\sqrt[4]{3^4}$ in the denominator.

$$\frac{6}{\sqrt[4]{27}} \cdot \frac{\sqrt[4]{\quad}}{\sqrt[4]{\quad}} = \frac{\quad}{\sqrt[4]{81}} = \frac{\quad}{3} =$$

4. Rationalize $\frac{5}{2\sqrt[3]{25}}$ by aiming for $2\sqrt[3]{125}$ or $2\sqrt[3]{5^3}$ in the denominator.

$$\frac{5}{2\sqrt[3]{25}} \cdot \frac{\sqrt[3]{\quad}}{\sqrt[3]{\quad}} = \frac{\quad}{2\sqrt[3]{125}} = \frac{\quad}{2 \cdot 5} =$$

We always want to aim for a radical in the denominator that has a perfect power in the radicand that corresponds with the index of the root.

Let's practice deciding on a good denominator to aim for:

5. Rationalize $\frac{12}{\sqrt[5]{4}}$ by aiming for ________ in the denominator.

6. Rationalize $\frac{\sqrt[3]{3}}{\sqrt[3]{4}}$ by aiming for ________ in the denominator.

7. Rationalize $\frac{6x^2}{\sqrt[3]{3x}}$ by aiming for ________ in the denominator.

8. Rationalize $\frac{\sqrt[4]{2}}{\sqrt[4]{9m}}$ by aiming for ________ in the denominator.

Finally, we'll put it all together. Rationalize the denominators in the following problems.

9. Rationalize: $\frac{40}{\sqrt[3]{10}}$

10. Rationalize: $\frac{x}{\sqrt[4]{8x^2}}$

11. Rationalize: $\sqrt[3]{\frac{25}{3}}$ (Hint: first write this with two separate radicals.)

12. Rationalize: $\frac{\sqrt[4]{6}}{\sqrt[4]{9x^2}}$

Student Activity D

Section 9.4: The Radical Reunion

Directions: Now that we've made it through addition, subtraction, multiplication, and division, it seems only fair to bring the whole gang back together for a reunion. Assume all variables represent positive numbers. Simplify each of the following radical expressions. If an expression cannot be simplified, say so.

$\sqrt{50}+\sqrt{18}$	$\sqrt{50-18}$	$\sqrt{50+18}$	$\sqrt{50}\cdot\sqrt{18}$	$\frac{\sqrt{50}}{\sqrt{18}}$

$\sqrt{16a^2b}+\sqrt{9a^2b}$	$\sqrt{16a^2b-9a^2b}$	$\sqrt{16a^2b}\cdot\sqrt{9a^2b}$	$\frac{\sqrt{16a^2b}}{\sqrt{9a^2b}}$	$\left(\sqrt{16a^2b}\right)^2$

$\sqrt[3]{135xy^4}-\sqrt[3]{40x^4y}$	$\sqrt[3]{135xy^4-40x^4y}$	$\sqrt[3]{135xy^4}\cdot\sqrt[3]{40x^4y}$	$\frac{\sqrt[3]{135xy^4}}{\sqrt[3]{40x^4y}}$	$\left(\sqrt[3]{40x^4y}\right)^3$

Assessment 9B

Chapter 9: Mid-chapter Assessment for Understanding

For each of the following, describe the type of problem and the strategies and key steps to remember while doing the problem. You do **not** have to complete the problems.

	Type of Problem	Strategies and Key Steps
1. Expand and simplify: $\left(4-\sqrt{2}\right)^2$		
2. Divide: $\frac{\sqrt{120x^4y}}{\sqrt{12y^5}}$		
3. Simplify: $\sqrt{3x}\cdot\sqrt{3x^3}$		
4. Simplify: $2\sqrt[3]{5}+\sqrt[3]{80}$		
5. Simplify: $\sqrt{\sqrt{81x^4}}$		
6. Simplify: $\sqrt{50}+\sqrt{18}$		
7. Rationalize the denominator: $\frac{12}{6-\sqrt{3}}$		
8. Is $\sqrt{8}$ rational or irrational?		
9. Simplify: $\sqrt{120x^5}$		
10. Evaluate: $4^{3/2}$		

Student Activity A

Section 9.5: Is it a solution?

Tic-tac-toe Directions: If the number in the square **IS** a solution of the equation, then circle the solution (placing an **O** on the square). If the given number **IS NOT** a solution, then put an **X** on the square.

$\sqrt[3]{7-x}=1-x$ -1	$(3-x)^{1/2}=x-3$ 3	$\sqrt{3-x}=x-3$ 0
$-3x=3\sqrt{-2x-1}$ 3	$-3x=3\sqrt{-2x-1}$ 0	$\sqrt{x}+\sqrt{x+3}=x-4$ 1
$x=\sqrt[4]{x^3+16}$ -1	$\sqrt{3x}+10=10$ 0	$-2=(3x-2)^{1/3}$ -2

Guided Learning Activity

Chapter 9.5: Radical Isolation

When we solve radical equations, it is important to **isolate the radical** before using the power rule of equality. *But why?* Consider the radical equation $\sqrt{x+3}+4=6$.

Squaring first: Attempting to square both sides without isolating the radical leaves another radical term. This is messier looking than what we started with and we're not even done!

$$\sqrt{x+3}+4=6$$
$$\left(\sqrt{x+3}+4\right)^2=(6)^2$$
$$\left(\sqrt{x+3}+4\right)\left(\sqrt{x+3}+4\right)=36$$
$$x+3+8\sqrt{x+3}+16=36$$

This is worse than the original problem!

Isolating first: If we isolate the radical term first, and then square both sides of the equation, we can quickly solve for x. Checking the solution shows that $x=1$ is a valid solution to this radical equation.

$$\sqrt{x+3}+4=6$$
$$\sqrt{x+3}=2$$
$$\left(\sqrt{x+3}\right)^2=(2)^2$$
$$x+3=4$$
$$x=1$$

Now let's try some together!

	Isolate the radical:	Solve by squaring both sides:	Check:
1.	$\sqrt{2x+3}+5=8$		
2.	$8\sqrt{x+5}-12=28$		
3.	$5+\sqrt{x+2}=0$		

4.

Isolate the radical:	Solve by cubing both sides:	Check:
$5+2x^{1/3}=13$		

5.

Isolate a radical:	Solve by cubing both sides:	Check:
$\sqrt[3]{3a+4}=\sqrt[3]{2a+7}$		

6.

Isolate the radical:	Solve by squaring both sides:	Check:
$\sqrt{x+7}-5=x$		

7.

Isolate a radical:	Squaring both sides:	
$\sqrt{x+4}+\sqrt{x-1}=5$		
Isolate the radical:	**Solve by Squaring both sides:**	**Check:**

Student Activity B

Section 9.5: Match Up on Solving Radical Equations

Match-up: Match each of the equations in the squares in the table below with its solution from the top. If the solution is not found among the choices A through D, then choose E (none of these).

A 5 **B** 3 **C** –2 **D** No real number solution **E** None of these

$3+\sqrt{7-x}=5$	$5\sqrt{x+6}=10$	$-3\sqrt{x+9}=-6$
$(2y-1)^{1/2}=3$	$8+\sqrt{x-1}=5$	$6\sqrt[3]{a-4}=6$
$x=2+\sqrt{14-x}$	$\sqrt[4]{3w}=\sqrt[4]{5+2w}$	$x+\sqrt[3]{3-12x}=x+3$
$-2\sqrt{3x-1}=10$	$x-1=(7-x)^{1/2}$	$x=\sqrt{4x-11}+2$

Student Activity C

Section 9.5: Working with the Language of Radicals

Fill in the table below. The first row has been done for you.

In words	Equation or Expression?	In Math Notation
1. the square root of the quantity $x+5$	Expression	$\sqrt{x+5}$
2. The quantity $3x+2$ squared is zero.		
3. The sum of 5 and the square root of x is 30.		
4.		$5(x+6)^2+8$
5.		$5\sqrt{x+6}+8$
6.		$7\sqrt{x}-7=42$
7. twice the square of the quantity $5x+9$		
8. 40 less the square root of x is 35.		
9.		$5\sqrt[3]{50-x}$
10. 5 times the cube root of the quantity $50-x$ is 25.		

Student Activity A
Section 9.6: Sail Into the Pythagorean Sunset

Directions: Almost all of the triangles in the figure below are right triangles. Using the sparse information you are given and the Pythagorean Theorem, work out the lengths of all the missing sides in the sailboat below. You should be able to work out the sides of the non-right triangles by piecing together the information you find from the right triangles. It might be helpful to start by finding the measure of a.

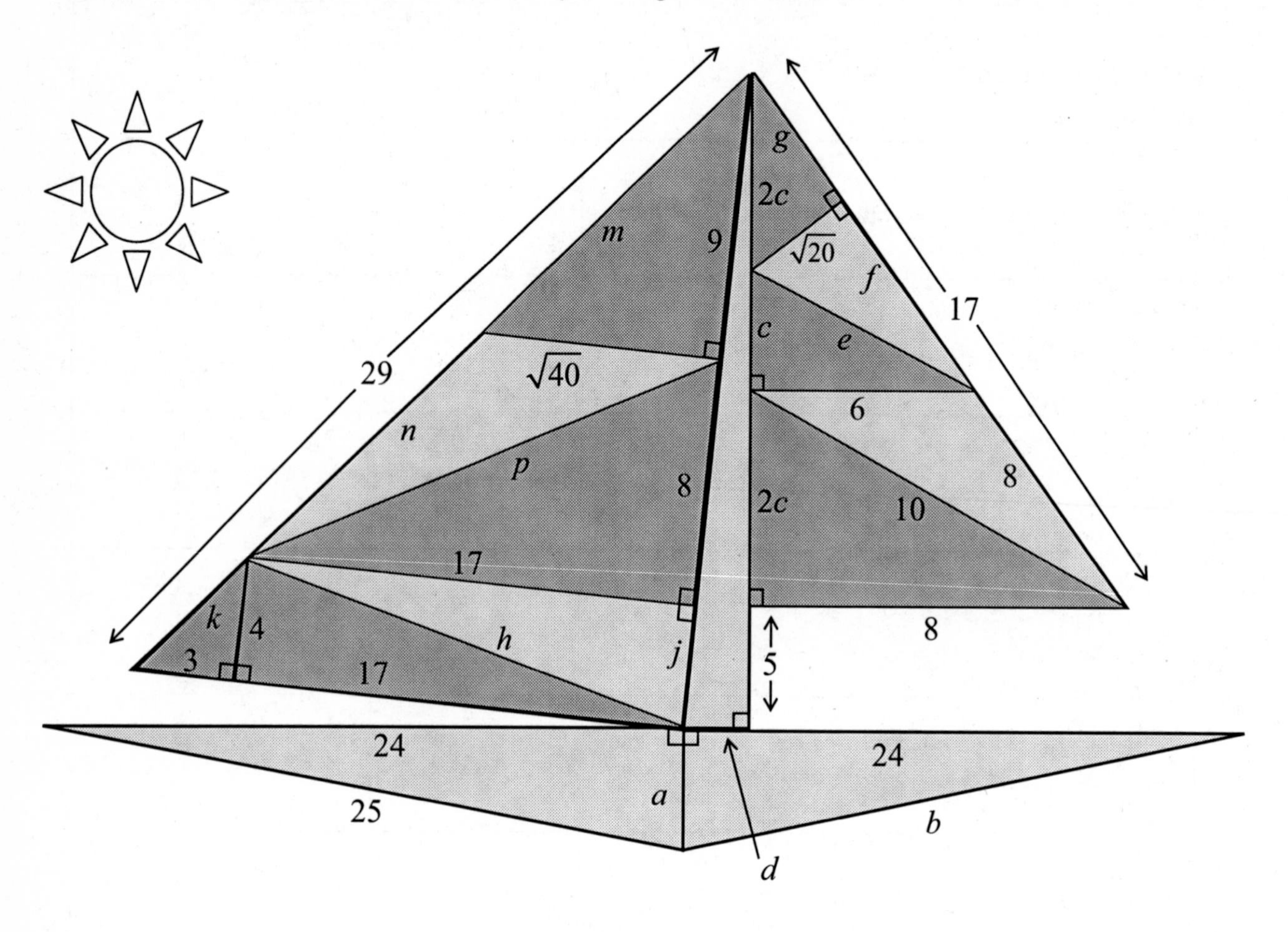

$a =$

$b =$

$c =$

$d =$

$e =$

$f =$

$g =$

$h =$

$j =$

$k =$

$m =$

$n =$

$p =$

Guided Learning Activity

Section 9.6: The Distance Formula

1. Find the length of side AB and side BC by measuring the distances on the graphing grid. Then use the Pythagorean Theorem to find the length of side AC.

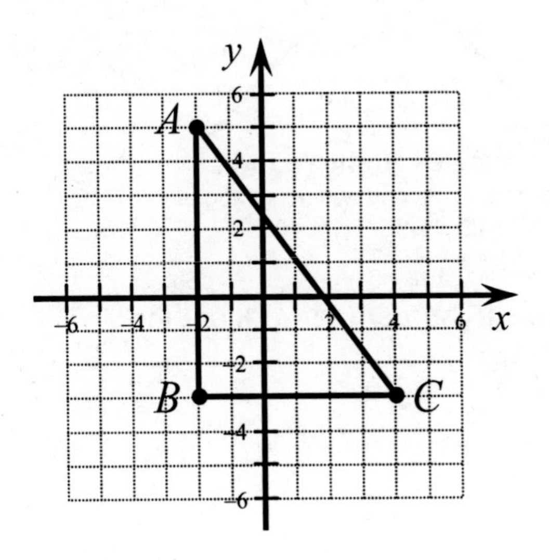

2. Find the length of line segment DE using a method similar to problem 1.

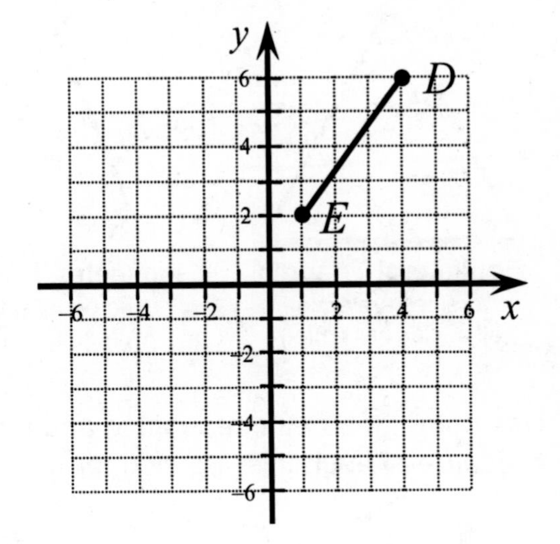

3. Find the length of the side d in the triangle below in terms of $x_1, x_2, y_1,$ and y_2.

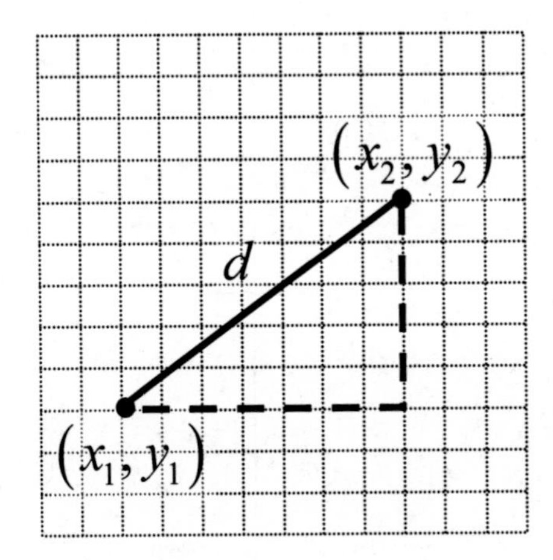

4. Now find the distance between $(4,-1)$ and $(7,-5)$ using the formula from Problem 3.

Student Activity B

Section 9.6: These Triangles are Just Special

Several triangles have special names. In every diagram, the equal sides are represented with thick lines.

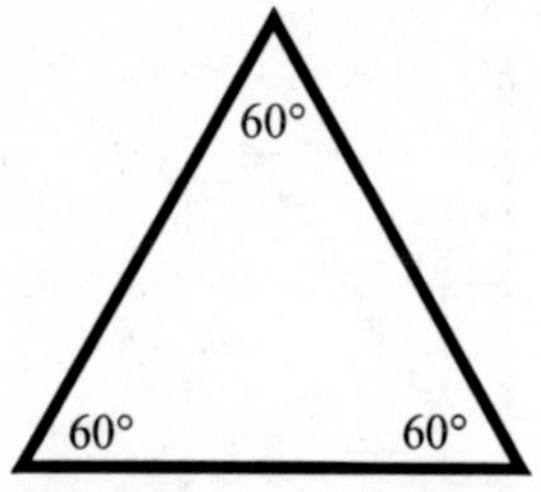

Equilateral Triangle
(all sides are equal,
all angles are equal)

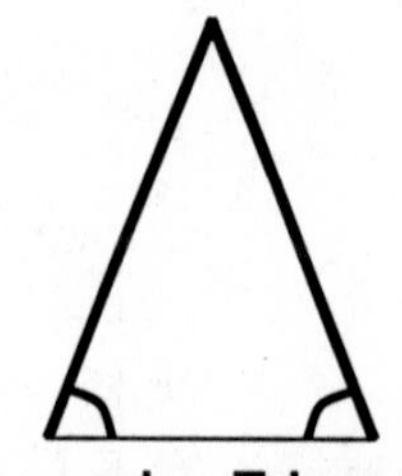

Isosceles Triangle
(two equal sides,
two equal angles)

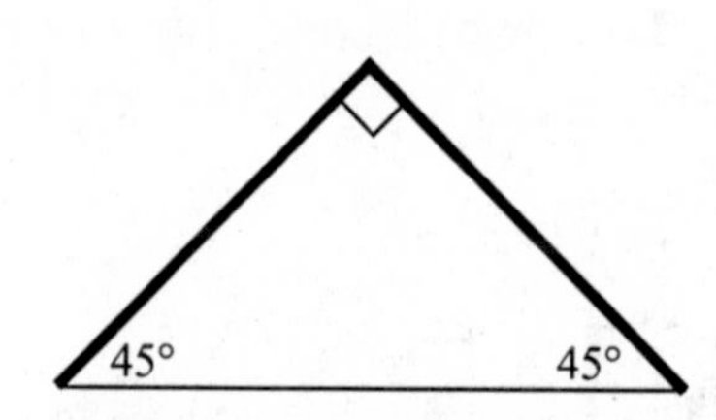

Isosceles Right Triangle
(one right angle and two equal angles)
45-45-90 Triangle

1. Use some geometric reasoning and the Pythagorean Theorem to work out the measure of each missing side for the triangles below.

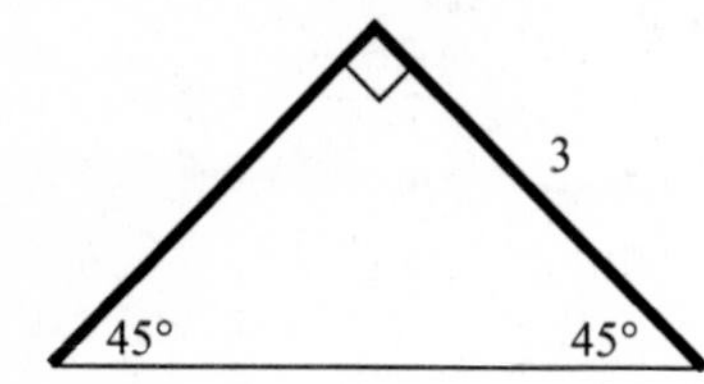

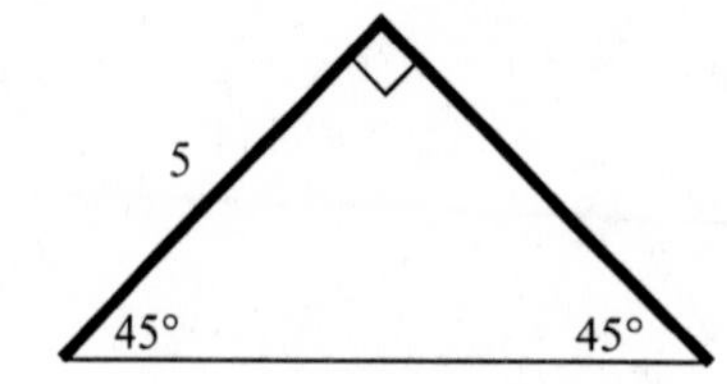

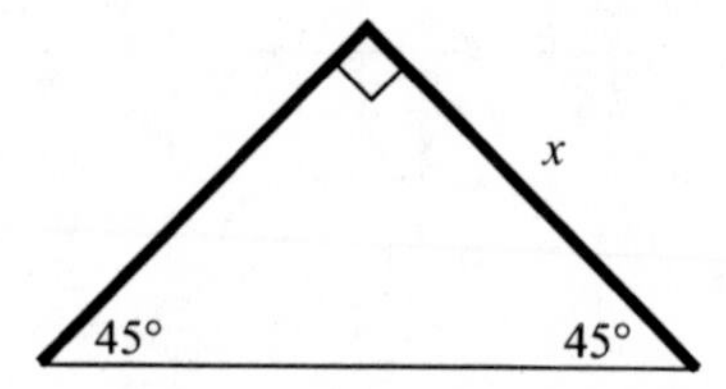

Based on your work above, complete the following conjecture: In an isosceles right triangle, if the legs have length l, then the hypotenuse has length ________.

If we take an equilateral triangle and cut it exactly down the middle, we get two congruent 30-60-90 triangles, demonstrated here:

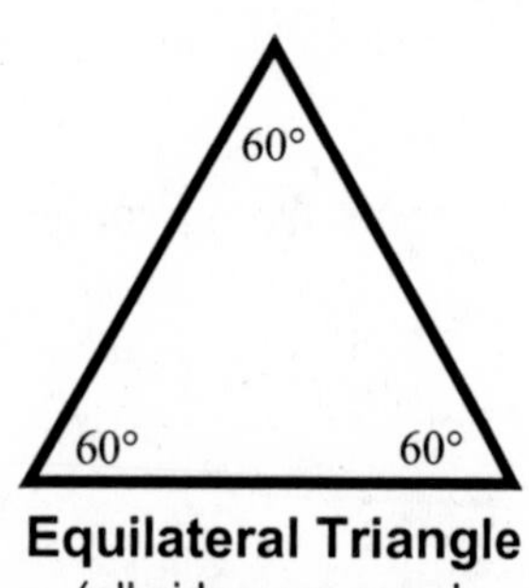

Equilateral Triangle
(all sides are equal,
all angles are equal)

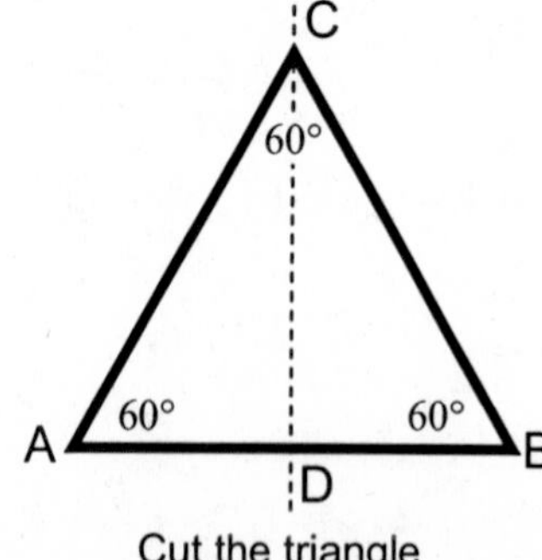

Cut the triangle
down the center.
$AD \cong DB = \frac{1}{2}AB$

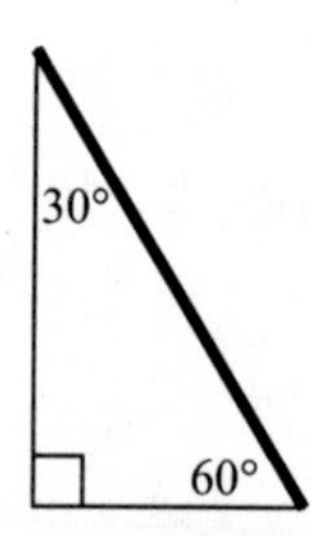

30-60-90 Triangle

2. Use some geometric reasoning and the Pythagorean Theorem to work out the measure of each missing side for the triangles below.

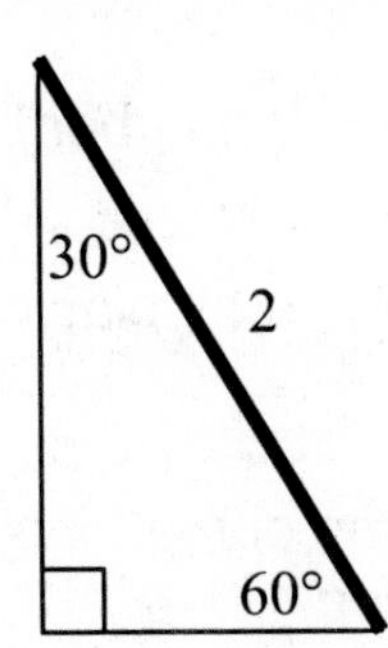

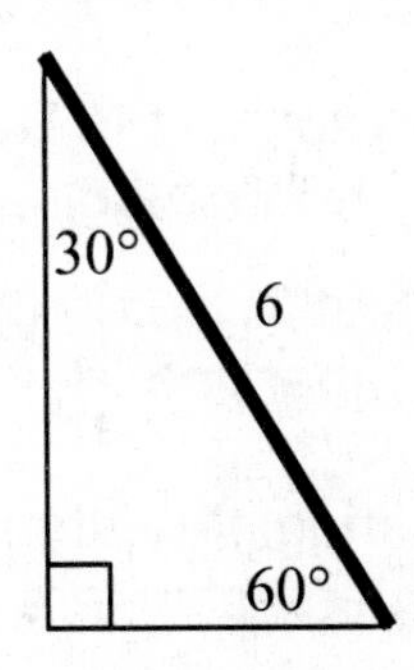

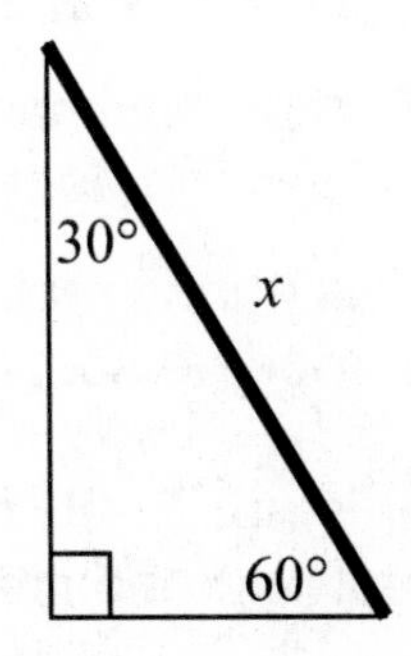

Based on your work above, complete the following conjecture: In a 30-60-90 triangle, if the shorter leg has length a, then the longer leg has length _______, and the hypotenuse has length _______.

3. Now use what you've learned about 30-60-90 and 45-45-90 triangles to work out the measure of the missing sides for the triangles below.

a.

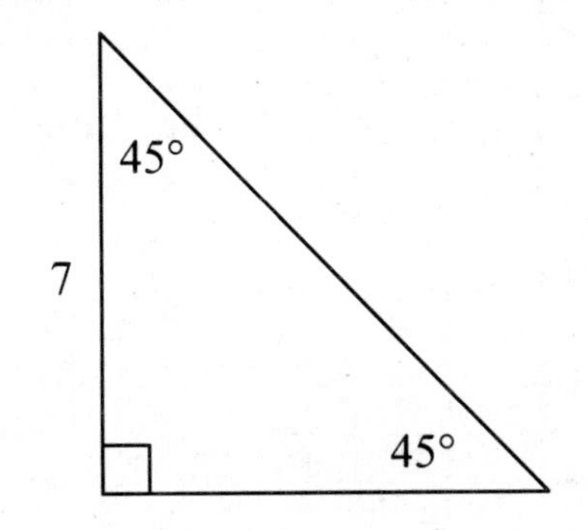

b.

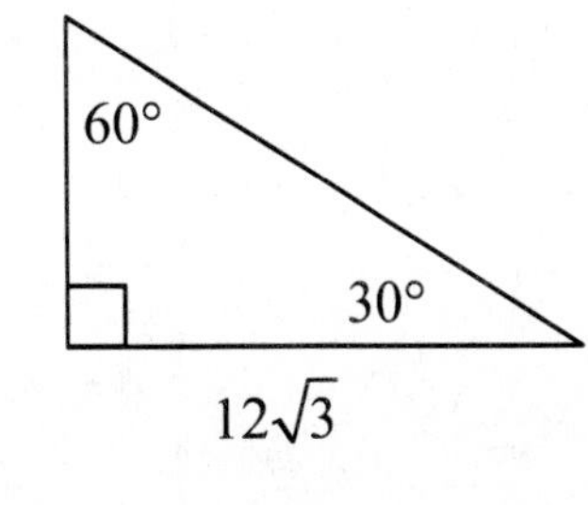

c.

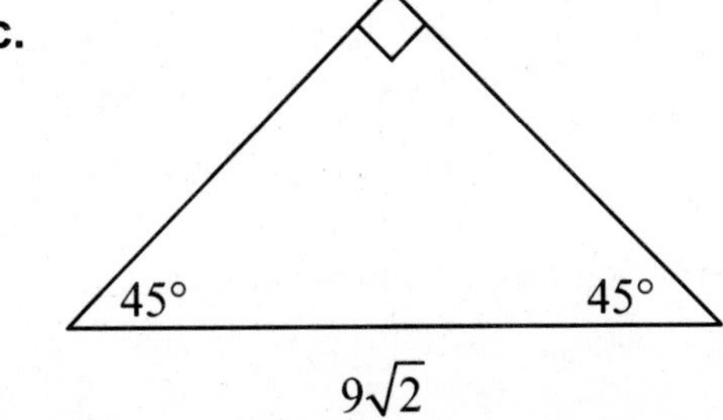

d.

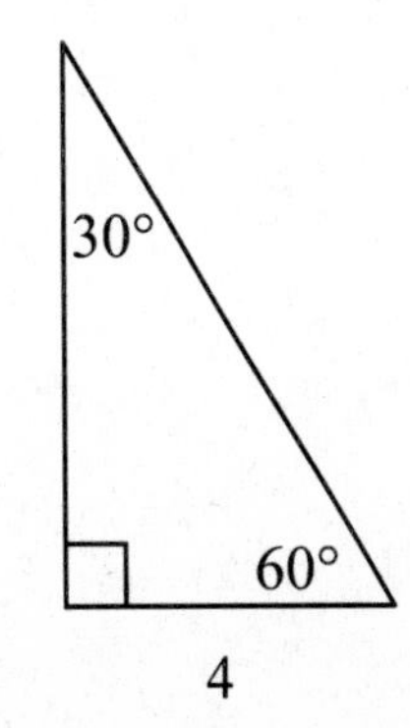

Student Activity A

Section 9.7: Complex Numbers on Trial

Directions: Each of the numbers below is "on trial" for being purely imaginary, purely real, or complex, and you are the judge.

If the number is purely imaginary, then sentence it to life on the imaginary number line by placing a dot in the appropriate place on the imaginary number line.

If the number is purely real, then sentence it to life on the real number line by placing a dot in the appropriate spot on the real number line.

If the number is not purely real or purely imaginary, then release it in the complex plane so that the real part and imaginary part intersect where the dot is placed.

As an example, ***A*** and ***B*** have been done for you.

A. $6i$

B. $-4-2i$

C. 7

D. $4+3i$

E. $-8i$

F. $-5i+4$

G. $-7+0i$

H. $\sqrt{-9}$

I. $-\sqrt{-25}$

J. $\sqrt{4}+\sqrt{-100}$

K. $-\sqrt{64}-\sqrt{-16}$

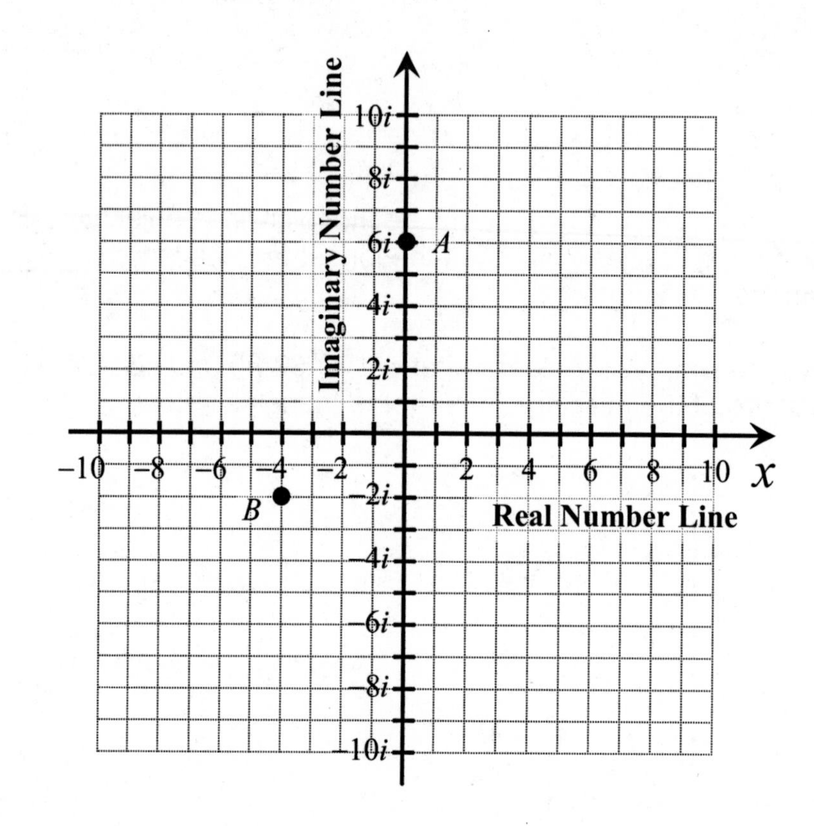

Student Activity B

Section 9.7: Match Up on Complex Numbers

Match-up: In each box, simplify the complex number expression given. Write the letter of the corresponding solution. If the solution does not appear in choices A – D, choose E (none of these).

A -2 **B** $-i$ **C** $5-5i$ **D** $5+5i$ **E** None of these

$(6+7i)+(-1-2i)$	$(2-i)(3-i)$	$-i(5+5i)$
$5(1-i)$	$\frac{2}{i}$	$\frac{-10+10i}{2i}$
$\frac{-6-4i}{3+2i}$	$\frac{15+5i}{2-i}$	$(6+7i)-(1+2i)$
$(2+i)(3+i)$	$\frac{1}{2}(1+i)(-1-i)$	$(1+i)(-1+i)$
$5(1-5i)$	$(10-i)-(5-6i)$	$\frac{-8-10i}{4+5i}$

Student Activity C

Section 9.7: Rotation Using Complex Numbers

Example 1: The complex number $4+3i$ is graphed in the complex plane. Also, you see a circle, centered at the origin, with radius 5. Notice that the point that represents $4+3i$ is a point on the circle.

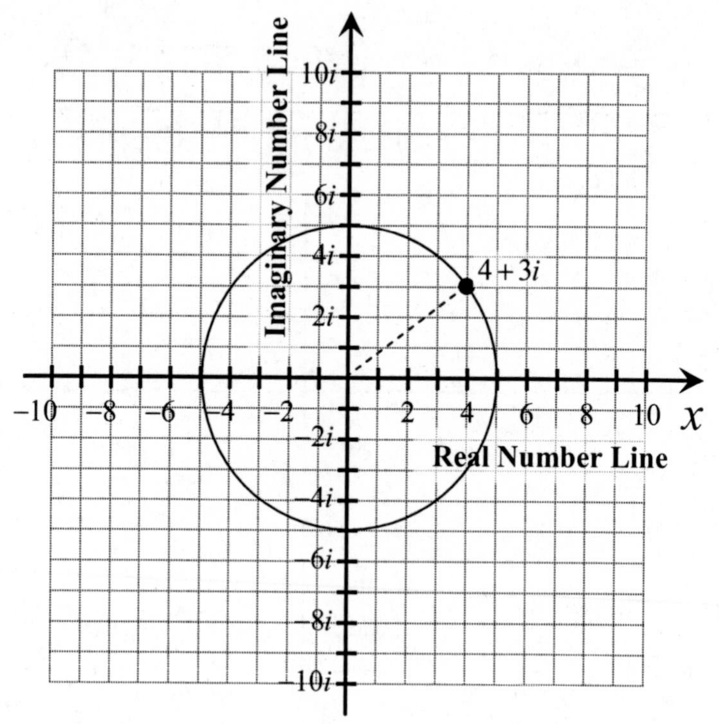

Multiply $4+3i$ by i and simplify. Graph this point.

Multiply the new number by i and simplify. Graph this point.

Multiply the new number by i and simplify. Graph this point.

Multiply the new number by i and simplify. Notice anything?

Draw more dashed lines that connect the **origin** to each of the points you have graphed. What do you notice about the lines?

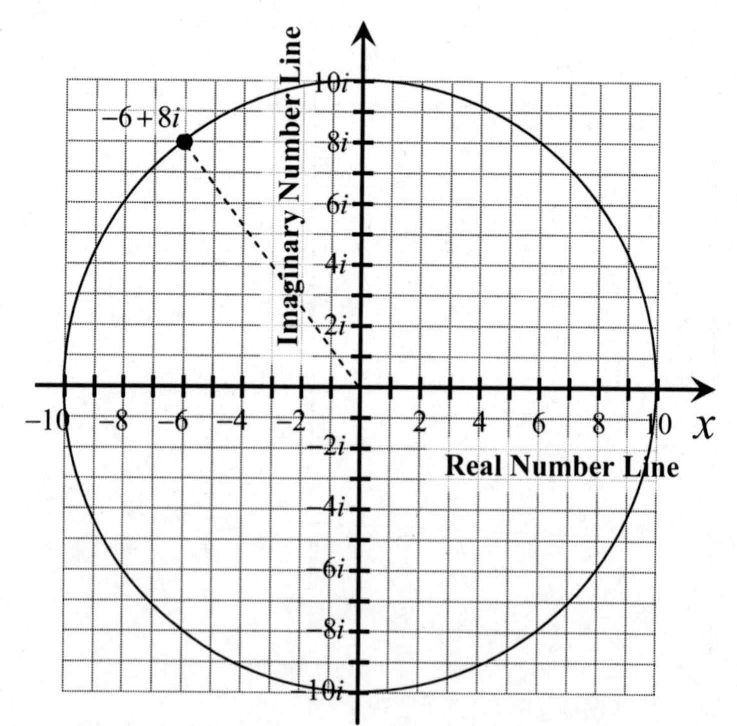

Example 2: Repeat this process of graphing and multiplying by i beginning with the number $-6+8i$, which has been graphed for you.

When we multiply by i, what affect does that have on the graph?

Student Activity D

Section 9.7: Parallel Expressions

Directions: Simplify each expression.

$(2+5x)+(4-3x)$	$(4+6x)-(3+x)$	$(3+x)(4+x)$
$(2+5\sqrt{3})+(4-3\sqrt{3})$	$(4+6\sqrt{5})-(3+\sqrt{5})$	$(3+\sqrt{2})(4+\sqrt{2})$
$(2+5i)+(4-3i)$	$(4+6i)-(3+i)$	$(3+i)(4+i)$

$(3x)^2$	$(3+x)^2$	$3(4-5x)$
$(3\sqrt{x})^2$	$(3+\sqrt{2})^2$	$3(4-5\sqrt{2})$
$(3i)^2$	$(3+i)^2$	$3(4-5i)$

$\sqrt{20}$	$2\sqrt{3}\cdot 5\sqrt{3}$	$\frac{4}{\sqrt{2}}$
$\sqrt{-20}$	$2i\cdot 5i$	$\frac{4}{i}$

$\frac{5}{2\sqrt{6}}$	$\frac{4}{2+\sqrt{3}}$	$\frac{2+\sqrt{3}}{3-\sqrt{3}}$
$\frac{5}{2i}$	$\frac{4}{2+i}$	$\frac{2+i}{3-i}$

Assessment 9C

Chapter 9: End-of-chapter Assessment for Understanding

For each of the following, describe the type of problem and the strategies and key steps to remember while doing the problem. You do **not** have to complete the problems.

	Type of Problem	Strategies and Key Steps
1. Find the distance between $(3,-6)$ and $(2,-4)$.		
2. Multiply. Write the answer in the form $a+bi$: $(-5+2i)(-5-2i)$		
3. Solve: $\sqrt{3x-1}+6=0$		
4. The smallest side of a 30-60-90 triangle is 6 cm. Find the lengths of the other two sides.		
5. Simplify: $\sqrt[4]{80x^4y^{10}}$		
6. Solve: $3\sqrt{2x-1}=6$		
7. Rationalize: $\frac{15}{\sqrt[3]{3}}$.		
8. Simplify: $x^{1/3}x^{2/5}$		
9. Simplify: $\sqrt{-9}+\sqrt{-36}$		
10. Add: $\sqrt{8}+\sqrt{50}$		

Assessment 9D

Chapter 9: Metacognitive Skills Assessment

Metacognitive skills refer to the ability to judge how well you have learned something and to effectively direct your own learning and studying. This is a self evaluation tool designed to help you focus your studying and to improve your metacognitive skills with regards to this math class.

Fill the 1st column out before you begin studying.
Fill the 2nd column out after you study and before you take the test.
Go back to this page after your test and circle any of the ratings that you would now change – this identifies the "disconnects" between what you think you know well and what you actually know well.

Use the scale below to assign a number to each topic.

5 I am confident I can do any problems in this category correctly.
4 I am confident I can do most of the problems in this category correctly.
3 I understand how to do the problems in this category, but I still make a lot of mistakes.
2 I feel unsure about how to do these problems.
1 I know I don't understand how to do these problems.

Topic or Skill	Before Studying	After Studying
Recognizing perfect squares (like 1, 4, 9, 16, 25…).		
Recognizing perfect cubes (like 1, 8, 27, 64, 125…), perfect fourths (like 1, 16, 81, …), and perfect fifths (like 1, 32, 243, …).		
Ability to use the terms radical, radicand, radical symbol, square, and square root, appropriately.		
Understanding why $\sqrt{a}$ is not a real number if $a < 0$.		
Understanding why $\sqrt[n]{a}$ is non-real if n is even and $a < 0$.		
Understanding why $\sqrt[n]{a}$ is a real number as long as n is odd.		
Categorizing a radical as rational or irrational, real or imaginary.		
Using a calculator to approximate an irrational radical.		
Simplifying radical expression involving numbers and variables that are perfect powers (like $\sqrt{16x^2}$).		
Rewriting a radical expression with rational (fractional) exponents or vice versa.		
Knowing which part of a rational (fractional) exponent represents the root and which part is the power.		
Evaluating numbers with rational exponent powers, like $8^{3/2}$ or $9^{-1/2}$.		
Using a calculator to evaluate radicals with higher-order roots or numbers written with rational exponents..		
Applying exponent rules correctly to simplify expressions involving rational exponents.		

Continued on next page.

Use the scale below to assign a number to each topic.

5 *I am confident I can do any problems in this category correctly.*
4 *I am confident I can do most of the problems in this category correctly.*
3 *I understand how to do the problems in this category, but I still make a lot of mistakes.*
2 *I feel unsure about how to do these problems.*
1 *I know I don't understand how to do these problems.*

Topic or Skill	**Before Studying**	**After Studying**
Applying the product and quotient rules for radicals correctly; using these rules to simplify radical expressions like $\sqrt{\frac{9x^2}{16}}$ or $\sqrt{4 \cdot 25}$.		
Simplifying radical expressions by finding perfect squares or perfect powers inside the radicand (for example: $\sqrt{8}$, $\sqrt[3]{x^4}$, or $\sqrt{50x^3y^2}$).		
Adding or subtracting radical expressions by simplifying each radical and using like terms.		
Simplifying radical expressions involving multiplication and/or distribution like $3\sqrt{6} \cdot 2\sqrt{2}$, $2\sqrt{3}(4\sqrt{2} - 5\sqrt{3})$, or $(4+\sqrt{3})(5-\sqrt{2})$.		
Simplifying a radical expression that is raised to a power, like $(2+\sqrt{3})^2$ or $(\sqrt{2x})^2$.		
Rationalizing the denominator when there is a single term involving a square root in the denominator.		
Rationalizing the denominator when there are two terms involving at least one square root.		
Rationalizing the numerator of an expression involving square roots.		
Rationalizing the denominator when there is a single term involving a higher-order root (like a cube or fourth root).		
Solving an equation containing a single radical.		
Solving an equation with a rational exponent power.		
Solving an equation with more than one radical expression.		
Checking the solution to a radical equation.		
Recognizing when a radical equation cannot possibly have a real number solution.		
Solving application problems involving radical equations.		
Knowing when the Pythagorean Theorem can be applied and correctly using it.		
Applying the distance formula correctly.		
Knowing the special properties of 45-45-90 and 30-60-90 triangles.		
Rewriting the square root of a negative number in terms of i.		
Performing operations (add, subtract, multiply, or divide) with complex numbers and writing the answer in the form $a+bi$.		
Simplifying powers of i using the pattern of powers of i.		

Student Workbook: Chapter 10

Table of Contents: *Quadratic Equations, Functions, and Inequalities*

Assessment 10A

Pretest and Diagnostic Tool: Quadratic Equations

Directions: Complete this assessment without looking back at your notes or your book. **Do not use a calculator for this assessment.**

1. Simplify: $\sqrt{\dfrac{49}{4}}$

2. Simplify: $\sqrt{20}$

3. Simplify: $\sqrt{150}$

4. Solve by factoring: $x^2 = 16$

5. Solve by factoring: $x^2 - 5x + 4 = 0$

6. Solve by factoring: $6x^2 = 4 - 5x$

7. Find ½ of 8 and square the result. What is the resulting number?

8. Find ½ of 5 and square the result. What is the resulting number?

For 9-10, fill in the blank to make the statement true.

9. $x^2 -$ _____ $+ 49 = (x-7)^2$

10. $x^2 +$ _____ $+ 36 = (x+6)^2$

11. Simplify: $-5 + \sqrt{25 + 96}$

12. Simplify: $11 - \sqrt{(-11)^2 - 4(30)}$

13. Simplify: $\dfrac{6 + \sqrt{40}}{2}$

14. Evaluate $x^2 + 2x + 1$ for $x = -3$.

15. Let $f(x) = 2x^2 - 1$. Find $f(-1)$.

16. Evaluate $(x-3)^2 + 5$ for $x = 2$.

17. Solve $\dfrac{2}{x} + 3 = 0$

18. Solve: $x + \dfrac{x}{x+1} = 0$

19. Does the point $(0,0)$ satisfy the inequality $y < x^2 - 6$?

20. Does the point $(0,0)$ satisfy the inequality $y \geq |x - 2|$?

Student Activity A

Section 10.1: Paint by Double-Signs

Directions: Simplify each of the expressions below. Then shade in a corresponding answer on the grid below. When you're done, you should see a picture!

1. -1 ± 6

2. -3 ± 5

3. $\dfrac{5 \pm 10\sqrt{2}}{5}$

4. $\dfrac{6 \pm 2\sqrt{2}}{2}$

5. $\dfrac{-8 \pm 24}{4}$

6. $\dfrac{12 \pm \sqrt{18}}{3}$

7. $\dfrac{-2 \pm \sqrt{25-9}}{2}$

$12-\sqrt{2}$	$12+\sqrt{2}$	$4+\sqrt{2}$	$4+\sqrt{18}$	$4-\sqrt{6}$
-2	$4-\sqrt{18}$	-7	7	$3-2\sqrt{2}$
2	$4-\sqrt{2}$	-3	-8	$3-\sqrt{2}$
$3+2\sqrt{2}$	$6+\sqrt{2}$	$3+\sqrt{2}$	-14	$6-\sqrt{2}$
$4+\sqrt{6}$	-6	$1+2\sqrt{2}$	$1+10\sqrt{2}$	3
8	-5	0	$1-10\sqrt{2}$	-4
-8	5	1	$1-2\sqrt{2}$	4

Guided Learning Activity

Section 10.1: The Square Root Property

I am thinking of a number. If I square the number, I get 9. What is the number?

The easy choice here is 3, since $3^2 = 9$.

What **other** number could satisfy the condition? $(\_\_\_\_)^2 = 9$

You could also set up the equation $x^2 = 9$ to solve this problem. We have learned how to solve this equation **by factoring**. Go ahead and do that. *Hint: Remember to get =0 on one side first.*

Solve: $x^2 = 9$

We can also solve this equation (often in less steps) by **taking a square root** on both sides of the equation.

Solve:
$$x^2 = 9$$
$$\sqrt{x^2} = \sqrt{9}$$
$$x = 3$$

But there is a problem with the process we have written above; we are missing one of the possible solutions! Recall that 9 has two square roots, -3 and 3. However, the notation $\sqrt{9}$ only gives the principal (positive) square root. Thus, we have to solve $x^2 = 9$ by listing both square root solutions, as follows.

Solve:
$$x^2 = 9$$
$$x = \sqrt{9} \quad \text{or} \quad x = -\sqrt{9}$$
$$x = 3 \quad | \quad x = -3$$

We now have both of the correct solutions using the square root property $\{\pm 3\}$

The Square Root Property of Equations:
For any nonnegative real number c, if $x^2 = c$, then $x = \sqrt{c}$ or $x = -\sqrt{c}$.

To solve with the Square Root Property, we must **isolate the squared part** of the equation first. Try these together with your instructor.

Example 1: Solve $5x^2 - 10 = 70$.

Example 2: Solve $\frac{(x-5)^2}{2} = 32$.

Example 3: Solve $3x^2 - 33 = 0$.

Example 4: Solve $4 = (4x-3)^2 - 9$.

Example 5: Solve $x^2 + 25 = 0$.

Student Activity B

Section 10.1: Square Isolation Tic-Tac-Toe

The square root property can be used to solve a quadratic equation if

1) you can isolate the part of the equation that is written as a square.
2) the only variables in the equation are in the squared part.

Directions: If the equation **can** be solved easily with the square root property using the rules above, then isolate the squared part, and circle the equation (placing an O on the square). If the equation cannot be easily solved with the square root property, then place an X on the square.

$3(x+2)^2 - 25 = 50$	$x^2 = 5x - 4$	$-5(3x-1)^2 = -45$
$3x^2 = 75$	$24 = (x+7)^2 + 8$	$x^2 + 8x - 20 = 0$
$16x^2 - 24x = -9$	$(x-3)^2 + 2x = 0$	$x^2 + 6x = 2(3x+2)$

Student Activity C

Section 10.1: Double The Fun on Quadratic Equations

Directions: When you solve some quadratic equations, you can choose to solve by factoring or by using the Square Root Property. **If possible**, solve the following equations both ways. If it is not possible to solve the problem using one of the methods, say so. The first one has been started for you.

	Quadratic Equation	Solve by Factoring	Solve by Square Root Property
1.	$x^2 = 25$	$x^2 = 25$ $x^2 - 25 = 0$	$x^2 = 25$ $x = \sqrt{25}$ or $x = -\sqrt{25}$
2.	$4x^2 - 81 = 0$		
3.	$(x-3)^2 = 4$	Hint: Expand the binomial first.	
4.	$x^2 - 8 = 0$		
5.	$x^2 + 16x + 64 = 25$		Hint: Do you see a perfect square?

Student Activity D

Section 10.1: Hatch the Missing Pieces

Directions: Fill in the blanks in each equation to complete the square and write the expression as a perfect square. All of the solutions are provided inside the "egg". When you've crossed out all the solutions, you have "hatched" the egg. The first one has been done for you.

$x^2 + 4x + \underline{4} = \underline{(x+2)^2}$

$x^2 + 3x + \_\_\_ = \_\_\_\_$

$x^2 - 4x + \_\_\_ = \_\_\_\_$

$x^2 - 16x + \_\_\_ = \_\_\_\_$

$x^2 + 5x + \_\_\_ = \_\_\_\_$

$x^2 + 8x + \_\_\_ = \_\_\_\_$

$x^2 + 12x + \_\_\_ = \_\_\_\_$

$x^2 - x + \_\_\_ = \_\_\_\_$

$x^2 - 10x + \_\_\_ = \_\_\_\_$

$x^2 + 2x + \_\_\_ = \_\_\_\_$

$\left(x - \frac{1}{2}\right)^2$ $\frac{25}{4}$ $\frac{9}{4}$ 16 $(x-2)^2$ $\frac{1}{4}$ 64 ~~$(x+2)^2$~~ 25 $\left(x + \frac{5}{2}\right)^2$ 4 $(x-8)^2$ $(x+4)^2$ $(x+6)^2$ $(x+1)^2$ $(x-5)^2$ 36 1 ~~4~~ $\left(x + \frac{3}{2}\right)^2$

Guided Learning Activity

Section 10.1: Learning to Complete

Recall that we can use the square root property to solve quadratic equations, if all of the variables appear within a quantity squared.

Example 1: Solve $(x+2)^2 = 9$

Now we will use the process of completing the square together with the square root property to help us solve quadratic equations.

Example 2: Solve $x^2 - 8x - 9 = 0$ by completing the square.

a) Get the constant term on the right side of the equation and leave a "blank" to complete the square in on the left side. (Note: Sometimes the equation is already set up this way.)

$$x^2 - 8x \_\_\_\_\_\_ = 9$$

b) Complete the square by adding the same constant to both sides.

$$x^2 - 8x \underline{+16} = 9 + 16$$

c) Write the left side as a perfect square, combine constant terms on the right side.

$$(x-4)^2 = 25$$

d) Now solve using the square root property.

e) And check your answers.

Example 3: Solve $x^2 - 10x + 5 = 0$ by completing the square.

a) Get the constant term on the right side of the equation and leave a "blank" to complete the square in on the left side.

b) Complete the square by adding the same constant to both sides.

c) Write the left side as a perfect square, combine constant terms on the right side.

d) Now solve using the square root property.

Example 4: Solve $a^2 = 24(a-1)$ by completing the square. *Hint: Simplify first!*

Example 5: Solve $x^2 + 5x + \frac{1}{4} = 0$ by completing the square.

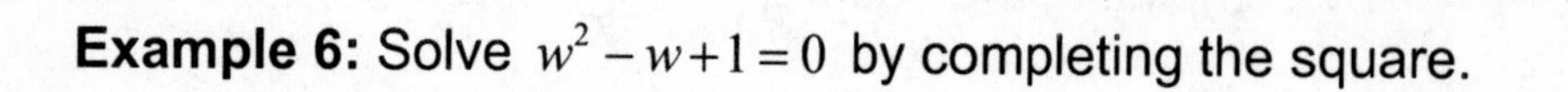

Example 6: Solve $w^2 - w + 1 = 0$ by completing the square.

What if the lead coefficient is not 1? We then begin by dividing both sides of the equation by the lead coefficient. Let's work through some of these problems together.

Example 7: Solve $4x^2 + 16x - 48 = 0$ by completing the square.

Example 8: Solve $3x^2 + 13x - 30 = 0$ by completing the square.

Student Activity A

Section 10.2: Recognizable Foe

Directions: Follow the directions for each step and fill in the boxes with the missing pieces as you go. Careful – as you move from line to line, the boxes might contain different expressions! At the end, you should get a formula that you might recognize!

Solve: $ax^2 + bx + c = 0$

Divide both sides by *a*: $x^2 + \square + \frac{c}{a} = \square$

Subtract $\square$ from both sides: $x^2 + \square \quad = -\frac{c}{a}$

Complete the square: $\frac{1}{2}\left(\square\right) = \square$ and $\left(\square\right)^2 = \square$

Add $\square$ to both sides: $x^2 + \frac{b}{a}x + \square = -\frac{c}{a} + \square$

Write the left side as a perfect square: $\left(x + \square\right)^2 = -\frac{c}{a} + \square$

Find a common denominator on the right: $\left(x + \square\right)^2 = -\frac{\square}{4a^2} + \frac{b^2}{4a^2}$

Combine the terms on the right side: $\left(x + \square\right)^2 = \frac{\square - 4ac}{4a^2}$

Apply the square root property: $x + \square = \pm\sqrt{\frac{\square - 4ac}{4a^2}}$

Simplify the denominator on the right side: $x + \square = \pm\frac{\sqrt{\square - 4ac}}{\square}$

Subtract $\square$ from both sides: $x = \square \pm \frac{\sqrt{\square - 4ac}}{\square}$

Combine terms with like denominators: $x = \frac{\square \pm \sqrt{\square - 4ac}}{\square}$

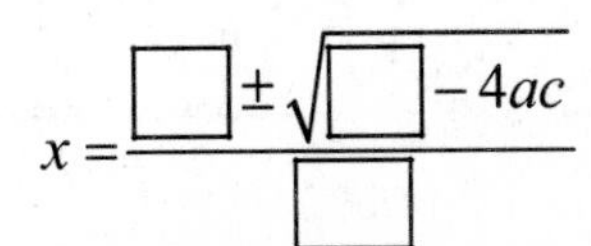

If you're really good… try it on your own on a blank sheet of paper without peeking!

Student Activity B

Section 10.2: Quadratic Formula with a Calculator

It can be tricky to evaluate the quadratic formula using your calculator because of all the order of operations that are implied by the symbols.

For example, suppose we wanted to calculate the value of:

$$\frac{-3 \pm \sqrt{(-3)^2 - 4 \cdot 1 \cdot (-2)}}{2 \cdot 1}.$$

There are many extra parentheses we would have to add to do the calculation directly. First, the numerator and denominator need parentheses, since a fraction bar is separates two distinct groups of terms:

$$\frac{\left(-3 \pm \sqrt{(-3)^2 - 4 \cdot 1 \cdot (-2)}\right)}{(2 \cdot 1)}$$

Then the radicand needs a set of parentheses, since the radicand must be simplified before the square root is taken:

$$\frac{\left(-3 \pm \sqrt{\left((-3)^2 - 4 \cdot 1 \cdot (-2)\right)}\right)}{(2 \cdot 1)}$$

Then we need to calculate two expressions because of the double-sign:

$$\frac{\left(-3 - \sqrt{\left((-3)^2 - 4 \cdot 1 \cdot (-2)\right)}\right)}{(2 \cdot 1)} \text{ or } \frac{\left(-3 + \sqrt{\left((-3)^2 - 4 \cdot 1 \cdot (-2)\right)}\right)}{(2 \cdot 1)}$$

Entering this into **my** calculator would look like this:

[(] [(-)] [3] [-] [2ND] √ [(] [(] [(-)] [3] [)] [^] [2] [-] [4] [×] [1] [×] [(-)] [2] [)] [)] [÷] [(] [2] [×] [1] [)] [ENTER]

The point is, **you will want to simplify your quadratic formula** to a more manageable expression **before** you enter it into your calculator.

$$\frac{-3 \pm \sqrt{(-3)^2 - 4 \cdot 1 \cdot (-2)}}{2 \cdot 1} = \frac{-3 \pm \sqrt{9+8}}{2} = \frac{-3 \pm \sqrt{17}}{2}$$ ←This is much more manageable!

Now we calculate the value of the two expressions: (your keystrokes may vary)

$\frac{\left(-3 - \sqrt{17}\right)}{2}$ [(] [(-)] [3] [-] [2ND] √ [1] [7] [)] [÷] [2] [ENTER] ≈ -3.56

$\frac{\left(-3 + \sqrt{17}\right)}{2}$ [(] [(-)] [3] [+] [2ND] √ [1] [7] [)] [÷] [2] [ENTER] ≈ 0.56

NOTE: On some calculators, you can edit an old entry. For example, on most TI graphing calculators, use ENTRY to pull up the last entry. You can use this to change the [-] to a [+] and then press ENTER to get the second answer.

Now try finding these with your calculator. The first row has been done for you. Be careful not to round values until the **end** of the problem. Use a ≈ symbol if the calculated result is rounded. Round to two decimal places if necessary.

Quadratic Formula Expression	Simplified Expression	Calculated Result
1. $\frac{-(-4)\pm\sqrt{(-4)^2-4(3)(1)}}{2(3)}$		
2. $\frac{-5\pm\sqrt{5^2-4(2)(-3)}}{2\cdot 2}$		
3. $\frac{-11\pm\sqrt{11^2-4(-2)(-6)}}{2(-2)}$		
4. $\frac{-(-3,850)\pm\sqrt{(-3,850)^2-4(550)(-5,500)}}{2(550)}$		
5. $\frac{-(-4.18)\pm\sqrt{(-4.18)^2-4(1.32)(-1.54)}}{2(1.32)}$		

Scrambled Answers

0.5	≈ 0.61	≈ −0.33	−3	≈ 8.22
1	3.5	≈ 0.33	≈ −1.22	≈ 4.89

Student Activity C

Section 10.2: Triple the Fun on Quadratic Equations

Directions: For many quadratic equations, you have the choice of solving by using the factoring method, the square root method (completing the square if necessary), or the quadratic formula. Solve each of the quadratic equations by the indicated methods (if possible). If it is **not** possible to use one of the methods, say so. We've given you some hints on the first problem. For each problem, put a ☺ by the method that seemed easiest.

1. Solve: $x^2 - 3x - 40 = 0$

Solve by Factoring (if possible):

$x^2 - 3x - 40 = 0$

$(\quad\quad)(\quad\quad) = 0$

Solve by the Square Root Property:
(completing the square if necessary)

$x^2 - 3x - 40 = 0$

$x^2 - 3x + \_\_\_\_ = 40 + \_\_\_\_\_$

Solve using the Quadratic Formula:

$x^2 - 3x - 40 = 0$

$a = \_\_\_\_,\ b = \_\_\_\_,\ c = \_\_\_\_$

2. Solve: $x^2 + 2x - 2 = 0$

Solve by Factoring (if possible):

$x^2 + 2x - 2 = 0$

Solve by the Square Root Property:
(completing the square if necessary)

$x^2 + 2x - 2 = 0$

Solve using the Quadratic Formula:

$x^2 + 2x - 2 = 0$

3. Solve: $4x^2 - 5x - 6 = 0$

Solve by Factoring (if possible):

$4x^2 - 5x - 6 = 0$

Solve by the Square Root Property:
(completing the square if necessary)

$4x^2 - 5x - 6 = 0$

Solve using the Quadratic Formula:

$4x^2 - 5x - 6 = 0$

4. Solve: $x^2 - 8 = 0$

Solve by Factoring (if possible):

$x^2 - 8 = 0$

Solve by the Square Root Property:
(completing the square if necessary)

$x^2 - 8 = 0$

Solve using the Quadratic Formula:

$x^2 - 8 = 0$

Student Activity A

Section 10.3: Match Up on the Discriminant

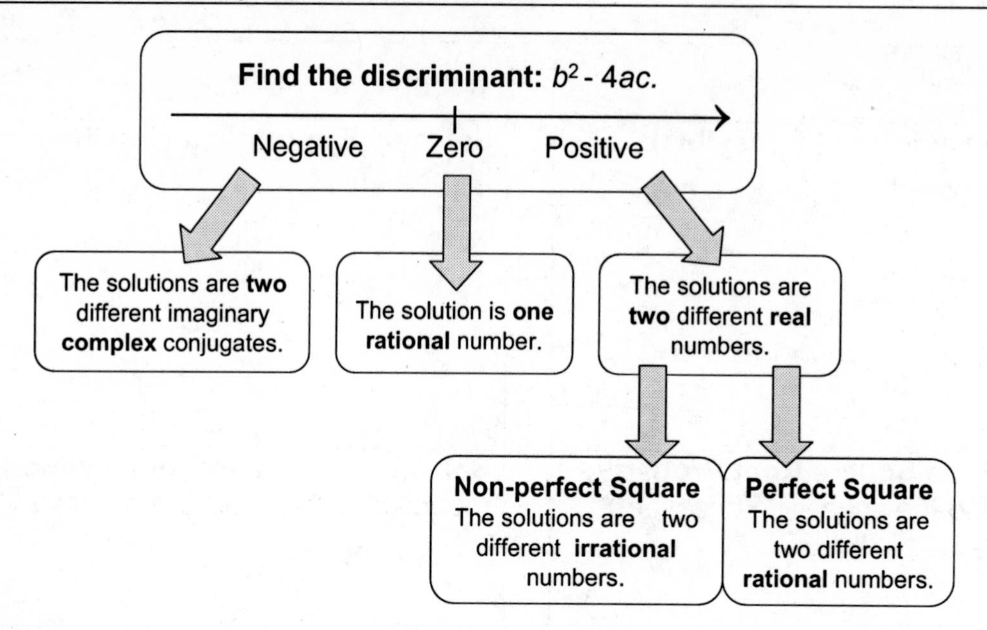

Directions: In each box, find the discriminant and then classify it.

A Two different rational numbers
B Two different irrational numbers
C One rational number
D Two different imaginary numbers

$25x^2 - 40x + 16 = 0$ $d = (-40)^2 - 4(25)(16) = 0$ C	$3x^2 + 14x - 24 = 0$	$x^2 - 2x + 2 = 0$
$9x^2 = 4$	$6x^2 - 3 = 0$	$x^2 + 3 = 5x$
$5(5x^2 + 2x) = -1$	$2x^2 + 3x = -3$	$3x^2 + 11x = 5$
$45x^2 + 24 = 67x$	$144x^2 - 264x + 121 = 0$	$4x^2 + 36x = -81$

Student Activity B

Section 10.3: Strategizing on Quadratic Equations

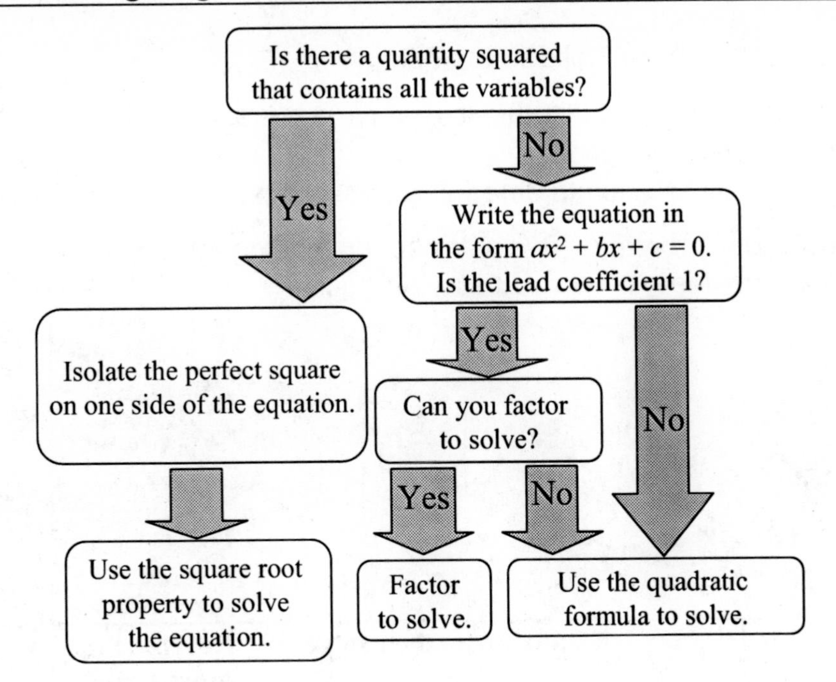

Directions: For each problem, use the flow chart above to help you make a decision about **how** to solve the quadratic equation (the square root property, factoring, or the quadratic formula). You do not need to solve the problems.

	Square root property	Factoring	Quadratic formula
1. Solve: $x^2 - 8x - 20 = 0$			
2. Solve: $4(x+3)^2 = 25$			
3. Solve: $4x^2 = 9$			
4. Solve: $x^2 + 5x - 3 = 0$			
5. Solve: $6x^2 + 7x = 20$			
6. Solve: $x^2 - 6x = 0$			
7. Solve: $x^2 - x + 6 = 0$			
8. Solve: $8x(x-2) = 2x + 5$			
9. Solve: $(x-2)^2 - 2x = 16$			
10. Solve: $(x-2)^2 = 16$			

Student Activity C

Section 10.3: Sink the Sub

Directions: For each problem, follow these steps:

- Find an appropriate y-substitution to make the equation quadratic in form.
- Solve the quadratic equation.
- Reverse the y-substitution to determine the solution set of the original equation.

The first problem is done for you. Two submarines holding the possible answers for $y=?$ and $x=?$ are provided to help you stay on track.

$y=?$

$y=x^{1/3}$ $y=2x-3$ $y=\sqrt{x}$

$y=\sqrt{x}$ $y=\frac{1}{x}$ ~~$y=x^2$~~

$x=?$

~~$x=\{-3,3,-i,i\}$~~ $x=\{-1,4\}$ $x=\{-6,5\}$

$x=\{36\}$ $x=\left\{64,\frac{1}{8}\right\}$ $x=\{25,36\}$

Solve.	$y=?$	Substitute, then solve for *y*.	Reverse the substitution	Solution set
1. $x^4-8x^2-9=0$	$y=x^2$	$y^2-8y-9=0$ $(y-9)(y+1)=0$ $y=9,-1$	$x^2=9$ $x^2=-1$ $x=\pm 3$ $x=\pm i$	$x=\{-3,3,-i,i\}$
2. $x-11\sqrt{x}+30=0$				
3. $x-\sqrt{x}-30=0$				
4. $2x^{2/3}-9x^{1/3}+4=0$				
5. $(2x-3)^2=25$				
6. $\frac{30}{x^2}-\frac{1}{x}-1=0$				

Assessment 10B

Chapter 10: Mid-chapter Assessment for Understanding

For each of the following, describe the type of problem and the strategies and key steps to remember while doing the problem. You do **not** have to complete the problems.

	Type of Problem	Strategies and Key Steps
1. Complete the perfect square trinomial: $x^2 - 14x +$ _____		
2. Is $x = -10$ a solution to the equation: $x^2 + 8x - 20 = 0$?		
3. Solve: $(x-9)^2 = 16$		
4. Solve $x^2 + 6x + 5 = 0$ by completing the square.		
5. Solve $x^2 - 10 = 0$ using the quadratic formula.		
6. Solve: $16(x-3)^2 = 9$		
7. Solve $15x^2 = 4 - 17x$ using the quadratic formula.		
8. Solve: $x - 8\sqrt{x} + 12 = 0$		
9. Solve $x^2 + 10 = 7x$ by factoring.		
10. How many solutions does $2x^2 + x - 21 = 0$ have? Are the solutions rational, irrational, or complex numbers?		

Guided Learning Activity

Section 10.4: Everything's Coming Up Parabolas

The graph of $f(x) = ax^2 + bx + c$, where $a \neq 0$, is a quadratic function, and its graph is a parabola. If *a* > 0, then the parabola opens up. If *a* < 0, then the parabola opens down. The vertex is the maximum or minimum point of the parabola (depending on whether it opens up or down. Each parabola is symmetric about the axis of symmetry.

1. On each parabola below, label the **vertex**, the **axis of symmetry**, whether the parabola **opens up** or **opens down** and whether ***a* > 0** or ***a* < 0**.

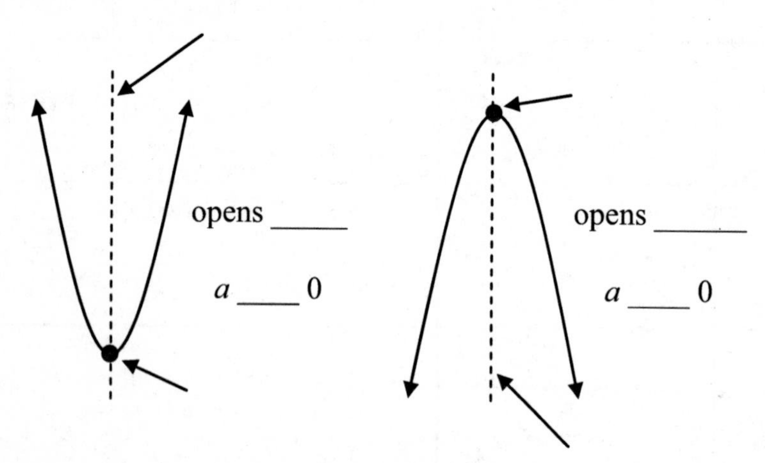

2. Draw the graph of $f(x) = -x^2 + 2x + 3$ by plotting points. On the graph, clearly indicate the point that is the vertex and a dashed line for the axis of symmetry.

x	y
−3	
−2	
−1	
0	
1	
2	
3	

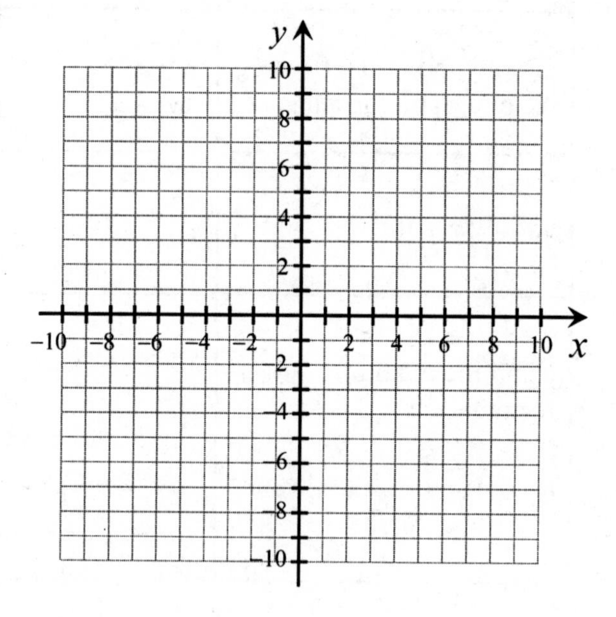

What is the vertex of the parabola? (___,___)

What is the equation for the axis of symmetry of the parabola? _________

What is the *y*-intercept of the parabola? (___,___)

What are the *x*-intercepts of the parabola? (___,___) and (___,___)

3.

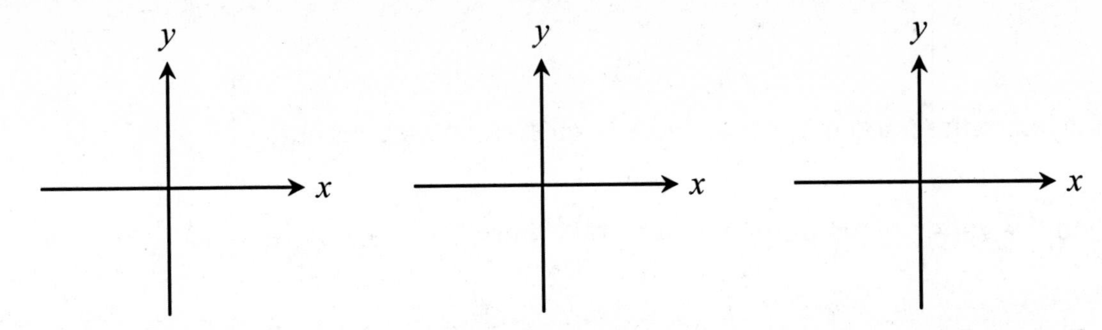

Draw a parabola with **two** x-intercepts.

Draw a parabola with **one** x-intercept.

Draw a parabola with **no** x-intercepts.

Will a parabola ever have **more** than one **y**-intercept? Why or why not?

Will a parabola ever have **less** than one **y**-intercept? Why or why not?

> To find the y-intercept, we find $f(0)$.
> Thus, the ***y*-intercept** is at a point given by $(0,\_\_\_)$.
>
> To find x-intercepts, we need to find $f(x)=0$.
> Likewise, the ***x*-intercepts** are at point(s) given by $(\_\_\_,0)$.

4. Find the y- and x-intercepts of the graph of $f(x)=x^2+4x-12$.

5. Find the y- and x-intercepts of the graph of $f(x)=2x^2+3$.

The **vertex** of the parabola $f(x) = ax^2 + bx + c$ has an x-coordinate of $x = -\frac{b}{2a}$.

6. Find the vertex of the parabola given by $f(x) = 2x^2 - 8x + 13$.

Find the x-coordinate: $x = \frac{-b}{2a} = \underline{\qquad} =$

Substitute this value of x into the given function to find the y-coordinate:

$f(x) = 2x^2 - 8x + 13$

$f(\quad) = 2(\quad)^2 - 8(\quad) + 13 = \underline{\qquad}$

Vertex: (____,____)

7. Find the vertex of the parabola given by $f(x) = -9x^2 - 6x - 8$.

8. Put it all together to graph $f(x) = -x^2 + 2x + 8$.

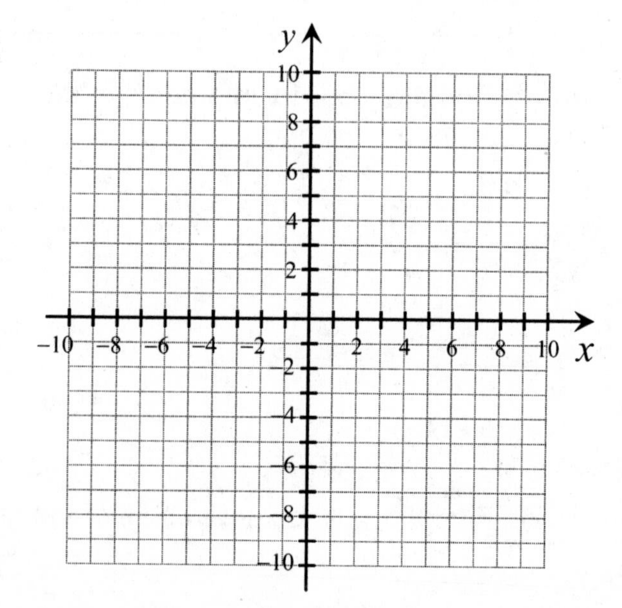

Does the parabola open up or down? _____

Find the y-intercept: $(0, \underline{\quad})$

Find the x-intercept(s), if there are any:

$(\underline{\quad}, 0)$ $(\underline{\quad}, 0)$

Find the vertex: (____,____)

Write the equation for the axis of symmetry: ________

Student Activity A

Section 10.4: Finding the Vertex and Intercepts Using a Calculator

Example: Consider the quadratic function $y = 8x^2 - 18x - 5$. First we will find all the characteristics of the graph using our **algebra** skills.

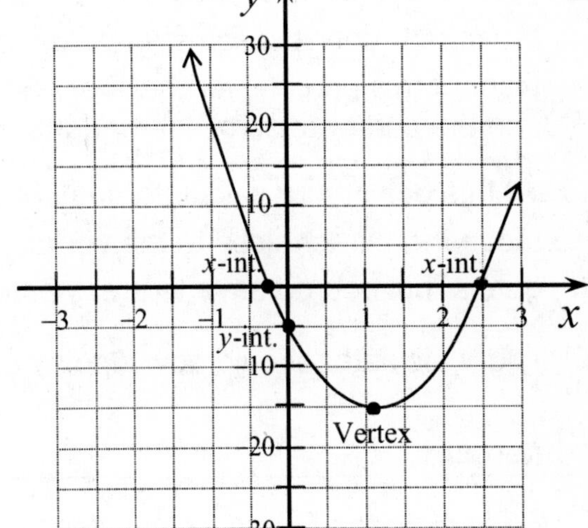

a. Find the vertex: (____,____)

b. Find the x-intercepts:
(____,0) and (____,0)

c. Find the y-intercept: (0,____)

Now we'll perform the same analysis using a **graphing calculator**.

☐ Using the $\boxed{y=}$ button or menu, graph the function $y = 8x^2 - 18x - 5$.

☐ Now use $\boxed{\text{WINDOW}}$ to find the window that best displays the parabola. Make sure you can see the vertex of the parabola in your graphing window.

Maximum or Minimum

Because the vertex of the parabola **is** the maximum or minimum point of the parabola, we can use this feature of the calculator to find the vertex.

☐ Find the graphing menu that has an option to find the MAX or MIN. You may find this in a MATH menu in the graphing panel, or under the CALC function. We will refer to this menu as the Graph Calculations Menu.

☐ Write down how you get to the Graph Calculations Menu on **your** calculator:

☐ Select the option to find the minimum of the graph (since in this case, the vertex is the minimum point).

☐ Your calculator will probably prompt you for the *Left Bound* or *Lower Bound*. Move the cursor until it is somewhere on the left of the vertex and then press ENTER. Repeat this process for the *Right Bound* or *Upper Bound*. Note: If you don't want to use the cursor to scroll left or right, you may just enter an *x*-value to be the bound.

☐ At this point, your calculator may prompt you to *Guess*. If this happens, you can move the cursor close to the vertex and press ENTER (note that you don't *have* to guess, but you do have to press enter).

☐ Your calculator should now display the coordinates of the vertex: (____,____).

y-intercept

☐ Go back to the Graph Calculations Menu. This time, select *Value* on the menu. Note: You can also use TRACE to do this.

☐ To find the *y*-intercept, we let $x = 0$. When prompted, enter 0 for the *x*-value.

☐ Your calculator should now display the coordinates of the y-intercept: $(0,$____$)$

x-intercept(s)

☐ Go back to the Graph Calculations Menu. This time, select *Zero* on the menu. It is called *Zero* because we let *y* be zero when we want to find the *x*-intercepts.

☐ Your calculator will probably prompt you for the *Left Bound* or *Lower Bound*. Move the cursor until it is somewhere on the left of the vertex and then press ENTER. Repeat this process for the *Right Bound* or *Upper Bound*.

☐ At this point, your calculator may prompt you to *Guess*. If this happens, you can move the cursor close to the *x*-intercept and press ENTER (note that you don't *have* to guess, but you do have to press enter).

☐ Your calculator should now display the coordinates of the x-intercept: (____,____).

☐ Repeat this process to find the coordinates of the second x-intercept: (____,____).

Possible Errors:

ERR: BOUND	You guessed a value that was not within the bounds you set.
ERR: INVALID	You tried to have the calculator find something with an *x*-value outside the viewing window.

Now use your calculator to find the vertex and the x – and y -intercepts of the following quadratic equations.

1. $y = 5x^2 + 15x + 10$

Vertex: (____,____)
y-intercept: (____,____)
x-intercept(s): (____,____) (____,____)

2. $y = -4x^2 + 20x - 24$

Vertex: (____,____)
y-intercept: (____,____)
x-intercept(s): (____,____) (____,____)

3. $y = x^2 - 4x + 14$

Vertex: (____,____)
y-intercept: (____,____)
x-intercept(s): (____,____) (____,____)

4. $y = \frac{1}{16}x^2$

Vertex: (____,____)
y-intercept: (____,____)
x-intercept(s): (____,____) (____,____)

Student Activity B

Section 10.4: Parabola Calisthenics

On each graph you are given the graph of the basic function $y = x^2$, a parabola. You will be asked to use a table of values to draw graphs of parabolas that are stretched or compressed from this basic graph. If possible, find each set of values in a different colored pencil and draw its graph in a corresponding color.

1. Draw the graphs of each quadratic function by making a table of values and plotting the points on the coordinate axes provided.

$f(x) = 2x^2$

x	$f(x)$
−2	
−1	
0	
1	
2	

$g(x) = \frac{1}{2}x^2$

x	$g(x)$
−3	
−2	
−1	
0	
1	
2	
3	

$F(x) = 10x^2$

x	$f(x)$
−2	
−1	
0	
1	
2	

$G(x) = \frac{1}{10}x^2$

x	$g(x)$
−3	
−2	
−1	
0	
1	
2	
3	

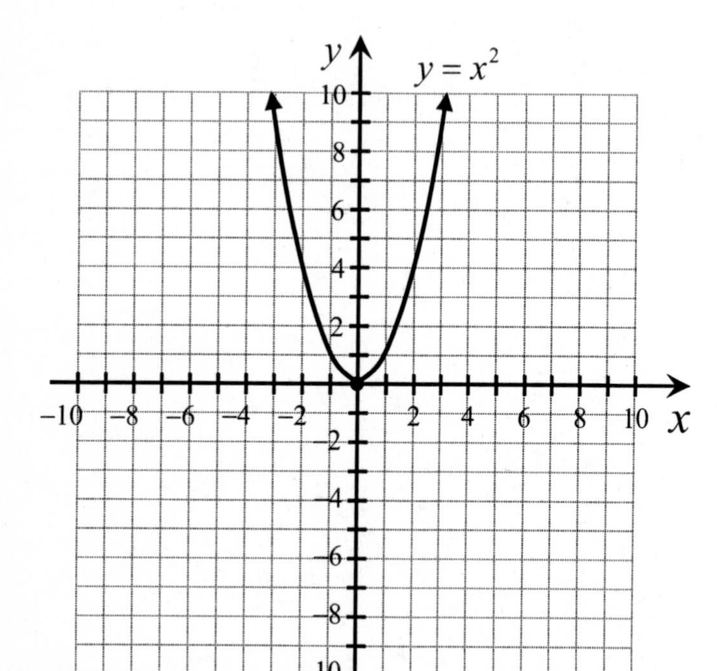

2. How do the graphs of f and F compare to the basic graph of $y = x^2$?

3. How do the graphs of g and G compare to the basic graph of $y = x^2$?

4. Using dashed lines, predict how $h(x) = 5x^2$ and $H(x) = \frac{1}{5}x^2$ would be graphed.

Student Activity C

Section 10.4: Parabola Transformations

1. Draw the desired transformation of $y = x^2$ on each graphing grid below

   a. Reflect the graph about the x-axis, then shift the graph up 3 units.

   b. Shift the graph up 3 units, then reflect the graph about the x-axis.

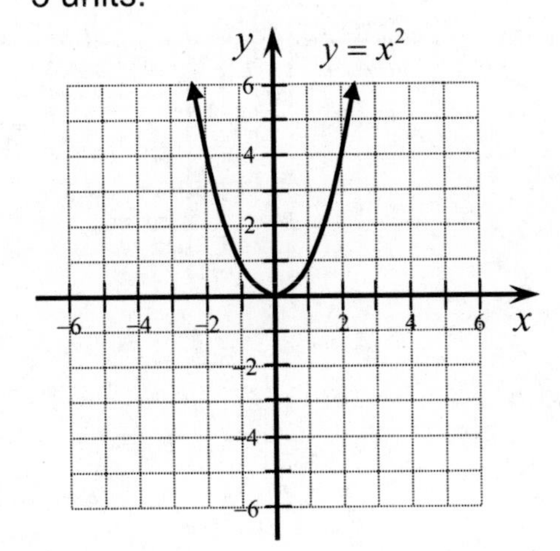

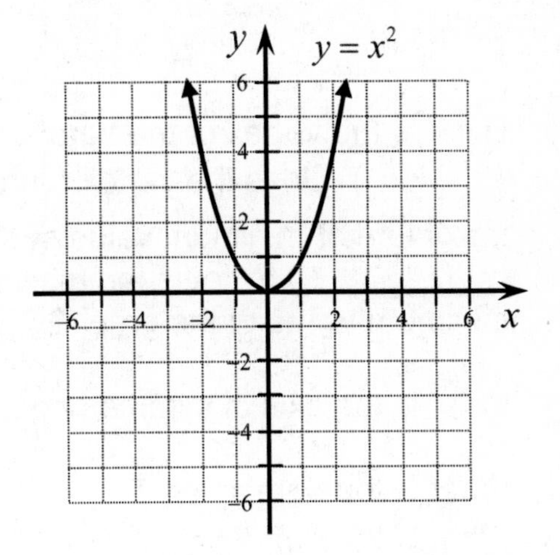

   c. Do you get the same result for both graphs? ____ Calculate a few points on the graph of $f(x) = -x^2 + 3$ to see which graph (**a** or **b**) is the appropriate graph of f. ____

2. Let's examine some functions that are of the form $y = ax^2$.

   $f(x) = 5x^2$ is narrower than $y = x^2$.

   $f(x) = \frac{1}{3}x^2$ is wider than $y = x^2$.

   $f(x) = -x^2$ is the graph of $y = x^2$ reflected about the x-axis.

   In all these cases, the stretch, compression, or reflection is caused by ________________ x^2 by a constant.

3. Let's examine some functions that include translations:

   Up 3 units: $f(x) = x^2 + 3$ Down 3 units: $f(x) = x^2 - 3$

   Right 3 units: $f(x) = (x - 3)^2$ Left 3 units: $f(x) = (x + 3)^2$

   In all these cases, the translation is caused by ___________ or ___________ a constant.

When we **graph**, the order of operations is the same as in arithmetic, multiplication first, then addition or subtraction.

Graphing Order of Operations:
1. Draw the graph of $y = ax^2$.
2. Identify the translations of $y = ax^2$ (up or down, left or right) and draw the final graph.

4. Graph $f(x) = -(x+1)^2 + 4$.

a. Draw the graph of $y = -x^2$ with a dashed line.

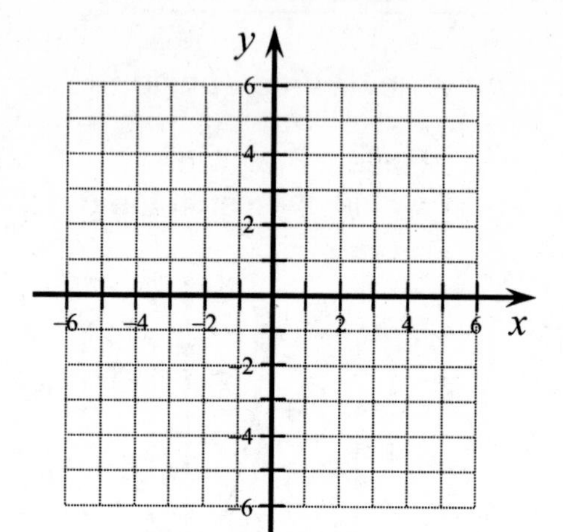

b. Identify any translations. Use the graph from part **a** to sketch the graph that includes these changes with a solid line.

c. What is the vertex of the parabola? (___,___) Does the vertex represent the maximum or minimum value of the graph? __________ What is the maximum or minimum **value** of the graph? ____

d. What is the axis of symmetry of the graph? _______

e. What is the domain of f? _________
What is the range of f? _________

5. Graph $f(x) = \frac{1}{2}(x+2)^2 - 3$.

a. Draw the graph of $y = \frac{1}{2}x^2$ with a dashed line.

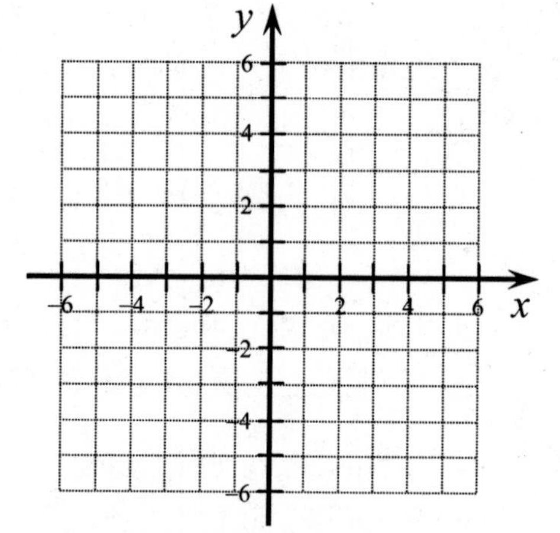

b. Identify any translations. Use the graph from part **a** to sketch the graph that includes these changes with a solid line.

c. What is the vertex of the parabola? (___,___) Does the vertex represent the maximum or minimum value of the graph? __________ What is the maximum or minimum **value** of the graph? ____

d. What is the axis of symmetry of the graph? _______

e. What is the domain of f? _________
What is the range of f? _________

Student Activity D

Section 10.4: Graphical Meaning of Complex Zeros

The *x*-intercepts of a graph can also be called the *zeros* (where the *y*-values are zero). When we try to find the *x*-intercepts of a quadratic function, and the result is a pair of complex numbers, does that give us any useful information for graphing? It turns out that the answer is **yes!**

Example 1: $f(x)=x^2-4x+29$

a) Solve $x^2-4x+29=0$ using the quadratic formula, as if you were finding the *x*-intercepts of the graph. The solution will be a pair of complex numbers.

b) If the zeros are given by $p \pm qi$, then for this problem, $p=$____ and $q=$____. We will also need the value of the leading coefficient: $a=$____.

c) Now we will graph the parabola, using only the values of p, q, and a.

- First we graph the vertex. The *x*-coordinate of the vertex is p and the *y*-coordinate of the vertex is aq^2. Write the vertex: (____,____). Then graph the vertex.

- We also need to locate the point $(p, 2aq^2)$. Since this is just a point of reference, just make a small **X** on the graph at this location: (____,____).

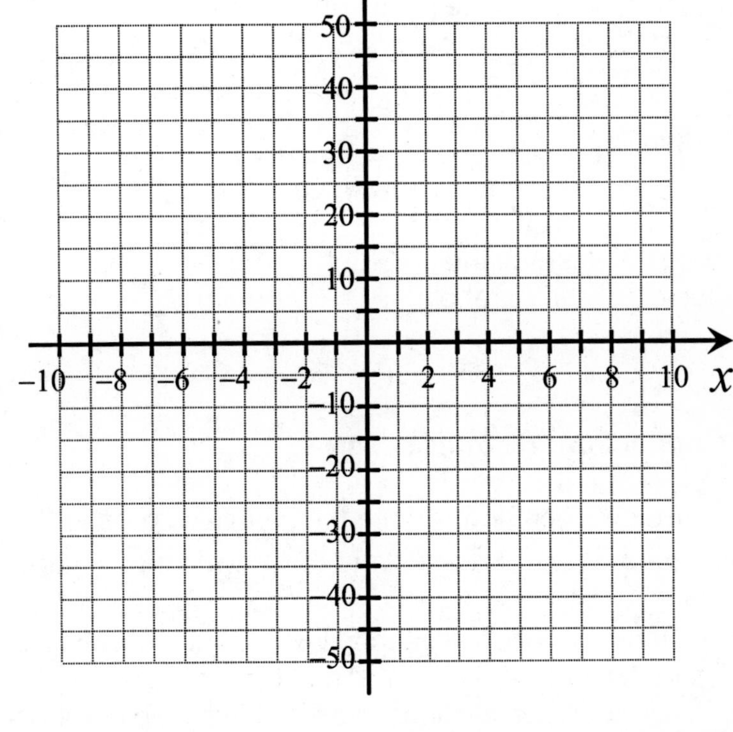

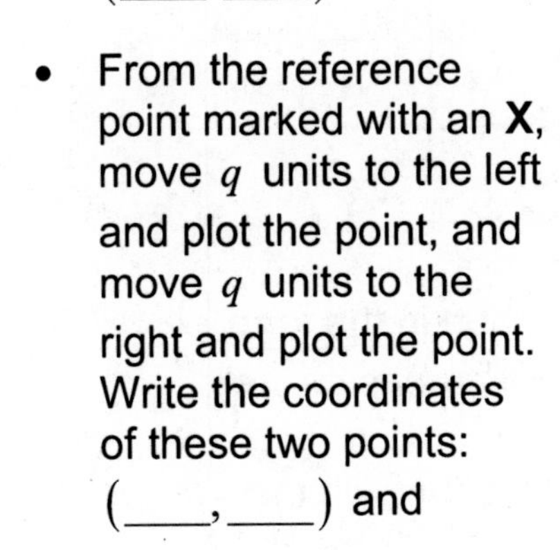

- From the reference point marked with an **X**, move q units to the left and plot the point, and move q units to the right and plot the point. Write the coordinates of these two points: (____,____) and (____,____).

- You should now have three points to use to sketch the graph of the parabola.

Example 2: $f(x) = -2x^2 + 4x - 20$

a) Solve $-2x^2 + 4x - 20 = 0$ using the quadratic formula, as if you were finding the x-intercepts of the graph. The solution will be a pair of complex numbers.

b) If the zeros are given by $p \pm qi$, then for this problem, $p =$ ____ and $q =$ ____. We will also need the value of the leading coefficient: $a =$ ____.

c) Now we will graph the parabola.

- Graph the vertex: $(p, aq^2) = ($____, ____$)$.

- Locate

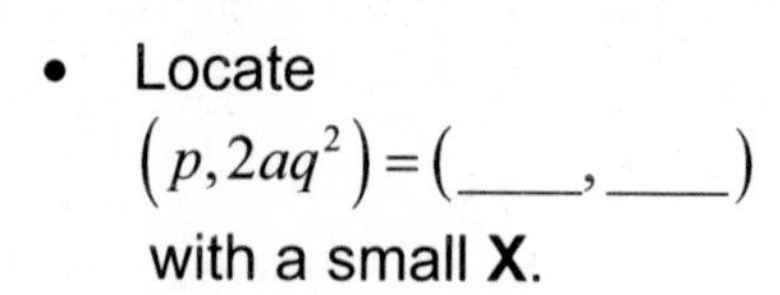

  $(p, 2aq^2) = ($____, ____$)$
  with a small **X**.

- From the **X**, plot points q units to the left and q units to the right. The points are

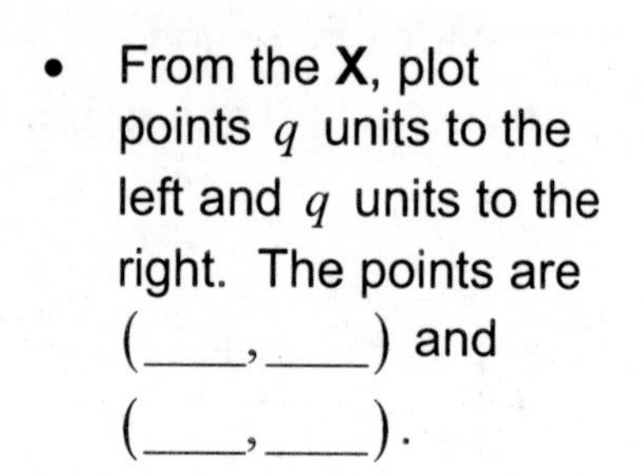

  (____, ____) and
  (____, ____).

- You should now have three points to use to sketch the graph of the parabola.

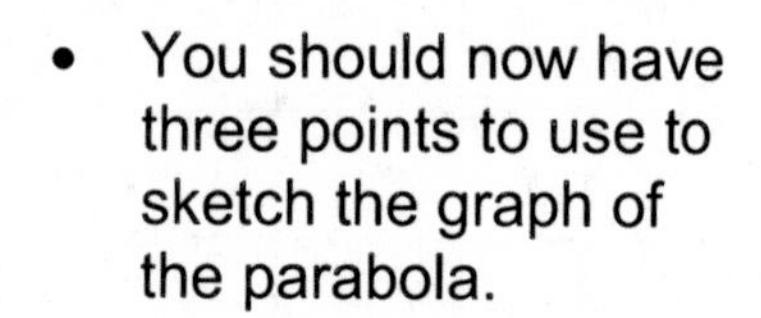

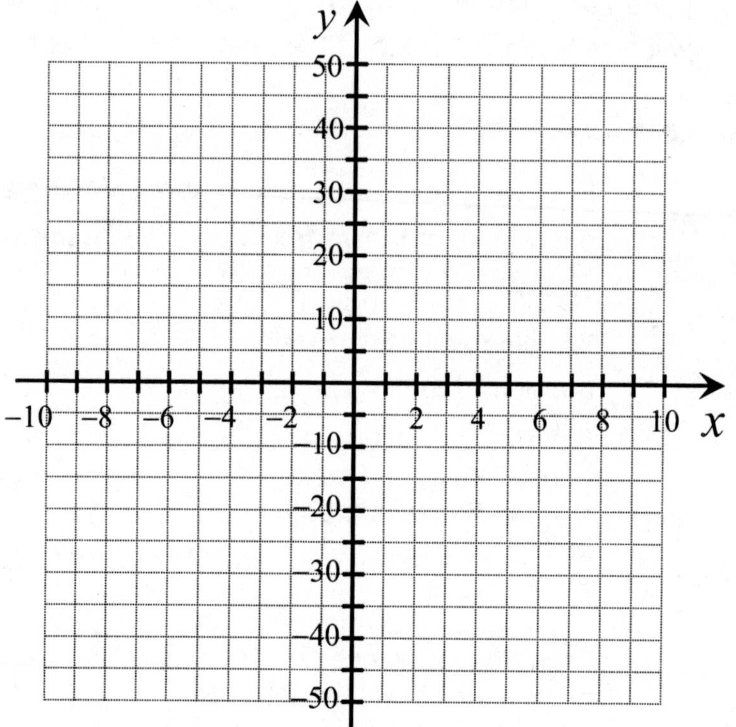

Connection to complex zeros:

a) Below each graph, write the complex zeros for the graph in the form $p \pm qi$. **Then calculate** $p \pm q$ and write those two values below the graphs as well.

b) Draw a horizontal dashed line on each graph that goes through the vertex of the parabola. Then draw a perfect reflection of the parabola over this dashed line. We'll call this new parabola the *imaginary parabola* for lack of a better name.

c) Where does the *imaginary parabola* cross the x-axis? In other words, what are its zeros? _____ and _____ Surprise!

Student Activity A

Section 10.5: Situation Critical

Directions: For each inequality, find the critical numbers. As you find the critical numbers, shade them in to paint the picture in the grid.

1. Solve: $x^2 - 8x < 20$

2. Solve: $9 - x^2 \le 0$

3. Solve: $2x^2 > 12 - 5x$

4. Solve: $x(x-5) \ge 0$

5. Solve: $x^2 \le 3$

6. Solve: $\frac{3}{x} + 6 < 0$

7. Solve: $x - \frac{5x}{x+1} \ge 0$

8. Solve: $\frac{x}{x+2} + \frac{3}{x+2} > 0$

9. Solve: $\frac{x+1}{(x+3)(x-3)} \le 0$

2	2	-2	$\frac{3}{2}$	$-\frac{3}{2}$	$\frac{5}{2}$
-5	$\frac{1}{2}$	-2	4	1	1
5	$-\frac{1}{2}$	-4	10	-1	-1
-3	-3	-3	0	3	3
$-\frac{3}{4}$	$\frac{3}{4}$	0	0	-10	-6
6	6	$-\sqrt{3}$	$\sqrt{3}$	$-\sqrt{5}$	$\sqrt{5}$

Student Activity B

Section 10.5: Wrap Up on Solving Quadratic and Rational Inequalities

Directions: For each problem, use test values to decide whether each interval makes the inequality true or false. Then write the solution to the inequality in interval notation. The first one has been **started** for you.

	Critical Numbers on a Number Line	Solution in Interval Notation
1. Solve: $x^2 - 8x - 20 < 0$	False | True | False; critical numbers: -2, 10	
2. Solve: $-x^2 + 9 \le 0$	-3, 3	
3. Solve: $2x^2 + 5x - 12 > 0$	-4, $\frac{3}{2}$	
4. Solve: $x^2 - 5x \ge 0$	0, 5	
5. Solve: $x^2 - 3 \le 0$	$-\sqrt{3}$, $\sqrt{3}$	
6. Solve: $\frac{3+6x}{x} < 0$	$-\frac{1}{2}$, 0	
7. Solve: $\frac{x^2 - 4x}{x+1} \ge 0$	-1, 0, 4	
8. Solve: $\frac{x+3}{x+2} > 0$	-3, -2	
9. Solve: $\frac{4}{x+2} \le 0$	-2	
10. Solve: $\frac{x+1}{x^2 - 9} \le 0$	-3, -1, 3	

Guided Learning Activity

Section 10.5: Shady Regions

1. For review, let's graph a linear inequality: $y < \frac{2}{3}x - 3$

   First graph the boundary **line**: $y = \frac{2}{3}x - 3$.

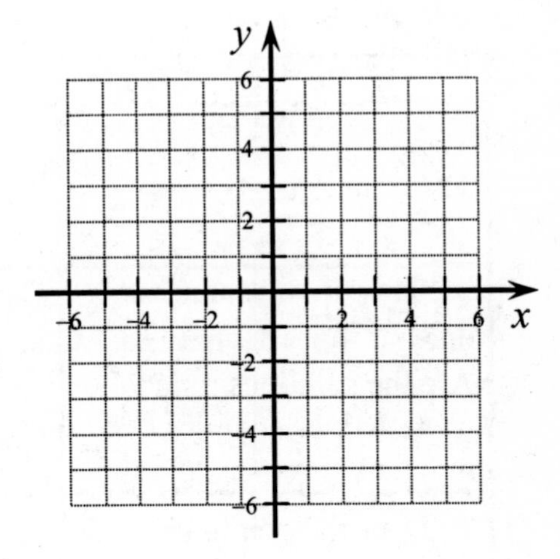

   a. Is this boundary solid or dashed? ________

   b. Choose a test point that is NOT on the boundary line: (____,____)

   c. Substitute the values of *x* and *y* into $y < \frac{2}{3}x - 3$ to see if the inequality is true or false.

   d. Should we shade the same side as the test point or the other side? ________

2. Now let's graph the inequality: $y \geq -x^2 + 3$

   First graph the boundary: $y = -x^2 + 3$

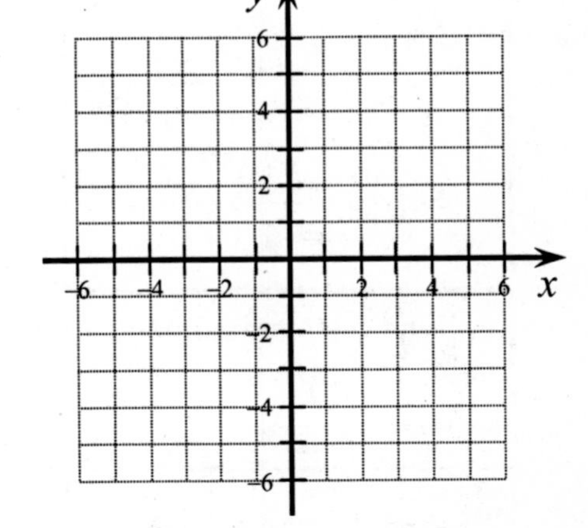

   a. Is this boundary solid or dashed? ________

   b. Choose a test point that is NOT on the boundary: (____,____)

   c. Substitute the values of *x* and *y* into $y \geq -x^2 + 3$ to see if the inequality is true or false.

   d. Should we shade the same side as the test point or the other side? ________

3. Now let's graph the inequality: $y > |x + 3|$

   First graph the boundary: $y = |x + 3|$

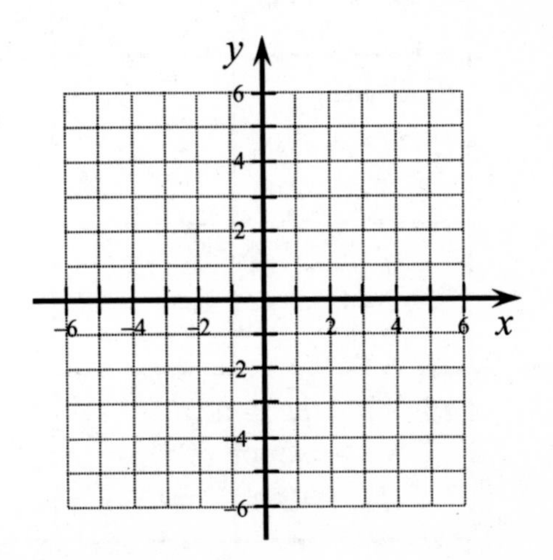

   a. Is this boundary solid or dashed? ________

   b. Choose a test point that is NOT on the boundary: (____,____)

   c. Substitute the values of *x* and *y* into $y > |x + 3|$ to see if the inequality is true or false.

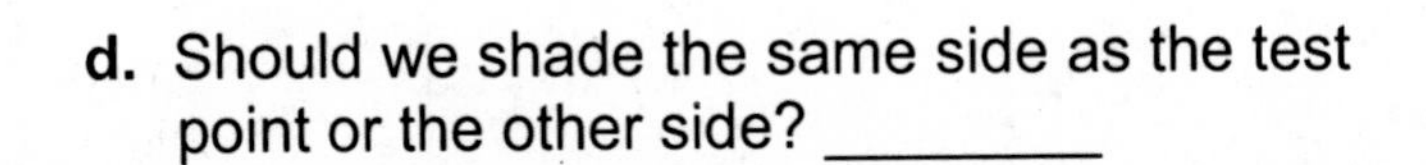

   d. Should we shade the same side as the test point or the other side? ________

Assessment 10C

Chapter 10: End-of-chapter Assessment for Understanding

For each of the following, describe the type of problem and the strategies and key steps to remember while doing the problem. You do **not** have to complete the problems.

	Type of Problem	Strategies and Key Steps
1. Solve: $x^2 - 8x = 20$		
2. How many solutions does $x^2 + 3 = 5x$ have? Are the solutions rational, irrational, or complex numbers?		
3. Does the graph of $y = 3 - x^2$ open up or down?		
4. Solve: $x^4 - 2x^2 - 35 = 0$		
5. Solve: $\frac{3}{x+4} < 1$		
6. Find the maximum value of $f(x) = -x^2 + 2x + 3$.		
7. Find the vertex of the graph of $f(x) = 3x^2 - 3x$..		
8. Write $f(x) = x^2 - 6x + 8$ in standard form.		
9. Graph $y < x^2 - 2$		
10. Find the x-intercept(s) of the graph of $f(x) = 3x^2 - 3x - 6$.		

Assessment 10D

Chapter 10: Metacognitive Skills Assessment

Metacognitive skills refer to the ability to judge how well you have learned something and to effectively direct your own learning and studying. This is a self evaluation tool designed to help you focus your studying and to improve your metacognitive skills with regards to this math class.

Fill the 1st column out before you begin studying.
Fill the 2nd column out after you study and before you take the test.
Go back to this page after your test and circle any of the ratings that you would now change – this identifies the "disconnects" between what you think you know well and what you actually know well.

Use the scale below to assign a number to each topic.

5 *I am confident I can do any problems in this category correctly.*
4 *I am confident I can do most of the problems in this category correctly.*
3 *I understand how to do the problems in this category, but I still make a lot of mistakes.*
2 *I feel unsure about how to do these problems.*
1 *I know I don't understand how to do these problems.*

Topic or Skill	Before Studying	After Studying
Solving a quadratic equation using factoring.		
Solving a quadratic equation using the square root property.		
Using the square root property and understanding where the double-sign comes from.		
Simplifying an expression written with a double-sign ($\pm$).		
Completing a perfect square trinomial.		
Solving a quadratic equation by completing the square.		
Knowing the quadratic formula.		
Solving a quadratic equation by using the quadratic formula.		
Finding the discriminant for a quadratic equation.		
Using the discriminant to determine the type and number of solutions.		
Looking at a quadratic equation and choosing the most appropriate method to solve the equation (factoring, square root property, completing the square, or the quadratic formula).		
Recognizing an equation that is quadratic in form.		
Solving an equation that is quadratic in form by making an appropriate y-substitution.		
Knowing the general shape of the graph of a quadratic function, including whether the parabola opens up or down.		
Finding the vertex of a parabola by first finding the x-coordinate.		
Finding the x- and y-intercepts of a parabola.		
Graphing a quadratic function of the form $f(x) = ax^2$.		
Graphing a quadratic function in standard form: $f(x) = a(x-h)^2 + k$.		

(continued on next page)

Use the scale below to assign a number to each topic.

5 I am confident I can do any problems in this category correctly.
4 I am confident I can do most of the problems in this category correctly.
3 I understand how to do the problems in this category, but I still make a lot of mistakes.
2 I feel unsure about how to do these problems.
1 I know I don't understand how to do these problems.

Topic or Skill	Before Studying	After Studying
Finding the vertex if the quadratic function is written in standard form.		
Writing a quadratic function in standard form by completing the square.		
Solving a quadratic inequality.		
Solving a rational inequality.		
Graphing a nonlinear inequality.		

Student Workbook: Chapter 11

Table of Contents: *Exponential and Logarithmic Functions*

Assessment 11A

Pretest and Diagnostic Tool: Exponential and Logarithmic Functions

Directions: Complete this assessment without looking back at your notes or your book. **Do not use a calculator for this assessment.**

1. Simplify: $(3x+5)-(x-4)$

2. Simplify: $(x+3)^2-4(x+3)$

3. Evaluate: $-(-3)^2+4(-3)-8$

4. Let $f(x)=9-\sqrt{3-x}$.
Evaluate $f(-1)$.

5. Simplify: $2\left(\dfrac{x-3}{2}\right)+3$

6. Solve for y: $x=\sqrt[3]{y+2}$

7. Solve for y: $x=(y-4)^2+5$

8. Evaluate: $\left(5^{1/2}\right)^2$

9. Evaluate: $2^{-2}+1$

10. Evaluate: $\left(\dfrac{1}{3}\right)^{-2}$

11. Evaluate: $2\left(3^2\right)$

Fill in the box to make a true statement.

12. $\square^3=27$

13. $\square^4=16$

14. $5^{\square}=\dfrac{1}{25}$

15. $10^{\square}=1$

16. Suppose we know a general rule that $\boxed{x\cdot y}=\boxed{x}+\boxed{y}$, then how can you rewrite $\boxed{2}+\boxed{3}$? __________

17. Suppose we know a general rule that $\boxed{\dfrac{x}{y}}=\boxed{x}-\boxed{y}$, then how can you rewrite $\boxed{\dfrac{4}{z}}$? __________

18. Given the equation $\boxed{y}=\boxed{x}+\boxed{x+2}$.
Is the equation
$\boxed{y}+1=\boxed{x}+1+\boxed{x+2}+1$
equivalent? _______

19. Solve for $\sqrt{x}$: $1000=500\sqrt{x}-4$

20. Solve for $|x-5|$: $2\cdot|x-5|+4=16$

Student Activity A

Section 11.1: Algebra of Domains

When we create a new function using the algebra of functions *f* and *g*, the domain of the new function is the set of real numbers that are in the domain of both *f* **and** *g* (as in, the intersection of the two domains). For a quotient *f* / *g*, there is the additional restriction that the new denominator, *g*, not be equal to zero.

Directions: Begin by writing the domain for the functions *f*, *g*, and *h*. Then complete the table below.

$f(x) = 2x - 1$ $\qquad$ $g(x) = \sqrt{x+4}$ $\qquad$ $h(x) = x + 5$

Domain: _______ $\qquad$ Domain: ________ $\qquad$ Domain: ________

	Find the new function and simplify it.	What is the domain of new function?
1.	$(f+g)(x)$	
2.	$(f-h)(x)$	
3.	$(f/h)(x)$	
4.	$(f \cdot h)(x)$	
5.	$(h/g)(x)$	
6.	$g(h(x))$	
7.	$g(f(x))$	

Student Activity B

Section 11.1: Graphical Algebra of Functions

When we perform an algebraic operation on functions (addition, subtraction, multiplication, division), we are performing the operation on the *output* of the functions.

1. Below you see the graphs of two functions, f and g. A table of values has been given for each of the functions. Fill in the table of values for $f+g$ and $f-g$.

x	$f(x)$	$g(x)$	$(f+g)(x)$ $f(x)+g(x)$	$(f-g)(x)$ $f(x)-g(x)$
–5	–5	–2		
–4	–4	3		
–3	–3	3		
–2	–2	3		
–1	–1	1		
0	0	1		
1	1	1		
2	2	3	5	–1
3	3	3		
4	4	–5		
5	5	–5		

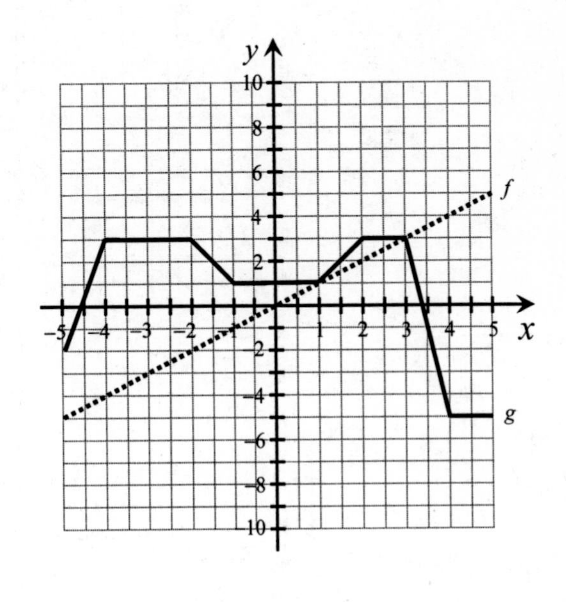

2. Now use the table or the graph to evaluate the expressions in the tables below.

$f(4)$	4
$g(4)$	–5
$f(4)+g(4)$	–1
$(f+g)(4)$	–1
$g(1)$	1
$g(3)$	3
$g(1+3)$	–5
$g(1)+g(3)$	4

$f(-3)$	–3
$g(-3)$	3
$f(-3)-g(-3)$	–6
$(f-g)(-3)$	–6
$g(1)$	1
$g(4)$	–5
$g(1-4)$	3
$g(1)-g(4)$	6

3. Why aren't $g(1+3)$ and $g(1)+g(3)$ equivalent?

4. Now we will graphically perform the addition and subtraction of the functions. For example, at $x = 2$, $f(2) = 2$ and $g(2) = 3$. Adding the values of f and g gives 5, so we plot a point at $(2,5)$ for the addition graph. Subtracting the values at two, $f - g$ gives -1, so we plot a point at $(2,-1)$ for the subtraction graph.

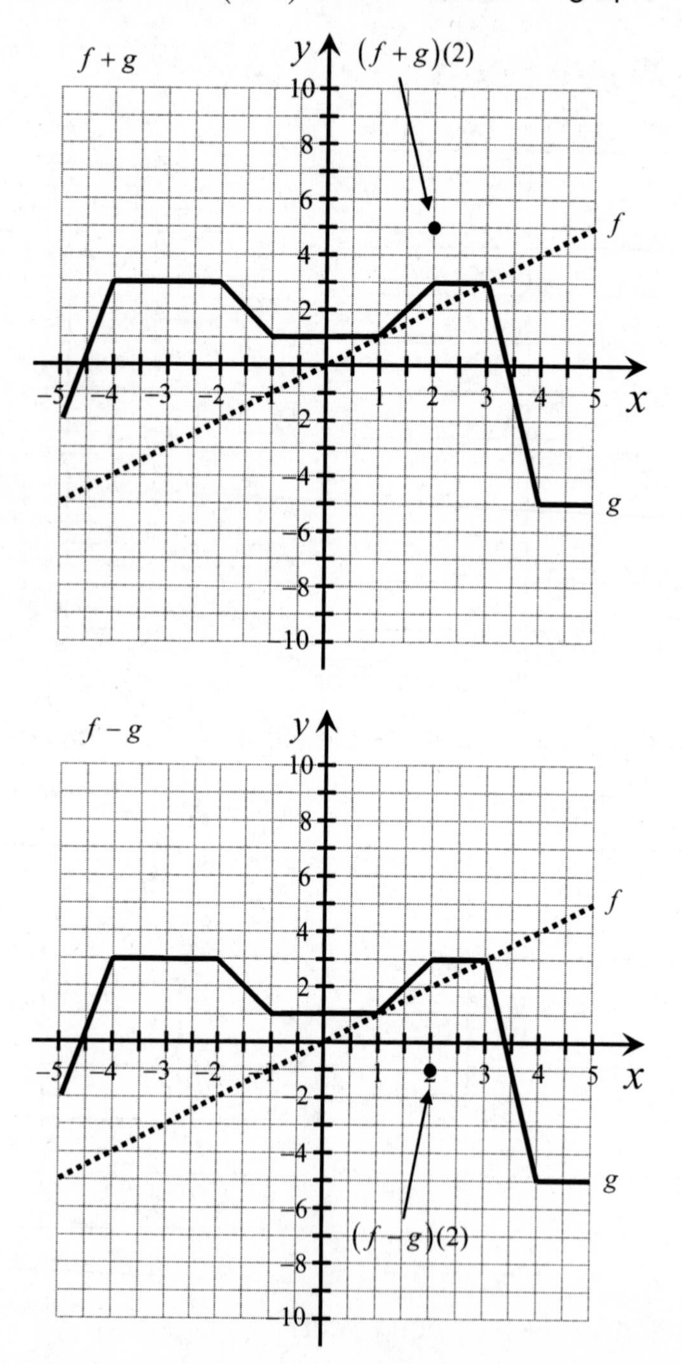

5. Now that you see the graphs of $f + g$ and $f - g$, try to explain what's happened to make the new curve for each one.

Student Activity C

Section 11.1: Composing Skeletons

Skeleton Method: To evaluate a function that uses function notation, it is helpful to "see" the function without the variables. To make the "skeleton" of the problem, replace the variables with an empty parentheses "skeleton".

For example, $f(x)=x^2+4x+3$ becomes $f(\ )=(\ )^2+4(\ )+3$.

For this activity, we define: $f(x)=3x-5$ $\quad g(x)=x-5$ $\quad h(x)=x^2-5x$

	Evaluate this expression	Write with parentheses notation	Make a parentheses skeleton of the outer function	Substitute the inner function and simplify
1.	$(f\circ g)(x)$	$f(g(x))$	$f(\quad)=3(\quad)-5$	$f(g(x))=3(x-5)-5$ $=3x-15-5=3x-20$
2.	$(g\circ f)(x)$			
3.		$h(g(x))$		
4.				$g(h(x))=(x^2-5x)-5$ $=x^2-5x-5$
5.	$(f\circ f)(x)$			
6.				$g(g(x))=(x-5)-5$ $=x-10$
7.		$h(f(x))$		

Student Activity D

Section 11.1: From Tables and Graphs

The functions *f*, *g*, and *h* are evaluated for the values of *x* in the table below.

For example, we can use the table to see that $f(0)=2$ or $g(-2)=7$.

x	$f(x)$	$g(x)$	$h(x)$
-2	0	7	4
-1	1	5	1
0	2	3	0
1	3	1	1
2	4	-1	4

x	$f(x)$	$g(x)$	$h(x)$
-2	0	7	4
-1	1	5	1
0	2	3	0
1	3	1	1
2	4	-1	4

$g(-2)=7$

$f(0)=2$

Directions: Evaluate each expression using the table of values.

1. $h(-1)=$
2. $(f+g)(2)=f(2)+g(2)=$
3. $(h-g)(1)=$
4. $(g/f)(-1)=$
5. $(f\cdot g)(0)=$
6. $g(3-1)=$
7. $h(f(0))=$
8. $g(h(-1))=$
9. $(g\circ f)(-2)=$
10. $f(f(-2))=$

Directions: Evaluate each expression below using the graphs of *f* and *g*.

11. $(f+g)(2)=f(2)+g(2)=$
12. $(g-f)(-2)=$
13. $(f\cdot g)(0)=$
14. $g(f(-1))=$
15. $(f\circ f)(-1)=$

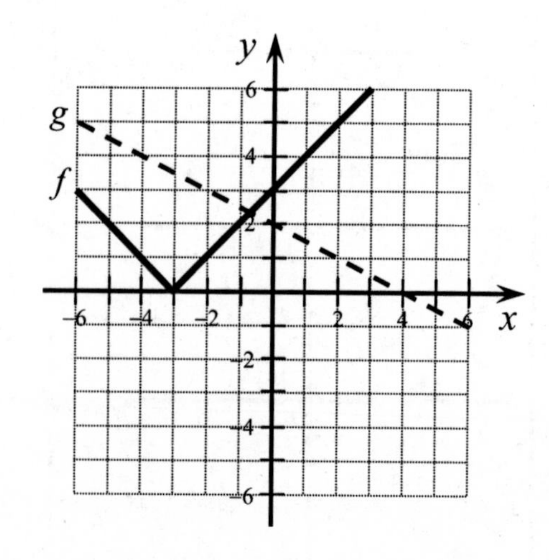

Student Activity A

Section 11.2: Why is 1-1 Important?

1. Below, you see a graph that represents f. Is f a function? _____
2. Is f one-to-one? _____
3. Make a table of solutions for f.
4. Swap the x- and y-coordinates of each solution to make second table.
5. Graph the solutions from the second table on the same set of axes as f.
6. Does the new graph represent a function? _____

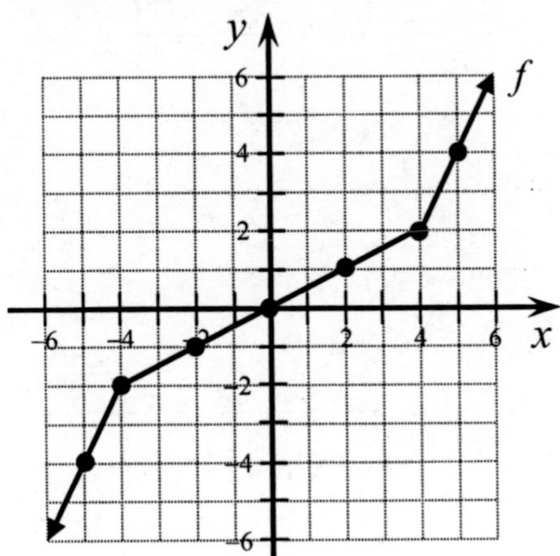

x	$f(x)$
–5	–4
–4	
–2	
0	
2	
4	
5	

x	y
–4	–5

7. Below, you see a graph that represents g. Is g a function? _____
8. Is g one-to-one? _____
9. Make a table of solutions for g.
10. Swap the x- and y-coordinates of each solution to make a second table.
11. Graph the solutions from the second table on the same set of axes as g.
12. Does this new graph represent a function? _____

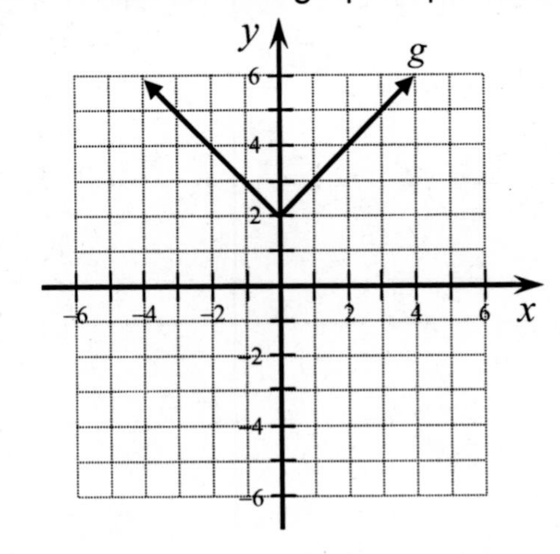

x	$g(x)$
–3	5
–2	
–1	
0	
1	
2	
3	

x	y
5	–3

13. If the original function is one-to-one, what does that mean for the inverse?

14. If the original function is NOT one-to-one, what does that mean for the inverse?

Guided Learning Activity
Section 11.2: The Magic of Inverse Functions

Operation	Inverse Operation
Add 5.	Subtract 5.
Multiply by 3.	
Subtract 7.	
Divide by -2.	
Cube it.	
Take a square root.	

We reverse any process by walking through the steps completely backwards.

f : **Putting on socks & shoes.**	f says: Put on a sock, put on a shoe, tie the shoe.
f^{-1} : **Removing socks & shoes.**	f^{-1} says: Untie the shoe, take off the shoe, take off the sock.

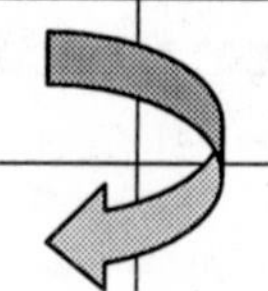

We can write inverse mathematical functions by reversing the function process, in the same way.

$f(x)=2x+3$	f says: Multiply by 2, then add 3.
$f^{-1}(x)=\dfrac{x-3}{2}$	f^{-1} says: Subtract 3, then divide by 2.

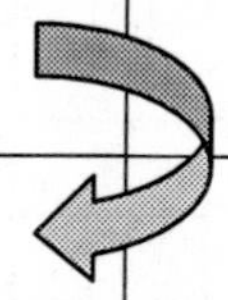

$f(x)=\dfrac{x-5}{3}$	f says:
$f^{-1}(x)=$	f^{-1} says:

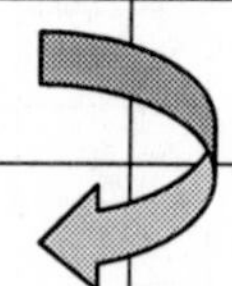

$f(x)=\sqrt[3]{x+4}$	f says:
$f^{-1}(x)=$	f^{-1} says:

$f(x)=4(x-3)$	f says:
$f^{-1}(x)=$	f^{-1} says:

Inverse functions have a special property that $f\left(f^{-1}(x)\right)=x$ and $f^{-1}\left(f(x)\right)=x$.
Why is that? Let's input a "value" into our Socks & Shoes function and see.

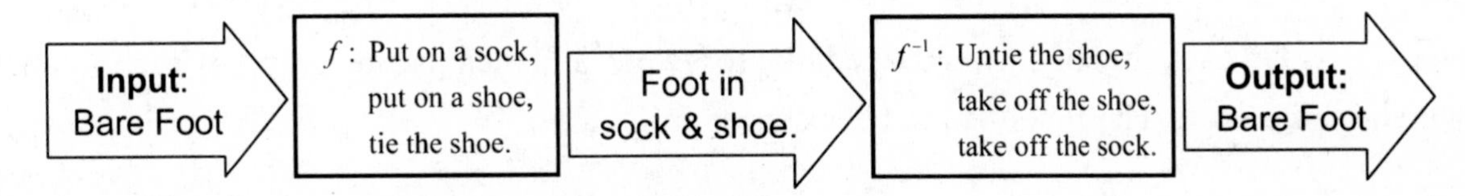

This represents the composition $f^{-1}\left(f(\text{bare foot})\right)=\text{bare foot}$.

Let's try a few mathematical compositions with numerical inputs.

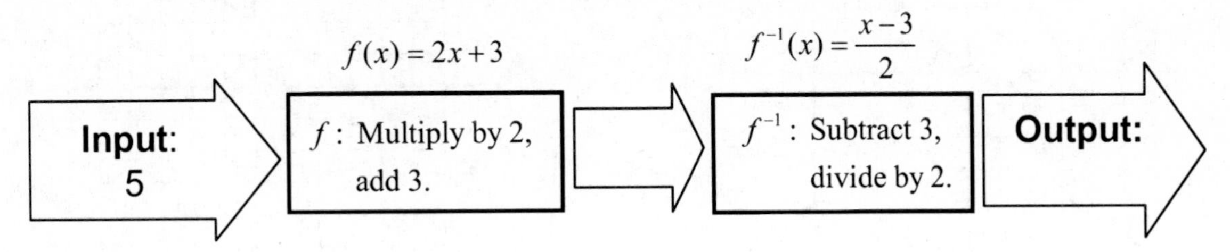

Here we illustrate $f^{-1}\left(f(x)\right)=x$:

Input	Apply the function	Output of function = Input to inverse	Apply the inverse function	Output
4	$f(x)=5x-2$		$f^{-1}(x)=\frac{x+2}{5}$	
3	$f(x)=\sqrt[3]{x+5}$		$f^{-1}(x)=x^3-5$	
8	$f(x)=\frac{x+4}{3}$		$f^{-1}(x)=3x-4$	

And here we illustrate $f\left(f^{-1}(x)\right)=x$:

Input	Apply the inverse function	Output of inverse = Input to function	Apply the function	Output
13	$f^{-1}(x)=\frac{x+2}{5}$		$f(x)=5x-2$	
2	$f^{-1}(x)=x^3-4$		$f(x)=\sqrt[3]{x+4}$	
5	$f^{-1}(x)=3x-4$		$f(x)=\frac{x+4}{3}$	

Student Activity B

Section 11.2: Mirror Images

Directions: Recall that if f and f^{-1} are inverse functions, then they should be symmetric over the line $y = x$. Also, if (a,b) is a point on the graph of f, then (b,a) will be a point on the graph of f^{-1}. On each graph, draw a dashed line for $y = x$ and then use the graph of f to graph f^{-1} on the same axes. The first one has been started for you.

1.

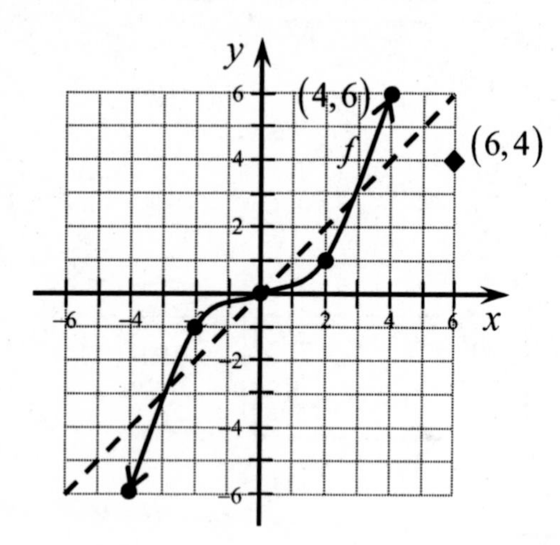

2.

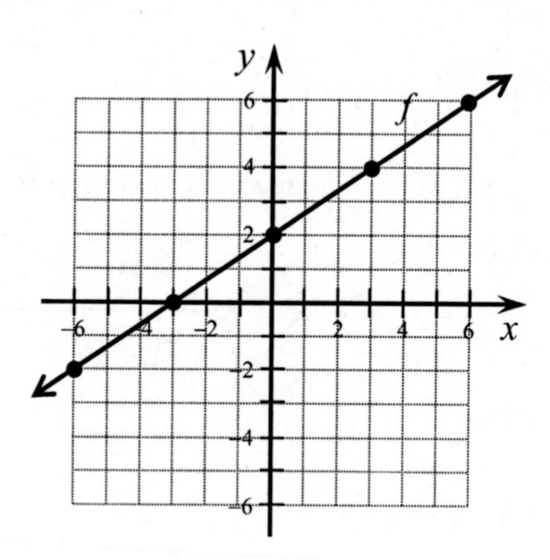

3.

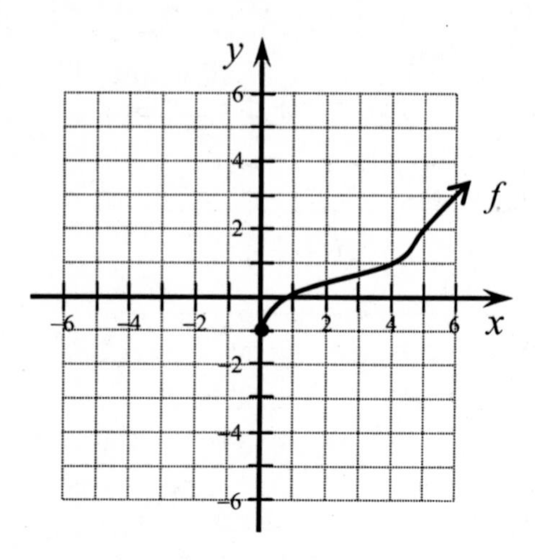

4.

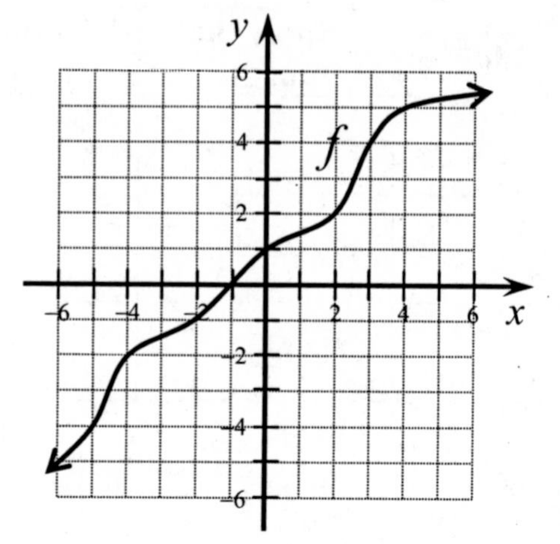

5. Now draw your own one-to-one function, and graph its inverse.

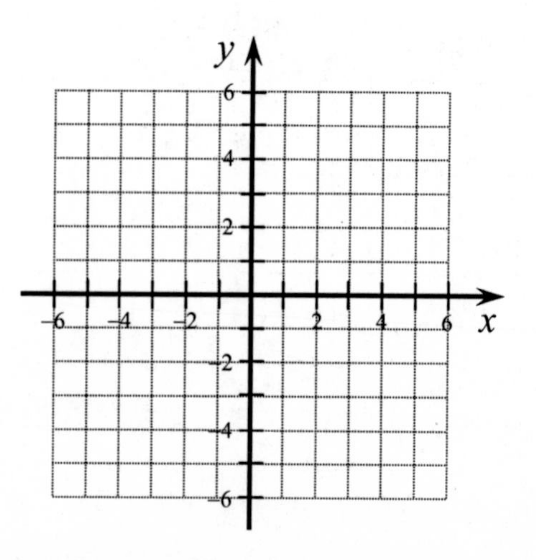

Student Activity C

Section 11.2: Put it All Together

If f and f^{-1} are inverse functions of each other then the following are true:

- The domain of f is the range of f^{-1} and the range of f is the domain of f^{-1}.
- Let (a,b) be a point on the graph of f, then (b,a) is a point on f^{-1}.
- The graphs of f and f^{-1} are symmetric over the line $y = x$.
- $f\left(f^{-1}(x)\right) = x$ and $f^{-1}\left(f(x)\right) = x$

Problem A: Let $f(x) = \frac{3}{2}x + 3$.

1. The point $(2,6)$ is on the graph of f. Without performing any calculations, what point must be on the graph of f^{-1}? (____,____)
2. Find $f^{-1}(x)$. Check this step with someone else before you go on.

3. Graph f and f^{-1} on the axes below. Label which is which.

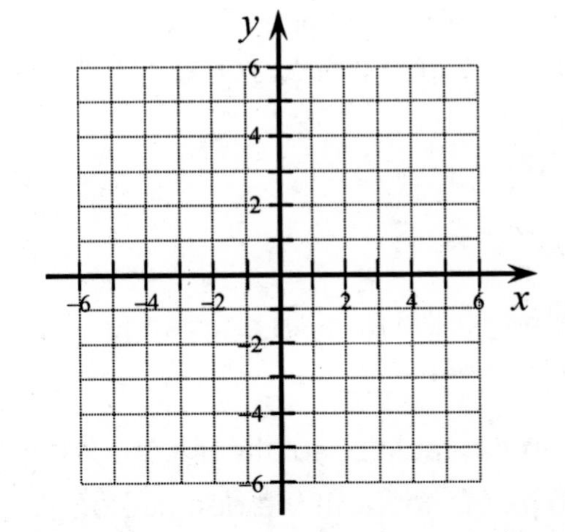

4. Draw the line $y = x$ on the same graph using a dashed line.

5. Fill in the domain and range for each:

   $f(x)$ Domain: ____________

   Range: ____________

   $f^{-1}(x)$ Domain: ____________

   Range: ____________

6. Find and simplify $f\left(f^{-1}(x)\right)$.

7. Find and simplify $f^{-1}\left(f(x)\right)$.

Problem B: Let $g(x) = \sqrt{x+3} + 1$.

1. Graph g on the axes below. Use the table of solutions to assist you. Label the graph with g.

x	$g(x)$
-5	
-4	
-3	
-2	
1	
6	

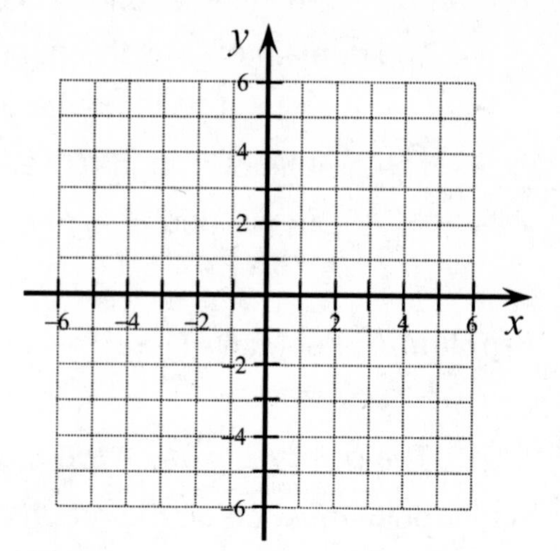

2. What is the domain of g? ________ What is the range? ________

3. Draw the line $y = x$ on the same graph using a dashed line. Draw a reflection of the graph of g over the line of symmetry. Label this g^{-1}.

4. Find the formula for $g^{-1}(x)$ using the formula for $g(x)$. Check this step with someone else before you go on.

5. Based on the domain and range of g, what is the domain of g^{-1}? __________ What is the range? _________

6. Fill in the table of solutions for g^{-1} using the formula you just found **and** the domain for g^{-1}. Then graph this function on the new axes provided.

x	$g^{-1}(x)$

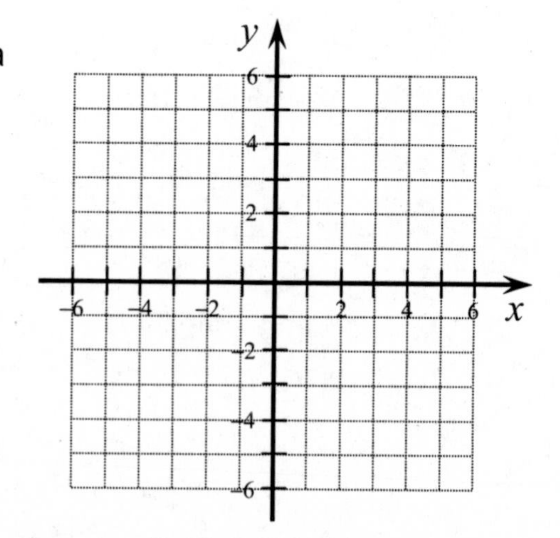

6. Include a condition with the formula for g^{-1} so that $g^{-1}(x)$ gives the true inverse of g. $g^{-1}(x) =$ ________________, $x \geq$ ____.

Student Activity A

Section 11.3: Paint by Exponential Expressions

Directions: For each expression, simplify the expression as much as possible. Shade in the squares containing the answers in the grid below. The first one has been done for you! Hint: It helps to write the problem number in the box that you shade. That way, if you mess up, it's easier to find the mistake.

1. $5^{\sqrt{2}} \cdot 5^{\sqrt{2}}$

$$= 5^{\sqrt{2}+\sqrt{2}} = 5^{2\sqrt{2}}$$

2. $\left(8^{\sqrt{3}}\right)^{\sqrt{3}}$

3. $a^4 \cdot a^{-\sqrt{4}}$

4. $a^{\sqrt{20}} \cdot a^{\sqrt{5}}$

5. $\sqrt{2}^{\sqrt{2}} \cdot \sqrt{2}^{-2\sqrt{2}}$

6. $x^{\sqrt{32}} \div x^{\sqrt{8}}$

7 $\left(a^{\sqrt{5}} b^{\sqrt{3}}\right)^{\sqrt{15}}$

8. $\left(x^{\sqrt{y}}\right)^{\sqrt{16y}}$

9. $x^{\sqrt{y}} \cdot x^{\sqrt{16y}}$

10. $(ab)^{\sqrt{6}} \div ab^{\sqrt{6}}$

11. $\sqrt{9}^{-2\sqrt{3}} \cdot 3^{\sqrt{3}}$

12. $\left(y^{\sqrt{x}}\right)^{\sqrt{x}} \cdot y^{-x}$

13. $\left(4^{\sqrt{9}}\right)^{1/3} \cdot \left(2^{\sqrt{3}}\right)$

ab	$a^{\sqrt{2}}$	$2^{3\sqrt{2}}$	a^9	xy	y^x
$x^{4\sqrt{y}}$	a^2	$2^{\frac{\sqrt{2}}{2}}$	64	512	$2^2 + 2^{\sqrt{3}}$
$a^{5\sqrt{3}}b^{3\sqrt{5}}$	$5^{\sqrt{2}}$	$2^{\frac{3\sqrt{2}}{2}}$	1	$x^{2\sqrt{2}}$	$2^{2+\sqrt{3}}$
$x^{5\sqrt{y}}$	$5^{2\sqrt{2}}$ (shaded)	$a^{\sqrt{6}-1}$	$a^{\sqrt{6}} - 1$	$a^{3\sqrt{5}}$	$a^{5\sqrt{3}}$
$3^{3\sqrt{3}}$	25	x^{4y}	x^{4^y}	$a^{\sqrt{15}}$	2
$3^{5\sqrt{3}}$	$25^{2\sqrt{2}}$	x^{y^4}	a	$\sqrt{2}^{\sqrt{2}}$	$\sqrt{2}$

Guided Learning Activity

Section 11.3: Graphing Exponential Functions

Example 1: Graph $f(x) = 2^x$.

a. Begin by filling in the table of solutions.

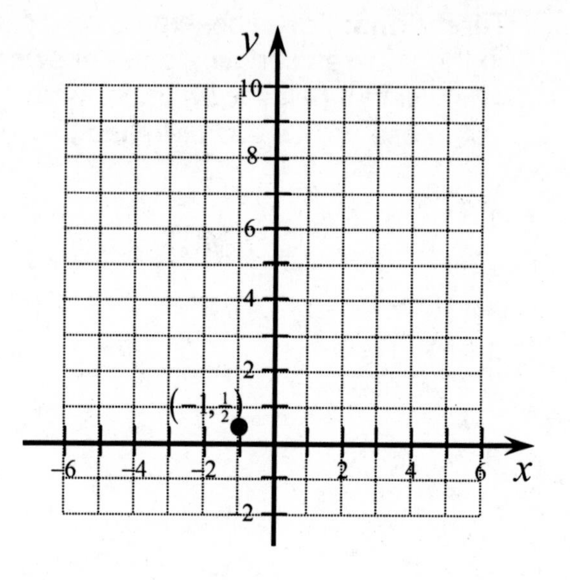

x	$f(x)=2^x$	$g(x)=2^x+2$	$h(x)=2^{x+2}$
-3			
-2			
-1	$2^{-1}=\frac{1}{2}$		
0			
1			
2			
3			

b. Plot the points on the axes provided. Then draw a smooth curve.

c. Notice that as x approaches $-\infty$, the y-values get closer and closer to zero. This is called an asymptote – a line that is approached by the graph. **Draw a dashed line** to represent the asymptote at $y = 0$.

d. What is the domain of $f(x) = 2^x$? __________ The range? __________

e. Fill in the table of solutions for $g(x) = 2^x + 2$, how do the y-values of g change from the y-values for f? ______________________________

f. The graph of $g(x) = 2^x + 2$ will be the graph of f shifted ________ units ________.

g. What is the equation of the asymptote of g? ________

h. What is the domain of g? __________ The range? __________

i. Fill in the table of solutions for $h(x) = 2^{x+2}$, how do the y-values of h change from the y-values for f? ______________________________

j. The graph of $h(x) = 2^{x+2}$ will be the graph of f shifted ________ units ________.

k. What is the equation of the asymptote of h? ________

l. What is the domain of h? __________ The range? __________

Example 2: Graph $f(x) = 3^x$.

a. Begin by filling in the table of solutions.

x	$f(x) = 3^x$	$g(x) = \left(\frac{1}{3}\right)^x$	$h(x) = -3^x$
-3			
-2			
-1			
0			
1			
2			
3			

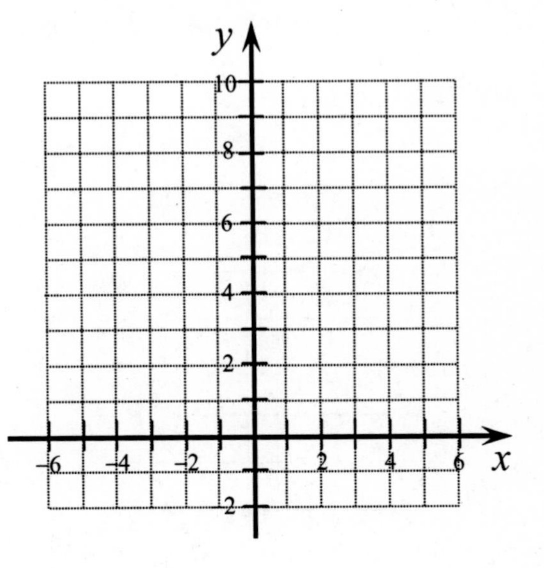

b. Plot the points on the axes provided. Then draw a smooth curve.

c. Again, as x approaches $-\infty$, the y-values get closer and closer to zero. **Draw a dashed line** to represent the asymptote at $y = 0$.

d. What is the domain of f? __________ The range? ____________

e. Fill in the table of solutions for $g(x) = \left(\frac{1}{3}\right)^x$ and graph it. How are the graphs of f and g related? ____

f. What is the domain of g? ___________ The range? ____________

g. Write 3^{-x} without negative exponents: ______ How is this related to the expression in the function g?

h. Fill in the table of solutions for $h(x) = -3^x$ and graph it. How are the graphs of f and h related? ____

i. What is the domain of h? ___________ The range? ____________

Follow Up Questions: Suppose we look at the function $f(x) = b^x$, where $b > 1$.

a. What point has to be a solution of f? ________

b. What is the equation for the asymptote of f? ________

c. What is the domain of f? __________ the range? __________

Student Activity B

Section 11.3: Graphing Exponential Functions with a Calculator

Let's first examine a graph by carefully finding and plotting a table of solutions.

By hand: $f(x)=\left(\frac{3}{2}\right)^x$

x	$f(x)$
-3	
-2	
-1	
0	
1	
2	
3	

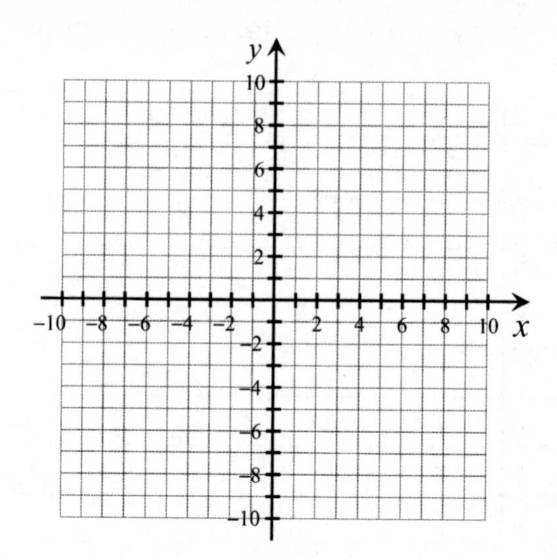

Graph the points and draw a smooth curve to represent the function. Use a dashed line to represent the asymptote.

What is the domain of f? ________ What is the range of f? ________

Is the function increasing or decreasing? ____________

What is the equation of the asymptote? ______

With Calculator: Use your calculator to graph $y=\left(\frac{3}{2}\right)^x$. Your calculator notation might look something like this: y = (3 / 2) ^ x .

- Use the WINDOW to look at a window that is the same size as the graph you drew in the previous problem.
- Use the TRACE feature to verify the values in the table above. For most graphing calculators, you can enter each x-value and press ENTER to see the corresponding y-value.
- Use the TABLE feature to see what happens as x approaches $-\infty$. That is, as we look at x values further and further to the left, what happens to the y-values? This is the kind of behavior you will look for to determine the location of a horizontal asymptote.

1. Use your calculator to graph $f(x) = -10^x$. Make sure to use the negative button (not the minus). No parentheses are necessary to enter this expression.

   a. Set the window so that x is $[-5,5]$ with a scale of 1 and y is $[-1000,1000]$ with a scale of 100.

   b. Use the TRACE feature to fill in the table of values below.

x	$f(x)$
-4	
-1	
0	
$\frac{1}{2}$	
2	

   c. Is the graph of f increasing or decreasing? ____________

   d. Use the TABLE feature to see look for the horizontal asymptote. Hint: You should look at what happens as x approaches ∞ or $-\infty$. What is the equation of the horizontal asymptote? _________

2. Use your calculator to graph $g(x) = 3 - 2^{x+1}$. Make sure to use the minus button between 3 and 2^{x+1}. You will need to put $x+1$ in parentheses.

   a. Set the window so that x is $[-5,5]$ with a scale of 1 and y is $[-20,20]$ with a scale of 2.

   b. Use the TRACE feature to fill in the table of values below.

x	$g(x)$
-3	
-2	
-1	
0	
1	
2	

   c. Is the graph of f increasing or decreasing? ____________

   d. Use the TABLE feature to see look for the horizontal asymptote. Hint: You should look at what happens as x approaches ∞ or $-\infty$. What is the equation of the horizontal asymptote? _________

Student Activity C

Section 11.3: Interesting Compounding

Paula is going to invest $10,000 in an account that pays 6% annual interest for 20 years. Use the compound interest formula to set up and answer each question below. Make sure you don't round decimals until you have reached the final answer.

Interest is compounded ...	$n =$	$nt =$	**Write the expression:** $A = P\left(1+\frac{r}{n}\right)^{nt}$	$A =$
Annually				
Quarterly	4	80	$A = 10,000\left(1+\frac{.06}{4}\right)^{80}$	$32,906.63
Monthly				
Weekly				
Daily				
Hourly				
Four times an hour				

Can Paula just keep making money at a greater rate if the number of compoundings per year were to increase?

Student Activity A

Section 11.4: Discovering e

Now we look at the expression $\left(1+\frac{x}{k}\right)^k$ ... Calculate the value of the expression for each of the blanks in the table. Round each result to the nearest hundredth.

	$x=1$	$x=2$	$x=3$
$k=10$			
$k=50$		$\left(1+\frac{2}{50}\right)^{50} \approx 7.11$	
$k=100$			
$k=500$			
$k=1{,}000$			
Now evaluate these expressions.	$(2.7183)^1 \approx$	$(2.7183)^2 \approx$	$(2.7183)^3 \approx$

So, as $k \to \infty$, the expression $\left(1+\frac{x}{k}\right)^k$ equals what? ______

Student Activity B

Section 11.4: Exponential Workout

Directions: For each expression, simplify and then find and shade the box with its equivalent form in the grid.

1. $\left(e^{\sqrt{2}}\right)^{\sqrt{2}}$

2. $\dfrac{2}{\sqrt{2}}$

3. $\dfrac{10^{3\sqrt{4}}}{10^{2\sqrt{4}}}$

4. e^{-1}

5. $\dfrac{6^7}{6^3 \cdot 36}$

6. $\left(\dfrac{1}{3}\right)^{-1}$

7. $\left(e^2\right)^{3/2}$

8. $\left(\sqrt{13}\right)^4$

9. $7^{36}7^{-34}$

10. $3460\left(\dfrac{1}{50}\right)^2$

11. $\dfrac{1}{\left(|-2|\right)^{-99}}$

12. $1-\dfrac{e^0}{9}$

13. $\left(\dfrac{1}{e}\right)^{-1}$

14. $\left(5^{2\sqrt{5}}\right)^3$

15. $\left(9^4\right)^{1/2}$

16. $0.001\left(10^3\right)$

17. $10^x \cdot 10^{3x}$

18. $(4e)^{1/2}$

19. $\dfrac{6^{\sqrt{3}+2\sqrt{2}}}{6^{\sqrt{8}}}$

20. $2(256)^{-1}2^8$

21. $3^x - \dfrac{1}{3^{-x}}$

22. $\dfrac{e^2}{e^{-3}}$

<table>
<tr><td>1</td><td>e</td><td>$2\sqrt{e}$</td><td>$2^{2\sqrt{2}}$</td><td>4^{4x}</td><td rowspan="2">27^{12}</td><td>14</td><td>99</td><td colspan="2">36</td><td>2</td></tr>
<tr><td rowspan="2">3</td><td>-2</td><td colspan="2">$9^{\sqrt{2}}$</td><td>e^2</td><td>49</td><td rowspan="3">$\frac{2}{3}$</td><td>$6^{\sqrt{3}}$</td><td>34</td><td>e^3</td></tr>
<tr><td>$5^{6\sqrt{5}}$</td><td>0</td><td rowspan="2">$\frac{1}{2}$</td><td>5</td><td>$\frac{8}{9}$</td><td>$\frac{1}{4}$</td><td>81</td><td colspan="2">1.384</td></tr>
<tr><td>2^{99}</td><td colspan="2">10</td><td>$\frac{1}{e}$</td><td>$\frac{6^4}{36}$</td><td>10^{4x}</td><td>169</td><td>3^{2-x}</td><td rowspan="2">9</td></tr>
<tr><td>$\sqrt{2}$</td><td colspan="2">100</td><td>8^x</td><td>9</td><td colspan="2">$-\frac{1}{3}$</td><td>-4</td><td>e^5</td><td>$2^{2\sqrt{2}}$</td></tr>
</table>

Student Activity C

Section 11.4: eGraphing

In this activity we will explore where the function $y = e^x$ fits in with the family of exponential curves we already know.

1. Graph $f(x) = 2^x$ and $g(x) = 3^x$ on the axes below.

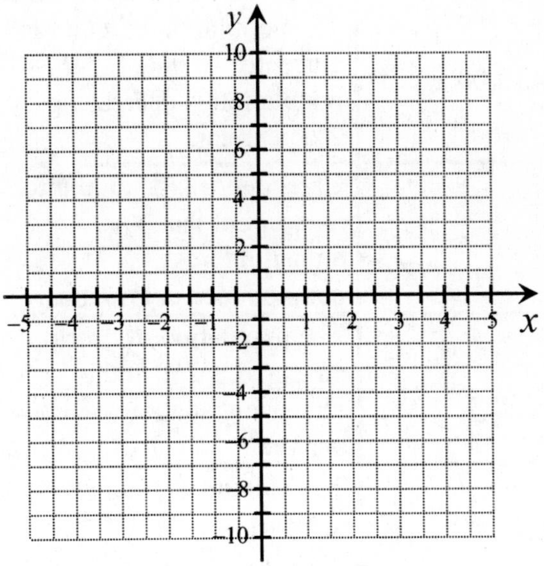

2. Fill in the table of solutions.

x	$h(x) = e^x$
−2	
−1	≈ 0.37
0	
1	
2	

3. Plot the solutions from the table on the same graph as the functions in the first problem. Draw a smooth curve to represent $h(x) = e^x$.

4. What is the numerical approximation for e? ________

5. Why is the resulting graph not surprising?

Student Activity D

Section 11.4: Modeling with Exponential Equations

Directions: Fill in the missing information in the table below. The first one has been done for you.

Exponential model	Description	Growth Rate (r)	Initial Amount (P)
$A = 10{,}000e^{-0.02t}$	The population of a city is currently 10,000, but the population is decreasing by 2% per year.	−0.02 (decreasing)	10,000
	Chris deposits $5,000 into an account that pays 7.5% annual interest, compounded continuously.		
	The population of deer on an island is 250 and is increasing at a rate of 4% per year.		
	The amount of radioactivity in a soil sample is R_0, and the radioactivity decreases at a rate of 18% per year.		

Set up an equation to solve each of the problems below. Do not solve them.

Equation	Description	Define the Missing Variable
	Jill's savings account just reached $10,000. The account pays 7.2% interest, compounded continuously. Jill made one deposit into the account 8 years ago. What was the initial deposit?	P = the initial deposit
	The population of Tigerlily is 4,234. The city council projects a decline of 5% annually due to poor economic conditions. In how many years will the population of Tigerlily be 4,000?	
	The number of bacteria in a petri dish is 500 at the beginning of an experiment. At the end of 1.5 hours, the population has doubled. What is the hourly growth rate?	

Assessment 11B

Chapter 11: Mid-chapter Assessment for Understanding

For each of the following, describe the type of problem and the strategies and key steps to remember while doing the problem. You do **not** have to complete the problems.

	Type of Problem	Strategies and Key Steps
1. Simplify: $\left(2^{\sqrt{3}}\right)^{2\sqrt{3}}$		
2. Let $f(x)=\sqrt{x+1}$ and $g(x)=x+2$. What is the domain of $(g/f)(x)$?		
3. A deposit of \$5,000 is left in an account that pays 10.5% annual interest, compounded continuously. How much will be in the account after 8 years?		
4. Let $f(x)=x-4$ and $g(x)=3x+8$. Find $(f\circ g)(5)$.		
5. Graph: $f(x)=5^x$		
6. Let $f(x)=x^2-5$ and $x\geq 0$. Find $f^{-1}(x)$.		
7. Is the function $f(x)=\lvert x-1\rvert$ one-to-one?		
8. An account has \$6,000 in it after 5 years. The interest rate on the account is 2.4% compounded monthly. How much was initially deposited?		
9. The initial population of Dogstar is 450, and it is increasing at an annual rate of 8%. What will the population be in 5 years?		
10. Graph $f(x)=e^x-5$ and write the equation for the asymptote.		

Student Activity A

Section 11.5: Why Logarithms?

Directions: Solve each of the equations. For each equation, note the operation that is performed on x and what the inverse of that operation is. There may be problems that you have not been taught how to solve. If you see one, say so.

<table>
<tr><td rowspan="2">1. Solve: $x+3=12$
$x+3-3=12-3$
$x=9$</td><td>Operation on x: Addition of 3</td></tr>
<tr><td>Inverse Operation on x: Subtraction of 3</td></tr>
<tr><td rowspan="2">2. Solve: $3x=12$</td><td>Operation on x:</td></tr>
<tr><td>Inverse Operation on x:</td></tr>
<tr><td rowspan="2">3. Solve: $x^3=12$</td><td>Operation on x:</td></tr>
<tr><td>Inverse Operation on x:</td></tr>
<tr><td rowspan="2">4. Solve: $3^x=12$</td><td>Operation on x:</td></tr>
<tr><td>Inverse Operation on x:</td></tr>
<tr><td rowspan="2">5. Solve: $\sqrt{x}=12$</td><td>Operation on x:</td></tr>
<tr><td>Inverse Operation on x:</td></tr>
<tr><td rowspan="2">6. Solve: $\frac{x}{2}=12$</td><td>Operation on x:</td></tr>
<tr><td>Inverse Operation on x:</td></tr>
<tr><td rowspan="2">7. Solve: $x^{3/4}=12$</td><td>Operation on x:</td></tr>
<tr><td>Inverse Operation on x:</td></tr>
<tr><td rowspan="2">8. Solve: $2^x=12$</td><td>Operation on x:</td></tr>
<tr><td>Inverse Operation on x:</td></tr>
</table>

Which problems are we stuck on? Why?

Student Activity B

Section 11.5: Paint by Equivalent Equations

Directions: Write each logarithmic equation as its corresponding exponential equation. Write each exponential equation as its corresponding logarithmic equation. Shade in your answers in the grid to reveal the picture. The first one has been done for you.

1. $5^3 = 125$
$\log_5 125 = 3$

2. $2^x = 8$

3. $\log x = 2$

4. $\log_4 64 = 3$

5. $10^x = \frac{1}{100}$

6. $16^x = 2$

7. $\log_3 x = -1$

8. $\log 1 = 0$

9. $4^{-x} = 2$

10. $\left(\frac{1}{2}\right)^x = 4$

11. $x^2 = 9$

12. $\log 1000 = x$

13. $\log_x 25 = 2$

14. $x^0 = 1$

15. $\left(\frac{1}{9}\right)^x = 3$

$\log_x 9 = 2$	$\log_2 9 = x$	$\log_x 3 = \frac{1}{9}$	$\log_5 125 = 3$	$\log_1 x = 0$	$\log \frac{1}{100} = x$
$10^2 = x$	$2^{10} = x$	$x = \log_8 2$	$\log_2 4 = -x$	$\log_{1/2} 4 = x$	$\log x = \frac{1}{100}$
$\log_{1/9} 3 = x$	$\log_x 1 = 0$	$\log_2 8 = x$	$\log_4 2 = -x$	$\log_4 x = 4$	$10^0 = 1$
$3^{-1} = x$	$x^{-1} = 3$	$4^3 = 64$	$3^4 = 81$	$\log_4 \frac{1}{2} = x$	$\log_3 x = \frac{1}{9}$
$10^x = 1000$	$x^2 = 25$	$\log_{16} 2 = x$	$\log_2 16 = x$	$25^x = 2$	$2^x = 25$

Guided Learning Activity

Section 11.5: Unraveling Exponential Functions

Let $f(x) = b^x$, where $b > 0$ and $b \neq 1$. Since f is a one-to-one function, we define notation for the inverse: $f^{-1}(x) = \log_b(x)$.

By the definition of inverses, $f\left(f^{-1}(x)\right) = x$, and so $b^{\log_b(x)} = x$.

Also by the definition of inverses, $f^{-1}\left(f(x)\right) = x$, and so $\log_b\left(b^x\right) = x$.

Part I: Write the corresponding function, inverse, or property to fill in the table below:

Function	Inverse	$b^{\log_b(x)} = x$	$\log_b\left(b^x\right) = x$
$f(x) = 3^x$	$f^{-1}(x) = \log_3(x)$	$3^{\log_3(x)} = x$	$\log_3\left(3^x\right) = x$
$f(x) = 5^x$			
		$10^{\log_{10}(x)} = x$	
	$f^{-1}(x) = \log_2(x)$		
			$\log_e\left(e^x\right) = x$
$f(x) = (1/2)^x$			

Part II: Circle the expressions below that simplify to be x:

$4^{\log_2(x)}$ $\quad 4^{\log_4(x)}$ $\quad 4\log_4(x)$ $\quad 4^{\log(4x)}$ $\quad \log_4(4x)$ $\quad \log_4\left(4^x\right)$

$\log_6\left(6^{2x}\right)$ $\quad \log_6(6+x)$ $\quad \log_6(6)$ $\quad \log_6\left(x^6\right)$ $\quad 6^{\log_6(x)}$ $\quad 6^{\log_x(6)}$

Just like we would never write 2^x as ${}^x2$, we also would never write $\log_2(x)$ as $(x)\log_2$.

$2^{(\ )}$ must operate on a number that follows the 2.

$\log_2(\ )$ must operate on a number that follows the $\log_2$.

Part III: Circle the expressions with correct mathematical notation below:

3^x $\quad x^3$ $\quad {}^3x$ $\quad {}^x3$ $\quad x\log_3$ $\quad \log_3 x$ $\quad \log_x 3$ $\quad 3\log_x$ $\quad 3^{\log_x}$

Sometimes, we can simplify expressions involving logs or exponentials by rewriting part of the expression. For example, $\log_9 81$ can be rewritten because 81 is the same as 9^2.

$$\log_9 81 = \log_9(81) = \log_9\left(9^2\right) = 2$$

Part IV: See if you can finish these off by being clever:

$\log_2 8 = \log_2(\quad) =$

$\log_4 16 = \log_4(\quad) =$

$\log_5\left(\frac{1}{5}\right) = \log_5(\quad) =$

$\log_3 \sqrt{3} = \log_3(\quad) =$

$\log_2\left(\frac{1}{4}\right) = \log_2(\quad) =$

> We have defined the inverse of the operation $b^{(\,)}$ (b to the ...) to be $\log_b(\,)$ (log base b of ...).

As with all the properties of equality, if we perform an operation on one side of an equation, we must also perform the same operation on the other side of the equation.

Example 1: To solve $2^x = 12$, we need to undo the operation $2^{(\,)}$.
So we need to use $\log_2(\,)$ on both sides:

$$2^x = 12$$
$$\log_2(2^x) = \log_2(12)$$
$$x = \log_2(12)$$

> Using your calculator, you can evaluate any logarithm using the LOG key and the **Change of Base Formula:** $\log_b(x) = \frac{\log(x)}{\log(b)}$

When the logarithm is written without a base it is called a **common log** and stands for $\log_{10}(\,)$.

So, to evaluate $\log_2(12)$, we calculate $\frac{\log(12)}{\log(2)}$ and get ≈ 3.585.

We can check this answer by finding $2^{3.585}$, which is $\approx 12.0003 \approx 12$.

Example 2: To solve $\log_3 x = 2$, we need to undo the operation $\log_3(\,)$.

So we need to use $3^{(\,)}$ on both sides:

$$\log_3 x = 2$$
$$3^{\log_3 x} = 3^2$$
$$x = 9$$

Part V: Decide which operation to use on both sides of the equation in order to solve for x. Then write out the step of applying the inverse operation. The first two have been done for you.

	Solve	Perform this Inverse Operation on both sides of the equation	First Step	Solve for x
a.	$3^x = 12$	$\log_3(\ )$	$\log_3(3^x) = \log_3(12)$	$x = \dfrac{\log 12}{\log 3} \approx 2.262$
b.	$\log_8 x = -1$	$8^{(\ )}$	$8^{\log_8 x} = 8^{-1}$	$x = \dfrac{1}{8}$
c.	$\log_3 x = 5$			
d.	$5^x = 9$			
e.	$(1/2)^x = 8$			
f.	$\log_6 x = \dfrac{1}{2}$			
g.	$0.432 = \log_{10} x$			
h.	$10 = 4^x$			
i.	$10^x = 140$			

Student Activity C

Section 11.5: Match Up on Log and Exponential Expressions

Match-up: Match each of the expressions in the squares in the table below with its simplified form from the top. If the simplified form is not found among the choices A through D, then choose E (none of these).

A 2 **B** $\frac{1}{2}$ **C** −2 **D** 1 **E** 0 **F** None of these

$\log_5 \sqrt{5}$ $\log_5 5^{1/2} = \frac{1}{2}$ B	$2^{\log_2 3}$	$\log \frac{1}{100}$	$10^{\log 10}$
$2^{\log 1}$	$\log_9 3$	$\log_3 9$	$10^{\log 2}$
$10^{\log_2 10}$	$\log_{1/2}\left(\frac{1}{4}\right)$	$\log 1$	$\log_4 2$
$\log_2 1$	$\log_4 16$	$\log_5 5$	$2^{\log_5 5}$
$(\log_5 2)^0$	$5^{\log_5 2} - \log_2 16$	$\log 100$	$\log_9 \frac{1}{81}$

Student Activity D

Section 11.5: Log Graphing

1. Graph $f(x)=2^x$ on the axes below using the table of values provided. Use a dashed line to represent the asymptote of the graph of f.

x	$f(x)=2^x$
−3	0.125
−2	0.25
−1	0.5
0	1
1	2
2	4
3	8

x	$f^{-1}(x)=\log_2 x$

2. The inverse function is written $f^{-1}(x)=\log_2 x$. We can make a table of solutions for f^{-1} by swapping the x- and y-coordinates of all the solutions. Fill in the table of solutions for $f^{-1}(x)=\log_2 x$.

3. Plot the points on the same graph as $f(x)=2^x$. Then draw a smooth curve.

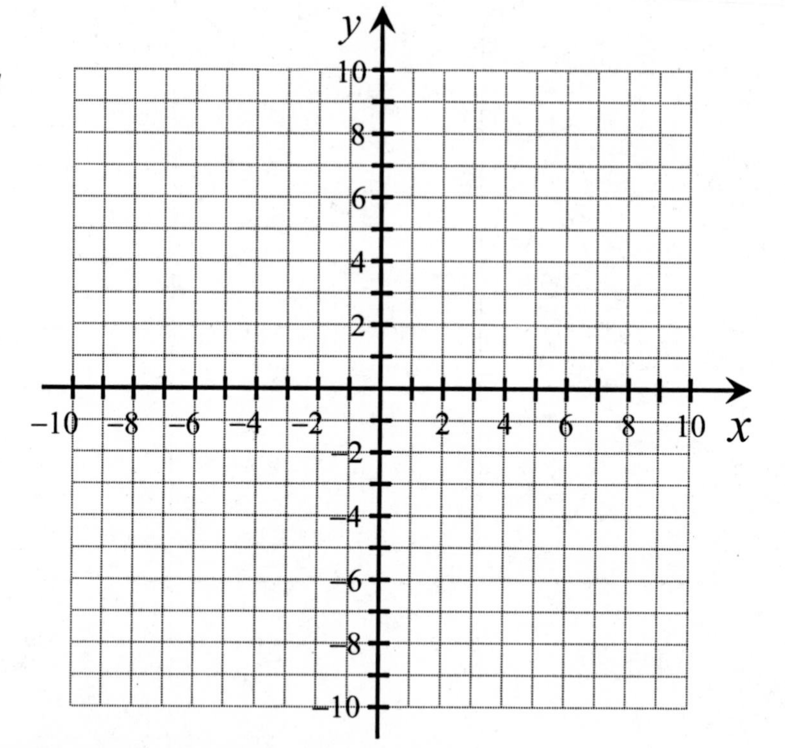

4. What is the equation for the asymptote of $f(x)$? ________

for $f^{-1}(x)$? ________

5. Write the domain and range for each function:

f: Domain ________

Range ________

f^{-1}: Domain ________

Range ________

6. If it is always true that the point $(0,1)$ is on the graph of $y=b^x$, then what point should always be on the graph of $y=\log_b x$? ________

7. If you wanted to graph $y=\log_5 x$, what would be an easy way to do it?

Student Activity A

Section 11.6: Shifty Logarithms

Complicated-looking variants of the $\ln(x)$ and e^x functions are often just simple *translations* of the function: that is, the original function shifted vertically or horizontally along the *y* or *x* axis, respectively.

Graph each set of functions below on the grid provided, and describe the translation(s).

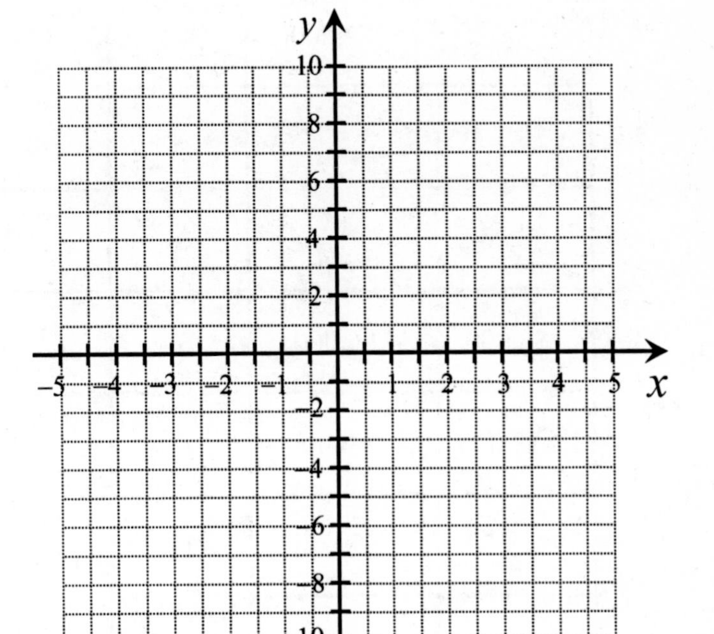

1. 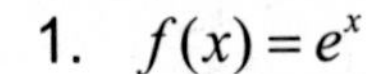 $f(x)=e^x$

2. $g(x)=e^x+2$

   Graph (2) is the same as $f(x)$

   shifted ______ 2 units.

3. 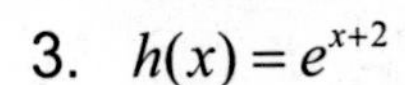 $h(x)=e^{x+2}$

   Graph (3) is the same as $f(x)$

   shifted ______ 2 units.

4. 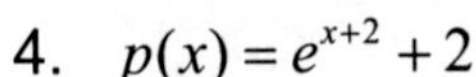 $p(x)=e^{x+2}+2$

   Graph (4) is the same as $f(x)$

   shifted ______ and ______ 2

   units.

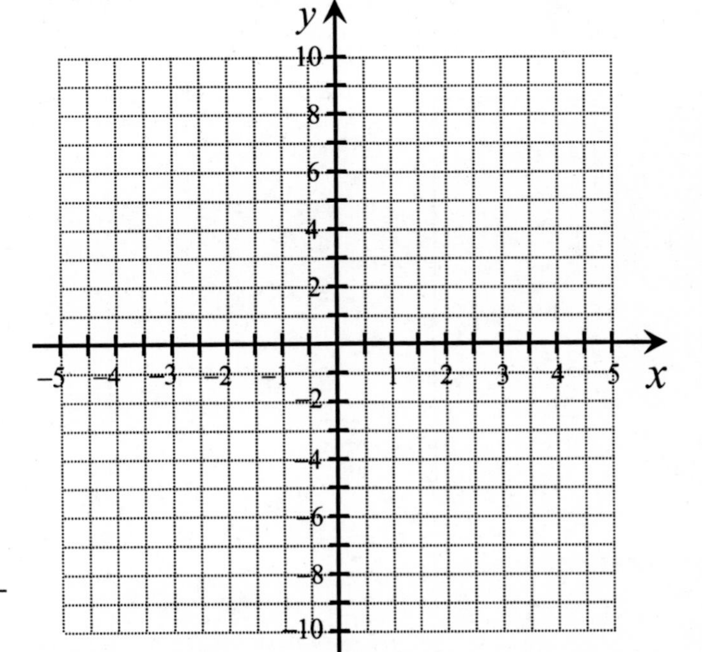

5. $f(x)=\ln(x)$

6. $g(x)=\ln(x-1)$

   $g(x)$ is the same as $f(x)$

   shifted _____ 1 unit.

7. $h(x)=3+\ln(x)$

   $h(x)$ is the same as $f(x)$

   shifted _____ 3 units

8. $p(x)=3+\ln(x+1)$

   $p(x)$ is the same as $f(x)$

   shifted up ___ units and left ___

   units.

Student Activity B

Section 11.6: Compounding the Problem

Directions: First set up an equation to help you solve for the missing information in the table below. Then solve each equation for the missing variable using the space provided at the bottom of the page. If k is infinity, use the formula for continuously compounded interest.

	A	P	k	r	t	**Equation**
1.		\$2,000	12	8%	15	
2.	\$9,500		4	5.9%	8	
3.	\$6,000		∞	6.4%	5	
4.	\$8,000	\$6,000	∞	8.2%		
5.	\$12,000	\$10,000	4		6	
6.	\$1,250	\$1,000	12	7.15%		
7.	$2P_0$	P_0	∞		10	
8.	$3A_0$	A_0	52	10.4%		

Student Activity C

Section 11.6: Persnickety Log Notation

Directions: Fill in the missing pieces to demonstrate your prowess in using logarithm language and notation!

	In words	Abbreviated Notation ($\ln x$ or $\log x$)	Notation with the base indicated
1.	Natural log of x	$\ln x$	$\log_e x$
2.	Log of the sum of 4 and x		
3.		$\ln\left(\frac{x}{y}\right)$	
4.			$\log_e(x-2)$
5.	10 to the log of 4		
6.		$\ln e^{-3}$	
7.			$e^{\log_{10} x}$
8.		$\log(5x)$	
9.	2 to the negative log of x		
10.	The quantity 10 to the natural log of x, squared.		

Student Activity D

Section 11.6: Interesting Comparison

1. Jake invests \$5,000 in Trustworthy Bank, which pays an annual interest rate of 5%, compounded **quarterly**, find the amount in his bank account **every ten years** by completing the table below:

$k =$

t	kt	Write the equation for the interest calculation:	A
10			
20	80	$A = 5{,}000\left(1 + \frac{0.05}{4}\right)^{80}$	\$13,507.42
30			
40			

2. Jake's friend Bob tells him that his bank, Dependable & Reliable Trust, pays interest compounded continuously. Jake is skeptical but decides to make the calculations to compare the difference between the two. Complete the table below for the same investment at the same interest rate, only this time compounding continuously. Should Jake switch?

t	kt	Write the equation for the interest calculation:	A
10			
20			
30			
40			

3. If Jake invests at Dependable & Reliable Trust, how long will it take for his investment to double?

Student Activity A

Section 11.7: Simply Logarithmic

Directions: Simplify each expression and shade in the equivalent simplified expression. **Solve each equation** and shade the solution. Only shade one square for each problem. The first one has been done for you.

1. $\log_3 3^2 = 2$
2. $6^{\log_6 4}$
3. $\ln e^9$
4. $2^{\log_2 2x} = 5$
5. $1 + \log 10 + 1$
6. $10^{\log 16x} = 16$
7. $\log_x 7 = 1$
8. $\log(\log 10)$
9. $\frac{1}{2}\ln e^{4x}$
10. $e^{\ln e}$
11. $5\log_8 8$
12. $\log_9 \frac{1}{729}$
13. $3\log_3 9$
14. $4x + \ln e^{-x}$
15. $\ln 1 + \log_3 \frac{1}{9} \log_3 \frac{6}{2}$
16. $\log_2 8^{y/3}$
17. $10^{\log e^2}$
18. $3 = \log_x 1000$
19. $(\log_3 9)^{-1}$

<table>
<tr><td rowspan="2">e</td><td>-4</td><td rowspan="2">$1+\log 11$</td><td>112</td><td>$\log_3 4$</td><td>100</td><td colspan="2">$e-2$</td><td colspan="2">5</td><td>1</td></tr>
<tr><td>11</td><td>8</td><td rowspan="2">$\frac{1}{2}$</td><td>3</td><td>2</td><td>-4</td><td>y</td><td>$\frac{7}{3}$</td><td>e^6</td></tr>
<tr><td>9</td><td>-8</td><td>$\frac{1}{2}$</td><td>-9</td><td>-10</td><td>1</td><td>-99</td><td>0</td><td>4</td><td>-3</td></tr>
<tr><td colspan="2">$2x$</td><td>6</td><td>$e^{1/6}$</td><td>7</td><td colspan="2">10</td><td colspan="2">$\ln 9$</td><td>$\frac{2}{5}$</td><td rowspan="2">$\frac{5}{2}$</td></tr>
<tr><td>100</td><td>$x-1$</td><td colspan="3">$\frac{1}{5}$</td><td>8</td><td>e^{2x}</td><td>12</td><td>$3x$</td><td>-2</td></tr>
</table>

Guided Learning Activity

Section 11.7: Become a Believer

Directions: Follow the directions for each step and fill in the boxes with the missing pieces as you go. Careful – as you move from line to line, the boxes might contain different expressions! At the end, you should get a formula that you might recognize!

Derive the Product Rule for Logarithms:

Let $M = b^x$ and $N = b^y$.

$MN = \square \cdot \square$ Solve $M = b^x$ for x. Solve $N = b^y$ for y.

$MN = b^{\square}$ $x = \square$ $y = \square$

Now we take $\log_b(\ )$ on both sides of the equation.

$$\log_b(MN) = \log_b(\square)$$

Simplify on the right side:

$$\log_b(MN) = \square$$

Replace x and y with the equivalent expressions found above:

$$\log_b(MN) = \square + \square$$

Derive the Quotient Rule for Logarithms:

Let $M = b^x$ and $N = b^y$. From above, we know: $x = \square$ and $y = \square$.

$$\frac{M}{N} = \frac{\square}{\square} \rightarrow \frac{M}{N} = b^{\square}$$

Now we take $\log_b(\ )$ on both sides of the equation.

$$\log_b\left(\frac{M}{N}\right) = \log_b(\square)$$

Simplify on the right side:

$$\log_b\left(\frac{M}{N}\right) = \square$$

Replace x and y with the equivalent expressions found above:

$$\log_b\left(\frac{M}{N}\right) = \square - \square$$

Student Activity B

Section 11.7: Pesky Product and Quotient Rules

A **product inside** the logarithm = A **sum outside** the logarithms.

$$\log(MN) = \log M + \log N \qquad \ln(MN) = \ln M + \ln N$$

A **quotient inside** the logarithm = A **difference outside** the logarithms.

$$\log\left(\frac{M}{N}\right) = \log M - \log N \qquad \ln\left(\frac{M}{N}\right) = \ln M - \ln N$$

Step 1: Use the product and/or quotient rules for logarithms to write each logarithm in expanded form, then simplify if possible. Write your answer in the second shaded column, and be careful not to leave exponents or radical symbols in your answer.

Step 2: Substitute $x = 2$ and $y = 3$ in both expressions, and use your calculator to verify that you have simplified correctly.

The first problem has been done for you.

Original Expression	Simplified Expression	Substitute $x = 2$ and $y = 3$ for original expression	Substitute $x = 2$ and $y = 3$ for simplified expression	Match?
$\log 9x$	$\log 9 + \log x$	$\log 9 \cdot 2 \approx 1.26$	$\log 9 + \log 2 \approx 1.26$	✓
$\log \frac{3x}{4}$				
$\ln xy^3$				
$\log \frac{10}{xy}$				
$\ln \frac{10^3}{2x}$				
$\log \sqrt{xy}$				

Student Activity C

Section 11.7: Escape the Properties Matrix

Directions: Begin at the box marked START. Without a calculator, look for pairs of expressions that are equivalent. Shade the matched pairs to reveal the path to escape the matrix! If you get really stuck, use your calculator to check the equivalence of the pairs you have already shaded.

Example: $\ln 5 + \ln 7 = \ln(5 \cdot 7) = \ln 35$

START $\ln 5 + \ln 7$	$\ln 35$	$\ln \frac{1}{4}$	$\ln 2 + \ln 8$	$\frac{\log 3}{\log 5}$
$\ln 12$	$\log \frac{5}{20}$	$\log 5 - \log 20$	$\ln 2^8$	$\log_3 5$
$\ln 3 + \ln 9$	$\frac{\log 5}{\log 20}$	$\log(5-20)$	$8\ln 2$	$5\ln 20$
$\log e^3$	$\log 3^2$	$\ln 1$	$\log 1$	$\ln 5^{20}$
3	$2 + \log 3$	$\log 100 + \log 3$	$\log 103$	$e^{\ln 4}$
$\ln 2$	$\ln 5 - \ln 10$	$-\ln 5$	$\log e$	$\log e^4$
$\log 50 - \log 5$	$\ln \frac{1}{2}$	$\log 2$	$\ln e$	$\ln \frac{1}{5}$
$\log 45$	$\log 4 + \log 4$	$2\log 4$	$\log 10$	$-\ln 5$
$\log 4 + \log 5$	$\log 4^4$	$\ln 3 + \ln 12$	$\ln 15$	**ESCAPE** the Properties Matrix

Student Activity A

Section 11.8: Escape the Logarithm Matrix

Directions: Begin at the box marked START. Either solve the equation or simplify the expression. Look for the result in the movement grid to the right. If your result is not in the movement grid, that's a BAD sign. For example, the answer to $3^x = 7$ is $x \approx 1.7712$, so we move to the right.
Next solve $\log(3-x) = 3$.

Left	Right	Down	
0.0025	~~1.7712~~	3.4190	1.0986
23.1331	10	0.7906	1.0986
	0.4474	2	8
	2.4022	−997	−2
	0.2091	4.3774	

START $3^x = 7$ →	$\log(3-x) = 3$	$e^0 + e^x$	
$\log_7(2x-3) = \log_7(x)$	$4^{x-2} = 27$	$\log_5 25 = x$	$4^x = 16$
$e^{-2.4t} = 14.2$	$e^{2.8x} = 3.5$	$\log_2(x-7) + \log_2 x = 3$	$\log 64 - \log 4x = 16$
	$\log 8 - \log x = 2$	$2^{x+2} = 3^x$	$7^x - 4$
$3^{x^2-3x+4} = 81$	$2^x = 16$	$\log 3x = \log 6$	
$\log 5 - \log 2x = 0.5$	$5^x = 7^{x-4}$	$\log \frac{2x}{5} = -3$	$3\log(y+3) = \log(y+4) - \log 12$
$e^{2x} = 9$	$\ln x = 5$	$\log_2 5x - \log_2 3 = 4$	$e^{3x} = 60$
$3^x - 11 = 3$	$5^{x+1} = 7$	$\log x^{900} = 900$	$6^{x+4} = 36$
$\frac{1}{3}\log(3x+5) = \log x$	$\ln 5x = 4$		$\ln e^{2x} = \ln 9$
$3^{9y-3} = 1$	$z + \ln e^{-z}$	$-2 + \log_4 z$	**ESCAPE** the logarithm matrix

Student Activity B

Section 11.8: Popular Applications

Population growth is a process that we can describe using the exponential growth model $P = P_0 e^{kt}$, where P is the population at some time t, P_0 is the initial population at $t = 0$, and *k* depends on the rate of growth.

Assume that a city San Populato had an initial population of 100, and that the city doubled its population in the first 5 years.

1. First, use this information to find *k (round to four decimal places).*

$P =$

$P_0 =$

$t =$

$k \approx$ ___________

2. Write the population growth model that gives the population at any given time.

3. Use the model to calculate the population of San Populato after 30 years.

4. Calculate the time needed for San Populato's population to grow to 1000 people.

5. Put it all together: In another remote part of the Country, the poor city of Lesgo is struggling to survive. A series of natural disasters is causing a steady flow of disgruntled citizens to leave the city. In fact, the situation is so bad that Lesgo lost half of its population in one year. If evacuation continues at its current negative exponential rate, and the city's initial population was about 1000 persons, approximately how much longer does the city have before only one lonely person is left?

Student Activity C

Section 11.8: Decaying Applications

Living organisms consume a variety of carbon isotopes in their supply of nutrients, including ^{14}C. The percentage of carbon that is ^{14}C in living organisms remains constant because it is always being renewed. However, when a living organism dies, it no longer consumes fresh carbon. At this point, the percentage of carbon that is made up of the isotope ^{14}C begins to decrease as the ^{14}C in the organism decays. By measuring the amount of Carbon-14 present in a dead organism, and comparing it with the percentage that would be in a live organism, scientists and archeologists can determine its approximate age. The half-life of Carbon-14 is 5,700 years; this means after 5,700 years, there will only be half of whatever amount of ^{14}C the sample had initially.

Radioactive decay is a process that we can describe using the exponential decay model $A = A_0 \cdot 2^{-t/h}$, where A is the amount of radioactive material present at time t, A_0 is the amount of radioactive material present at $t = 0$, and h is the material's half-life.

1. Write a model for the decay of Carbon-14.

2. If 2/3 of the initial estimated amount of Carbon-14 is present in a sample, how many years old is the sample? Hint: Replace A with $\frac{2}{3}A_0$ and solve for t.

3. If 75% of the initial estimated amount of Carbon-14 has **decayed**, how many years old is the sample? Hint: Don't replace A with $0.75A_0$.

4. A professor is examining a museum skeleton that he suspects might be a fake, so he takes a measurement of the Carbon-14 remaining in one of the suspect bones of the specimen. From this measurement, he determines that 65% of the original ^{14}C remains in the sample. The skeleton is estimated to be 5,000 years or older. Perform your own calculation of the age of the skeleton, and then decide if the skeleton might be a fake.

Assessment 11C

Chapter 11: End-of-chapter Assessment for Understanding

For each of the following, describe the type of problem and the strategies and key steps to remember while doing the problem. You do **not** have to complete the problems.

	Type of Problem	Strategies and Key Steps
1. Write $\log_2 x + \log_2 5$ as a single logarithm.		
2. Simplify: $10^{\log(4x)}$		
3. Solve: $\log(0.05x) = 2.1$		
4. After 10 years, Nancy removes the amount in her trust fund, \$12,298. If Nancy initially invested \$5,000, and the fund is compounded continuously, what was the annual interest rate?		
5. Expand the expression $\ln\left(\frac{e}{x}\right)^2$ and then simplify.		
6. Simplify: $\ln\frac{1}{e^3}$		
7. If $f(x) = 3^x$, what is $f^{-1}(x)$?		
8. Solve: $800 = 1000e^{-0.04x}$		
9. How are the graphs of $f(x) = \log x$ and $g(x) = \log(x+3)$ related?		
10. Solve: $\log_3 x + \log_3 6 = 4$		

Assessment 11D

Chapter 11: Metacognitive Skills Assessment

Metacognitive skills refer to the ability to judge how well you have learned something and to effectively direct your own learning and studying. This is a self evaluation tool designed to help you focus your studying and to improve your metacognitive skills with regards to this math class.

Fill the 1st column out before you begin studying.
Fill the 2nd column out after you study and before you take the test.
Go back to this page after your test and circle any of the ratings that you would now change – this identifies the "disconnects" between what you think you know well and what you actually know well.

Use the scale below to assign a number to each topic.

5 *I am confident I can do any problems in this category correctly.*
4 *I am confident I can do most of the problems in this category correctly.*
3 *I understand how to do the problems in this category, but I still make a lot of mistakes.*
2 *I feel unsure about how to do these problems.*
1 *I know I don't understand how to do these problems.*

Topic or Skill	Before Studying	After Studying
Finding the domain of a function.		
Combining functions using addition, subtraction, multiplication, or division.		
Finding the composition of two functions.		
Determining whether a function is one-to-one.		
Given the graph of a function, drawing the inverse.		
Given the equation of a function, finding the equation for the inverse.		
Using the properties that relate functions and their inverses.		
Correctly using exponent rules to simplify expressions.		
Graphing basic exponential functions, including $y = e^x$.		
Graphing exponential functions with translations.		
Identifying the asymptote of the graph of an exponential function, its domain and its range.		
Solving application problems involving the compound interest formula or the continuously compounded interest formula.		
Evaluating exponential expressions with irrational exponents.		
Evaluating exponential expressions involving base-*e*.		
Writing an exponential model to represent a real-world growth or decay problem.		
Understanding why and when it is necessary to use logarithms.		
Simplifying logarithmic expressions.		
Simplifying expressions in the form $\log_b b^x$ or $b^{\log_b x}$.		

(continued on next page)

Use the scale below to assign a number to each topic.

5 *I am confident I can do any problems in this category correctly.*
4 *I am confident I can do most of the problems in this category correctly.*
3 *I understand how to do the problems in this category, but I still make a lot of mistakes.*
2 *I feel unsure about how to do these problems.*
1 *I know I don't understand how to do these problems.*

Topic or Skill	**Before Studying**	**After Studying**
Graphing basic logarithmic functions, including $y = \ln x$.		
Graphing logarithmic functions with translations.		
Identifying the asymptote of the graph of an logarithmic function, its domain and its range.		
Applying the product, quotient, and power rules for logarithms to expand a logarithmic expression.		
Applying the product, quotient, and power rules for logarithms to write an expanded log expression as a single logarithm.		
Solving a logarithmic equation.		
Solving an exponential equation.		
Solving a logarithmic equation where the product or quotient rule for logarithms needs to be used first.		
Solving application problems involving population growth or decay.		
Solving application problems involving the Richter scale or sound.		
Solving application problems involving pH.		
Solving application problems involving radioactive decay.		
Solving application problems involving continuously compounded interest, including problems with doubling times.		

Student Workbook: Chapter 12

Table of Contents: *More on Solving Systems of Equations*

Assessment 12A

Pretest and Diagnostic Tool: More on Systems of Equations

Directions: Complete this assessment without looking back at your notes or your book. **Do not use a calculator on this assessment.**

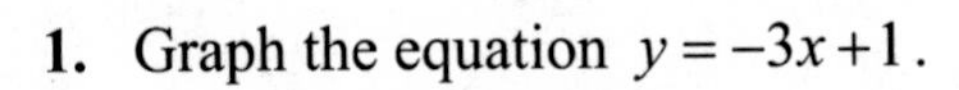

1. Graph the equation $y = -3x + 1$.

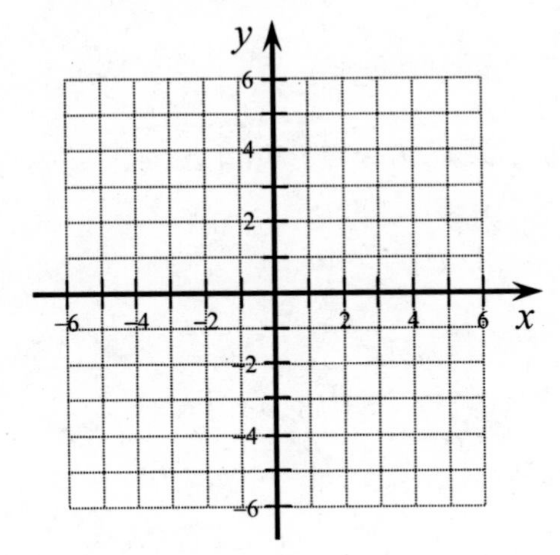

2. Write the equation $x - 2y = 8$ in slope-intercept form.

3. Solve for x: $3x - 2(x - 4) = 8$.

4. Solve for y: $z - (4z + 10) = -1$.

5. Solve for y: $12\left(\frac{3}{4}y\right) - 4 = y$.

6. If $y = -1$, solve the equation $3x - 2y = 6$ for x.

7. Add the two expressions using the vertical method:
$$\begin{array}{r} 5x - 3y \\ +\ 2x + 3y \\ \hline \end{array}$$

8. Add the two expressions using the vertical method:
$$\begin{array}{r} -12x - 4y + z \\ +\quad 8x - 4y + z \\ \hline \end{array}$$

9. Multiply both sides of the equation $6x + 5y = -1$ by 4.

10. Multiply both sides of the equation $x - 4y + 2z = 7$ by -2.

11. Multiply both sides of the equation $\frac{x}{3} - \frac{1}{2} = -2$ by 6.

12. Multiply both sides of the equation $0.05x + 0.06y = 210$ by 100.

13. Find an expression for the interest earned after one year if x dollars earns 4.5% annual interest.

14. Find an expression for the distance traveled by an airplane that travels for 4 hours at a rate of $x + 15$ miles per hour.

15. Find an expression for the revenue made by a ticket office if x tickets are sold for \$12.00 each and y tickets are sold for \$6.00 each.

16. Below you see two rows of numbers. Add the two rows together.

5   2   −1

3   −2   4

17. Below you see two rows of numbers. Multiply the first row by −2 and add it to the second row.

1   0   −2   5

2   4   0   −3

18. Simplify: $3(-2)-(-4)(5)$

19. Evaluate: $\frac{0}{-2}$

20. $a=4$ $b=-1$ Find $ad-bc$.

$c=0$ $d=8$

Student Activity A

Section 3.2: Triple Play on Systems of Two Equations

Directions: For each of the problems below, solve the system of equations by graphing, by substitution, and by elimination – your solution should be the same for all three methods methods. If they are not, go back and look for a mistake in your work.

1.

System of Equations:

$$\begin{cases} x + 2y = -2 \\ -2x - 3y = 5 \end{cases}$$

Solve by Graphing:

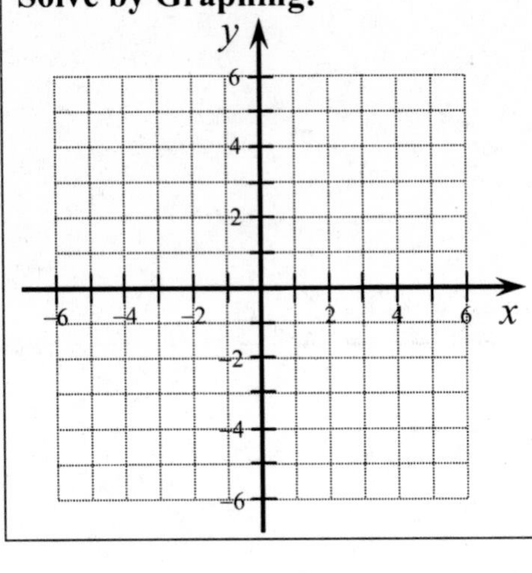

Solve by Substitution

Solve by Elimination

2.

System of Equations:

$$\begin{cases} y = 3x - 5 \\ 2y - 6x = -10 \end{cases}$$

Solve by Graphing:

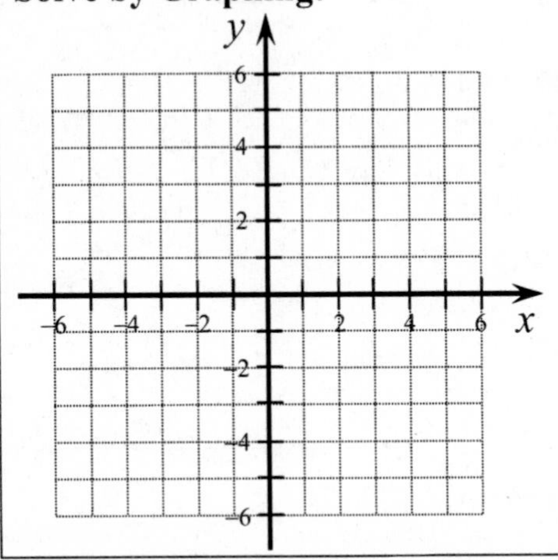

Solve by Substitution

Solve by Elimination

3.

System of Equations:

$$\begin{cases} 4x+3y=0 \\ 2x-9y=7 \end{cases}$$

Solve by Graphing:

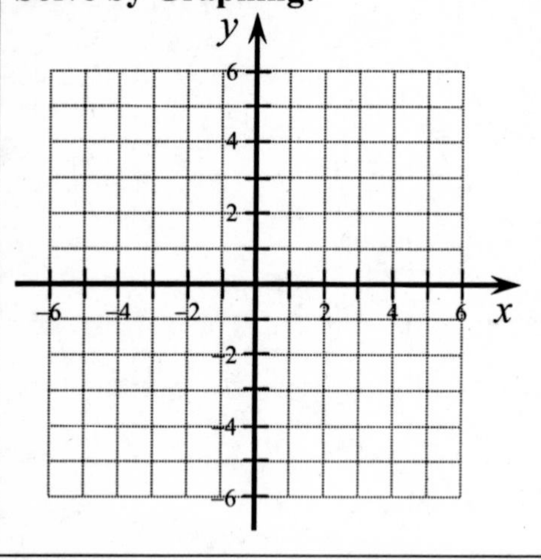

Solve by Substitution

Solve by Elimination

4.

System of Equations:

$$\begin{cases} 3y-2x+12=0 \\ \frac{x}{2}-\frac{3y}{4}=-\frac{3}{4} \end{cases}$$

Solve by Graphing:

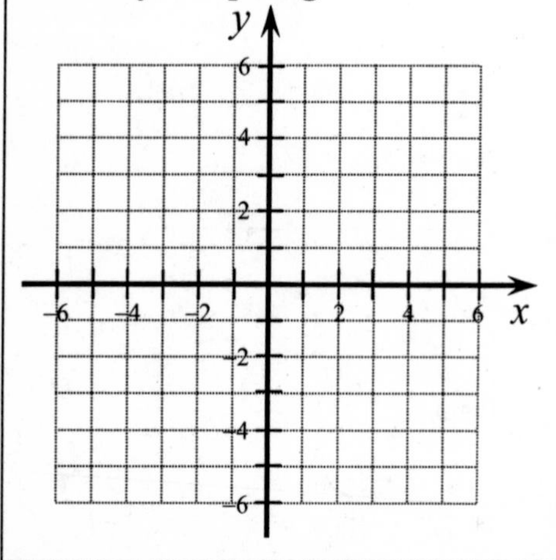

Solve by Substitution

Solve by Elimination

Student Activity B

Section 12.1: Tactical Choices

Directions: For each system of equations below, decide whether it will be easier to solve using substitution or elimination (choose your tactic).

	System of Equations	**Tactic** (substitution or elimination)	**Tactical Plan** **Substitution** – which variable in which equation will you solve for? **Elimination** – which variable will you eliminate and how?
1.	$\begin{cases} 3x+2y=37 \\ y=5x-1 \end{cases}$		
2.	$\begin{cases} -2x+y=4 \\ 5x-3y=-14 \end{cases}$		
3.	$\begin{cases} y=-x+1 \\ 4x+4y=4 \end{cases}$		
4.	$\begin{cases} 5x+6y=-53 \\ 5x+3y=-29 \end{cases}$		
5.	$\begin{cases} 12x-2y=48 \\ 3x=2y+21 \end{cases}$		
6.	$\begin{cases} 0.03x+0.01y=-0.11 \\ -3x+4y=16 \end{cases}$		
7.	$\begin{cases} -\frac{x}{7}+\frac{y}{2}=\frac{17}{14} \\ -\frac{6x}{5}+\frac{y}{5}=-\frac{9}{5} \end{cases}$		
8.	$\begin{cases} x=10y+7 \\ -x+10y=18 \end{cases}$		

Student Activity C

Section 12.1: System Setup

Directions: Read each problem carefully.
a) **Define the variables** you would use to solve the problem.
b) **Write a system** of two equations that could be used to solve each problem. You do not need to solve the application problems.

1. At a hotdog stand, plain hotdogs cost $2.50 and hotdogs with chili cost $3.25. One day the stand brought in $480 in revenue from selling 165 hotdogs. How many plain hotdogs were sold? How many hotdogs with chili were sold?

   $p =$

   $c =$

   Equations:

2. First, Christine rode a bicycle for a while at a rate of 10 mph. Then, she got off the bike and ran at a rate of 6 mph. During the 5 hours she was biking and running, she traveled a distance of 44 miles. How long did Christine spend on her bike, and how long did she run?

3. An investment club split $12,000 between an account that ended up paying 9% annual interest and one that paid 11%. If one year of income from these investments is $1,240, how much was invested at each rate?

4. Sheila has stock solutions of 7% salt and 12% salt on the shelf of her lab. She wants to mix them together obtain 50 ounces of a 10% salt solution. How many ounces of each of the stock solutions should she use?

Guided Learning Activity

Section 12.2: Pairing Off

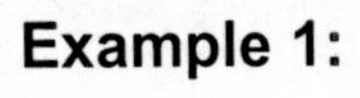

Example 1:

HOW TO SOLVE:
A system of three equations with three variables.

→ **GOAL:** A system of two equations with two variables.

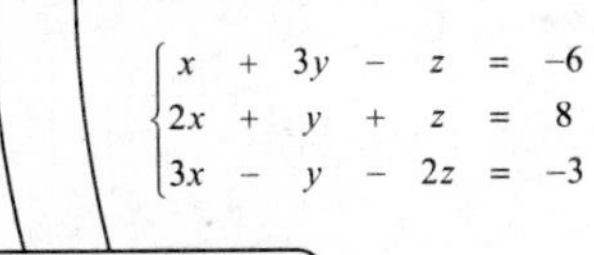

$$\begin{cases} x + 3y - z = -6 \\ 2x + y + z = 8 \\ 3x - y - 2z = -3 \end{cases}$$

Choose a variable to eliminate.

$$\begin{cases} x + 3y - z = -6 \\ 2x + y + z = 8 \\ 3x - y - 2z = -3 \end{cases}$$

Let's eliminate z.

Pick any two equations and eliminate the chosen variable.

$$\begin{cases} x + 3y - z = -6 & A \\ 2x + y + z = 8 & B \\ 3x - y - 2z = -3 & C \end{cases}$$

Add equations A and B.

$$3x + 4y = 2$$

We eliminated z.

Pick a different pair of equations and eliminate the **same** variable.

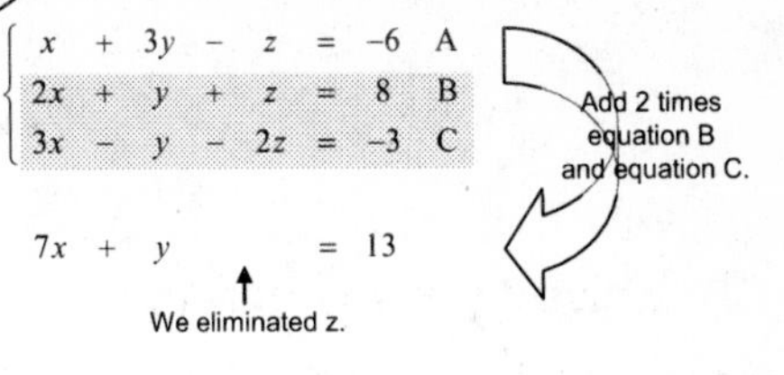

$$\begin{cases} x + 3y - z = -6 & A \\ 2x + y + z = 8 & B \\ 3x - y - 2z = -3 & C \end{cases}$$

Add 2 times equation B and equation C.

$$7x + y = 13$$

We eliminated z.

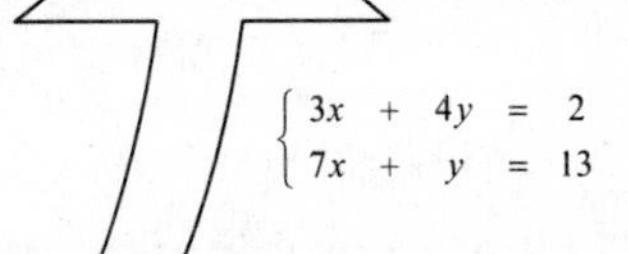

$$\begin{cases} 3x + 4y = 2 \\ 7x + y = 13 \end{cases}$$

Example 2:

HOW TO SOLVE:
A system of three equations with three variables when there are missing variables.

→ **GOAL:** A system of two equations with two variables.

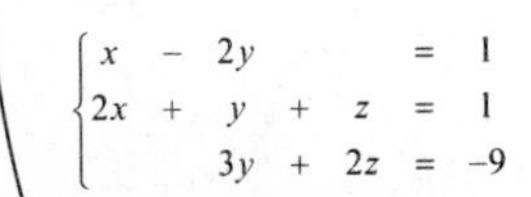

$$\begin{cases} x - 2y = 1 \\ 2x + y + z = 1 \\ 3y + 2z = -9 \end{cases}$$

Choose a variable to eliminate. If the system is missing a variable in one of the equations, that would be an easy choice.

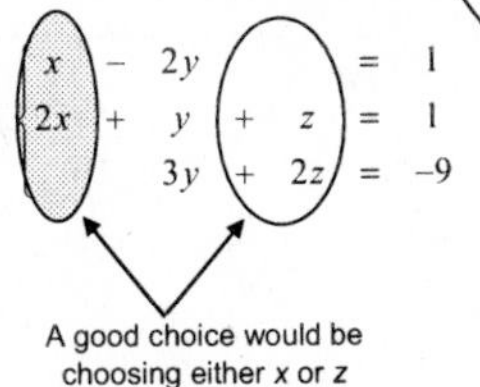

$$\begin{cases} x - 2y = 1 \\ 2x + y + z = 1 \\ 3y + 2z = -9 \end{cases}$$

A good choice would be choosing either *x* or *z* to eliminate. Let's choose *x*.

Notice that one of the equations already has the chosen variable eliminated.

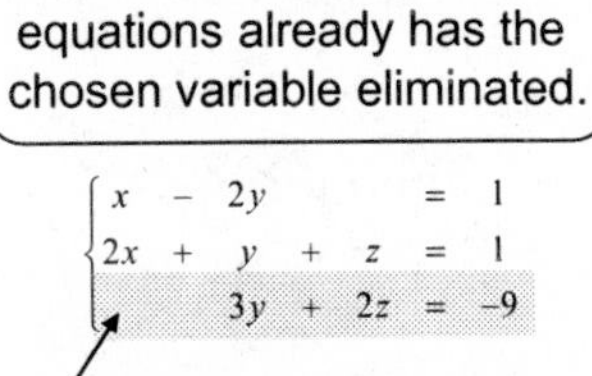

$$\begin{cases} x - 2y = 1 \\ 2x + y + z = 1 \\ 3y + 2z = -9 \end{cases}$$

x was already eliminated.

$$3y + 2z = -9$$

Pick a pair of equations and eliminate the **same** variable.

$$\begin{cases} x - 2y = 1 & A \\ 2x + y + z = 1 & B \\ 3y + 2z = -9 & C \end{cases}$$

Add -2 times equation A and equation B.

$$5y + z = -1$$

We eliminated x.

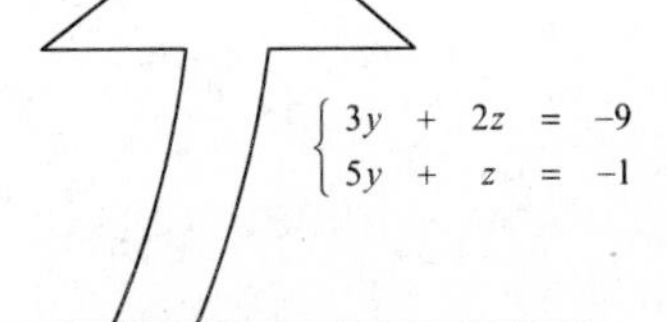

$$\begin{cases} 3y + 2z = -9 \\ 5y + z = -1 \end{cases}$$

Directions: For each system of three equations, use linear combinations to get a system of two equations by eliminating the chosen variable. **Write the system of two equations in the box for each problem.**

1. Solve by eliminating **z**:

$$\begin{cases} x - 3y + z = 2 & \text{A} \\ 2x + 4y - z = 10 & \text{B} \\ 3x - 4y + 2z = 8 & \text{C} \end{cases}$$

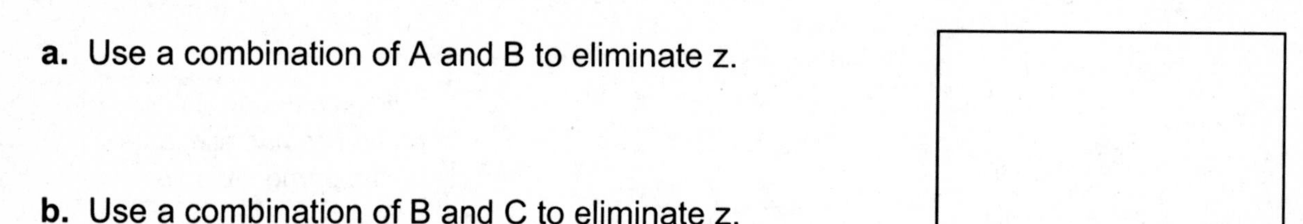

a. Use a combination of A and B to eliminate z.

b. Use a combination of B and C to eliminate z.

2. Solve by eliminating **y**:

$$\begin{cases} 2x + y + z = -1 & \text{A} \\ x + 2y + 3z = 5 & \text{B} \\ x - y - 4z = -2 & \text{C} \end{cases}$$

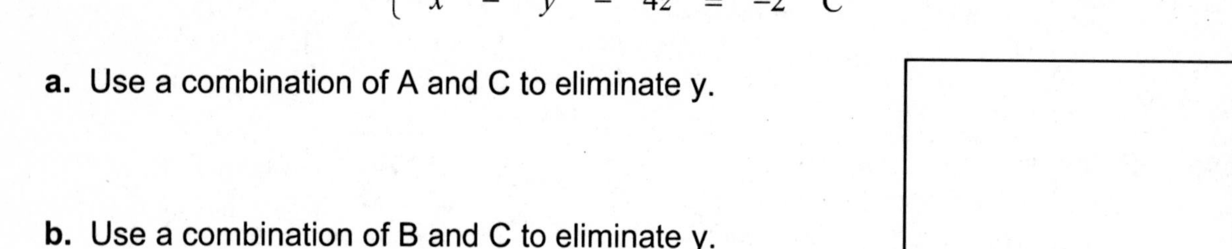

a. Use a combination of A and C to eliminate y.

b. Use a combination of B and C to eliminate y.

3. Solve by eliminating **x**:

$$\begin{cases} 2x + y + z = -1 & \text{A} \\ 4x - y - z = 6 & \text{B} \\ y + 3z = -2 & \text{C} \end{cases}$$

a. Write the equation that already has x eliminated.

b. Use a combination of A and B to eliminate x.

Still have time? Finish solving each system of equations on another sheet of paper.

Student Activity A

Section 12.2: Plane Facts

The graph of a linear equation in three variables, such as $2x+3y-5z=8$, is a flat surface called a **plane**. It is very difficult to imagine graphing in 3-D on 2-dimensional paper, but we'll give it a try for this activity. When we solve a system of three linear equations with three variables, there are many possible ways the three planes could be positioned in 3-dimensional space (each equation represents a plane).

Recall that the solution of a system of three equations is an ordered triple of numbers that satisfies each equation. Graphically, a point is a solution of a system of three equations in three variables if it lies on **each** of the corresponding planes. Below we show five ways that three planes could be positioned in space.

Directions: For each case, decide on the **number** of possible solutions (one, none, or infinitely many) and then use a colored pencil to indicate where the solution point (or points) lies.

1.

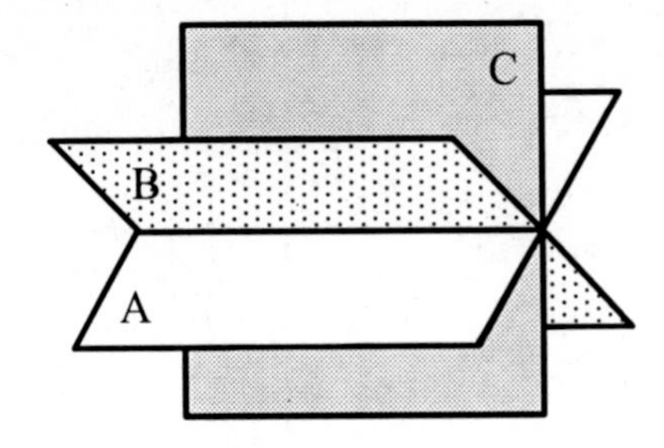

2.

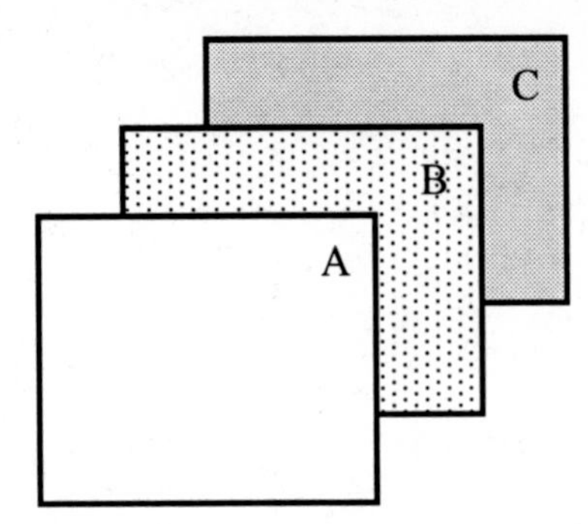

3.

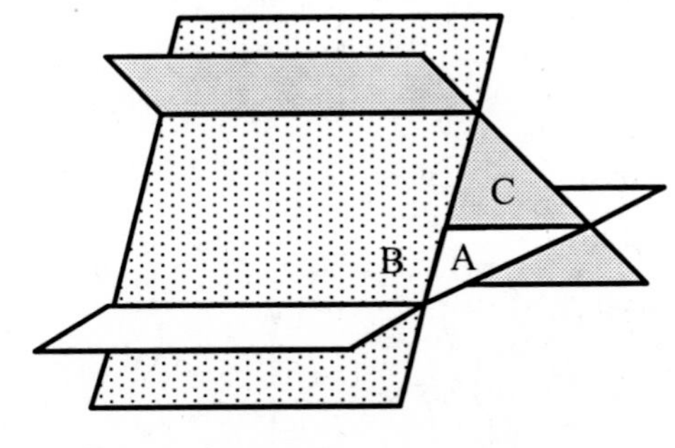

4.

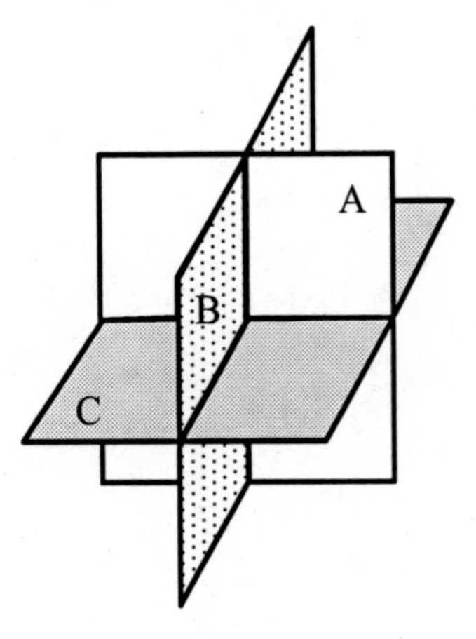

5.

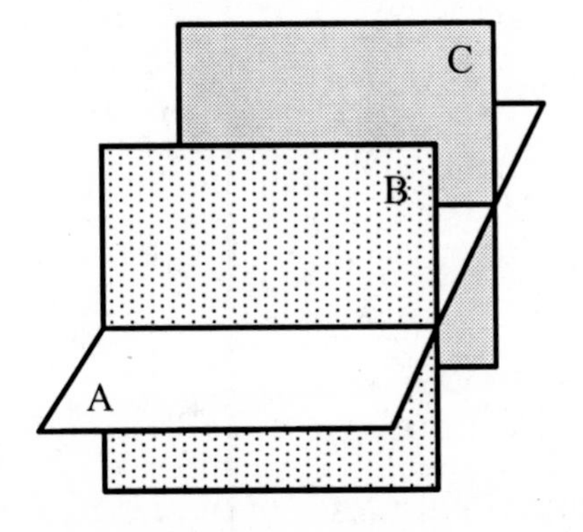

6. Which of the graphs represent a system that is consistent?

7. Which of the graphs represent a equations that are dependent?

Student Activity B

Section 12.2: Strange Systems

Directions: For each system of equations, **decide** what makes the system "strange" and then decide on a strategy for solving the system. **Carry out enough steps** to transform the system from a "strange" one to one that looks standard.

1. Solve: $\begin{cases} 3x + 4y - 2z = 2 \\ 6x - 4y + z = 0 \\ 5z = 5 \end{cases}$

2. Solve: $\begin{cases} 0.02x + 0.1y + 0.02z = 0 \\ x + 10y + 3z = 1 \\ \dfrac{x}{2} + \dfrac{z}{8} = -\dfrac{1}{2} \end{cases}$

3. Solve: $\begin{cases} x + z = -y \\ y - 3 = z \\ 10 + 3z = 2y \end{cases}$

4. Solve: $\begin{cases} x + y = 0 \\ y + 2z = 3 \\ x - z = -1 \end{cases}$

Still have time? Go ahead and solve the systems using another sheet of paper.

Assessment 12B

Chapter 12: Mid-chapter Assessment for Understanding

For each of the following, describe the type of problem and the strategies and key steps to remember while doing the problem. You do **not** have to complete the problems.

	Type of Problem	Strategies and Key Steps
1. Use substitution to solve the system: $\begin{cases} x-3y=-1 \\ 3x-4y=7 \end{cases}$		
2. Solve the system: $\begin{cases} x+y-z=8 \\ 2x-y+3z=-4 \\ x+2z=0 \end{cases}$		
3. Is the system consistent or inconsistent? $\begin{cases} x-4y=6 \\ 2x=8y+10 \end{cases}$		
4. Is $(1,-2,0)$ a solution to the system? $\begin{cases} 3x+y+4z=1 \\ x-y+z=-1 \\ 2x+y+3z=0 \end{cases}$		
5. Rafi paddles his canoe upstream 24 miles in 4 hours. If he can make the downstream trip in 3 hours, find the rate that Rafi can paddle and the rate of the current.		
6. Use elimination to solve the system: $\begin{cases} x+3y=-2 \\ x+4y=0 \end{cases}$		
7. Solve the system by graphing: $\begin{cases} y=-2x+4 \\ y=2x \end{cases}$		
Solve the system: $\begin{cases} 5x+4y+9=z \\ x=-(y+z) \\ 2x+y=2z+5 \end{cases}$		

Student Activity A

Section 12.3: Retirement Allocation

Last year, at her annual appointment with her financial advisor, Jackie agreed to reallocate a portion of the $500,000 that she had saved for retirement so that less money was allocated to high-risk investments. The high-risk account was projected to make 8% annual interest, a medium-risk account was projected to earn 6%, and the "safe" low-risk account was projected to earn 4.5%. Jackie and her advisor split the $500,000 into these three accounts. Looking at the performance statements at the end of the year, Jackie sees that her retirement funds now total $530,250. Jackie also notices that the interest earned by the money in the medium-risk account was twice the interest earned by the money in the low-risk account. How much of Jackie's retirement fund was allocated into each account at the beginning of last year?

Define three variables:

Let H = ____.

Let M = ____.

Let L = ____.

Write three equations:

Total invested:

Interest earned:

Mid-risk / low-risk relationship:

Solve the system:

Write a sentence to describe the solution:

Student Activity B

Section 12.3: Blog Readers

Patrick is looking through the analytics on the number of readers for his popular math blog "From the Origin." In the blog statistics, Patrick can his readers from blog hits from inside the U.S., blog hits from outside the U.S., and blog hits from an RSS reader. For the month of January, he had a total of 4400 readers. The number of readers with blog hits from the United States was equal to the sum of the blog hits from readers outside the U.S. and the readers who used an RSS feed to read the blog. The number of RSS readers is 200 less than twice the number of blog readers from outside the U.S. Find the number of readers with blog hits from the U.S., the number of readers with blog hits from outside the U.S., and the number of readers who used an RSS feed.

Define three variables:

Let $x =$ ______________________________.

Let $y =$ ______________________________.

Let $z =$ ______________________________.

Write three equations:

Solve:

Write a sentence to describe the solution:

Guided Learning Activity

Section 12.3: Fitting a Quadratic Curve

The equation of a parabola is of the form $y = ax^2 + bx + c$.

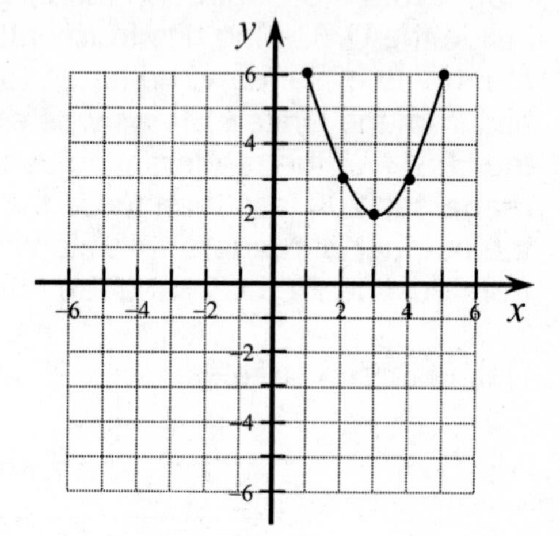

If we are given a graph of the parabola, we can use points from the graph to determine the equation of the parabola by setting up a system of three equations and three unknowns.

Begin by writing down the coordinates of **three** of the points that are indicated on the graph to the right.

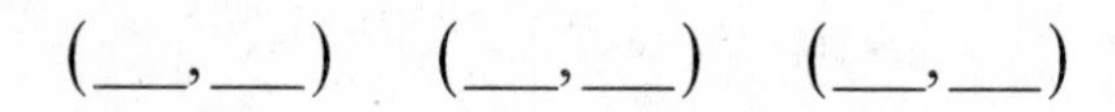

Now we write a "skeleton" of parentheses for the equation of a parabola:

$$y \;=\; ax^2 \;+\; bx \;+c$$
$$(\quad) = a(\quad)^2 + b(\quad) + c$$

Use the *x*- and *y*-coordinates of each point to complete an equation

$(\_\_,\_\_)$ $\quad (\quad) = a(\quad)^2 + b(\quad) + c \qquad \_\_\_\_a + \_\_\_\_b + \_\_\_\_c = \_\_\_\_$

$(\_\_,\_\_)$ $\quad (\quad) = a(\quad)^2 + b(\quad) + c \qquad \_\_\_\_a + \_\_\_\_b + \_\_\_\_c = \_\_\_\_$

$(\_\_,\_\_)$ $\quad (\quad) = a(\quad)^2 + b(\quad) + c \qquad \_\_\_\_a + \_\_\_\_b + \_\_\_\_c = \_\_\_\_$

Solve the system of three equations and three variables to find *a*, *b*, and *c*:

Write the equation of the parabola: $y = \_\_\_\_x^2 + \_\_\_\_x + \_\_\_\_$

Student Activity A

Section 12.4: Tic-Tac-Toe on Row Echelon Form

A matrix is in **row echelon form** if it has 1's down its main diagonal and a zero below its main diagonal.

Directions: If the matrix in the square is in row echelon form, then circle it (put an **O** on the square). If the matrix is **not** in row echelon form, then mark it with an **X**.

$\left[\begin{array}{cc\|c} 1 & 3 & 4 \\ 0 & 2 & 6 \end{array}\right]$	$\left[\begin{array}{cc\|c} 1 & 3 & 5 \\ 0 & 1 & 4 \end{array}\right]$	$\left[\begin{array}{ccc\|c} 1 & 2 & 5 & -7 \\ 0 & 1 & 2 & 15 \\ 0 & 0 & 1 & \frac{1}{2} \end{array}\right]$
$\left[\begin{array}{ccc\|c} 1 & 0 & 5 & 2 \\ 0 & 1 & 0 & 3 \\ 0 & 0 & 1 & 6 \end{array}\right]$	$\left[\begin{array}{ccc\|c} 1 & 0 & 4 & 5 \\ 0 & -1 & 2 & 6 \\ 0 & 0 & 1 & 0 \end{array}\right]$	$\left[\begin{array}{ccc\|c} 1 & 0 & 4 & 5 \\ 0 & 1 & 2 & 0 \\ 0 & 0 & 1 & 0 \end{array}\right]$
$\left[\begin{array}{cc\|c} 1 & 0 & \frac{3}{4} \\ 0 & 1 & \frac{1}{2} \end{array}\right]$	$\left[\begin{array}{cc\|c} 1 & 2 & 3 \\ 1 & 0 & 4 \end{array}\right]$	$\left[\begin{array}{cc\|c} 4 & 1 & 2 \\ 1 & 0 & -3 \end{array}\right]$

Student Activity B

Section 12.4: Rowing to Freedom

Directions: Begin with the matrix in the box marked "Start". Follow each of the given row operations and shade the box that contains your result for each one (each row operation is carried out on the previous answer). At the "pit stop" you'll get a new matrix to use. You can escape this matrix by "rowing" your way to freedom!

1. $-2R_2 \to R_2$
2. $R_2 + R_3 \to R_3$
3. $2R_1 + R_2 \to R_2$
4. $-\frac{3}{4}R_2 \to R_2$
5. $R_2 + R_3 \to R_3$
6. $-\frac{1}{3}R_2 \to R_2$

You should now go to the PIT STOP and proceed using the new matrix.

7. $R_1 \leftrightarrow R_3$
8. $R_1 + (-2)R_2 \to R_2$
9. $R_2 + 2R_3 \to R_3$
10. $-1R_3 \to R_3$
11. $-\frac{1}{2}R_2 \to R_2$
12. $\frac{1}{4}R_1 \to R_1$

START $\begin{bmatrix} 1 & 1 & 1 & 0 \\ 1 & -1 & 1 & -6 \\ 2 & 1 & 3 & -2 \end{bmatrix}$	$\begin{bmatrix} 1 & 1 & 1 & 0 \\ -2 & 2 & -2 & 12 \\ 2 & 1 & 3 & -2 \end{bmatrix}$	$\begin{bmatrix} 1 & 1 & 1 & 0 \\ -2 & 2 & -2 & 12 \\ 0 & 3 & 1 & 10 \end{bmatrix}$	$\begin{bmatrix} 1 & 1 & 1 & 0 \\ 0 & 4 & 0 & 12 \\ 0 & 3 & 1 & 10 \end{bmatrix}$
$\begin{bmatrix} 1 & 1 & 1 & 0 \\ 1 & -1 & 1 & -6 \\ -4 & -2 & -6 & 4 \end{bmatrix}$	$\begin{bmatrix} -1 & 3 & -1 & 12 \\ -2 & 2 & -2 & 12 \\ 2 & 1 & 3 & -2 \end{bmatrix}$	$\begin{bmatrix} 1 & 1 & 1 & 0 \\ 0 & -1 & 0 & -3 \\ 0 & 0 & 1 & 1 \end{bmatrix}$	$\begin{bmatrix} 1 & 1 & 1 & 0 \\ 0 & -3 & 0 & -9 \\ 0 & 3 & 1 & 10 \end{bmatrix}$
$\begin{bmatrix} 4 & 2 & 1 & 5 \\ 2 & 2 & 2 & 2 \\ 0 & 1 & 1 & 1 \end{bmatrix}$	PIT STOP $\begin{bmatrix} 0 & 1 & 1 & 1 \\ 2 & 2 & 2 & 2 \\ 4 & 2 & 1 & 5 \end{bmatrix}$	$\begin{bmatrix} 1 & 1 & 1 & 0 \\ 0 & 1 & 0 & 3 \\ 0 & 0 & 1 & 1 \end{bmatrix}$	$\begin{bmatrix} 1 & 1 & 1 & 0 \\ 0 & -3 & 0 & -9 \\ 0 & 0 & 1 & 1 \end{bmatrix}$
$\begin{bmatrix} 4 & 2 & 1 & 5 \\ 0 & -2 & -3 & 1 \\ 0 & 1 & 1 & 1 \end{bmatrix}$	$\begin{bmatrix} 0 & 1 & 1 & 1 \\ 2 & 2 & 2 & 2 \\ 4 & 3 & 2 & 6 \end{bmatrix}$	$\begin{bmatrix} 1 & 1 & 1 & 0 \\ 0 & 1 & 0 & 0 \\ 0 & 0 & 1 & 1 \end{bmatrix}$	$\begin{bmatrix} 1 & 1 & 1 & 0 \\ 0 & 3 & 0 & 9 \\ 0 & 0 & 1 & 1 \end{bmatrix}$
$\begin{bmatrix} 4 & 2 & 1 & 5 \\ 0 & -2 & -3 & 1 \\ 0 & 0 & -1 & 3 \end{bmatrix}$	$\begin{bmatrix} 4 & 2 & 1 & 5 \\ 0 & 1 & \frac{3}{2} & -\frac{1}{2} \\ 0 & 0 & 1 & 0 \end{bmatrix}$	$\begin{bmatrix} 1 & \frac{1}{2} & \frac{1}{4} & \frac{5}{4} \\ 0 & 1 & \frac{3}{2} & 0 \\ 0 & 0 & 1 & 0 \end{bmatrix}$	$\begin{bmatrix} 1 & \frac{1}{2} & \frac{1}{4} & 0 \\ 0 & 1 & \frac{3}{2} & 0 \\ 0 & 0 & 1 & 0 \end{bmatrix}$
$\begin{bmatrix} 4 & 2 & 1 & 5 \\ 0 & -2 & -3 & 1 \\ 0 & 0 & 1 & -3 \end{bmatrix}$	$\begin{bmatrix} 4 & 2 & 1 & 5 \\ 0 & 1 & \frac{3}{2} & -\frac{1}{2} \\ 0 & 0 & 1 & -3 \end{bmatrix}$	$\begin{bmatrix} 1 & \frac{1}{2} & \frac{1}{4} & \frac{5}{4} \\ 0 & 1 & \frac{3}{2} & -\frac{1}{2} \\ 0 & 0 & 1 & -3 \end{bmatrix}$	**You've successfully "rowed" to freedom!**

Guided Learning Activity

Section 12.4: Parallel Operations

Directions: Follow the steps to solve the system of equations, first by elimination, then using matrix operations. Note that the shaded directions are for the matrix operations column (on the right).

$x - y + 4z = 5 \quad (1)$ $2x + y + z = 4 \quad (2)$ $-x + y - z = -5 \quad (3)$	**1.** Solve this system of equations. **1.** This system can be represented by this coefficient matrix.	$\left[\begin{array}{ccc\|c} 1 & -1 & 4 & 5 \\ 2 & 1 & 1 & 4 \\ -1 & 1 & -1 & -5 \end{array}\right]$
	2. Add equation #1 to equation #3. **2.** Add row 1 to row 3 (replacing row 3).	
	3. Solve the resulting equation for z by dividing both sides by 3. **3.** Multiply row 3 by $\frac{1}{3}$.	
	4. Multiply equation #1 by -2 and add to equation #2. **4.** Multiply row 1 by -2 and add to row 2 (replacing row 2).	
	5. Divide both sides of the resulting equation by 3. **5.** Multiply row 2 by $\frac{1}{3}$.	
	6. Gather together equations found in steps 3 and 5, as well as equation 1. **6.** Rewrite matrix as a system of equations.	
	7. Solve using back-substitution. **7.** Solve using back-substitution.	

Student Activity C

Section 12.4: Strategy Comparison

In this activity we look at two different strategies for row-reducing the same matrix.

Strategy 1 will attempt to get the zeros in the matrix in this order:

$$\begin{bmatrix} 4 & 3 & 2 \\ ⑤ & 1 & 3 \\ ① & ② & 0 \end{bmatrix}$$

$2^{nd} \rightarrow$ row 2, column 1 (5); $1^{st} \rightarrow$ row 3, column 1 (1); 3rd ← row 3, column 2 (2)

Strategy 2 will attempt to get the zeros in the matrix in this order:

$$\begin{bmatrix} 4 & 3 & 2 \\ ⑤ & 1 & 3 \\ ① & ② & 0 \end{bmatrix}$$

$2^{nd} \rightarrow$ row 2, column 1 (5); $3^{rd} \rightarrow$ row 3, column 1 (1); 1st ← row 3, column 2 (2)

Strategy 1	Strategy 2
Step 1: $R_2 + (-5)R_3 \rightarrow R_3$	**Step 1:** $-2R_2 + R_3 \rightarrow R_3$
Step 2: $-5R_1 + 4R_2 \rightarrow R_2$	**Step 2:** $-5R_1 + 4R_2 \rightarrow R_2$
Step 3: $9R_2 + (-11)R_3 \rightarrow R_3$	**Step 3:** $9R_1 + 4R_3 \rightarrow R_3$ What happens here?
Step 4: $\frac{1}{4}R_1 \rightarrow R_1$ $-\frac{1}{11}R_2 \rightarrow R_2$ $-\frac{1}{15}R_3 \rightarrow R_3$	**Step 4:** $27R_2 + 11R_3 \rightarrow R_3$
	Step 5: $\frac{1}{4}R_1 \rightarrow R_1$ $-\frac{1}{11}R_2 \rightarrow R_2$ $-\frac{1}{12}R_3 \rightarrow R_3$

Was one strategy be easier than the other? If so, which one? If not, why?

Student Activity A

Section 12.5: Match Up on Determinants

Match-up: In each box of the grid below, you will find **mostly** determinants to evaluate. However, there are a few expressions that are absolute values! Match the result with the appropriate choice. If the result is not found among the choices A through D, then choose E (none of these).

A 4 **B** –7 **C** 5 **D** 0 **E** None of these

$\begin{vmatrix} 3 & 5 \\ 5 & 6 \end{vmatrix}$	$\begin{vmatrix} -2 & 5 \\ 1 & -5 \end{vmatrix}$	$\begin{vmatrix} 2 & 1 & 3 \\ 0 & -1 & 2 \end{vmatrix}$	$\begin{vmatrix} \frac{3}{4} & \frac{3}{2} \\ \frac{1}{3} & \frac{2}{3} \end{vmatrix}$
$\begin{vmatrix} -2 & 5 \\ 3 & 1 \\ 1 & -5 \end{vmatrix}$	$\begin{vmatrix} 3 & 5 \\ 1 & 3 \end{vmatrix}$	$2 \cdot \begin{vmatrix} -2 & 6 \\ 1 & -4 \end{vmatrix}$	$\begin{vmatrix} 4 & 2 \\ 6 & 3 \end{vmatrix}$
$\begin{vmatrix} 5 & 0 \\ -1 & 0 \end{vmatrix}$	$\lvert 6-11 \rvert$	$\begin{vmatrix} 1 & 2 \\ -2 & 0 \end{vmatrix}$	Find k if $\begin{vmatrix} 3 & 1 \\ k & 3 \end{vmatrix} = 5$.
Evaluate $\begin{vmatrix} x & 8 \\ x & x \end{vmatrix}$ if $x = 1$.	$\begin{vmatrix} 5 & -3 \\ -10 & 6 \end{vmatrix}$	$\left\lvert \frac{1}{4} + \frac{2}{3} \right\rvert$	Find k if $\begin{vmatrix} 4 & -8 \\ 3 & k \end{vmatrix} = -4$.

Guided Learning Activity

Section 12.5: The Minor League

It is easy to find the determinant of a 2×2 matrix: $\begin{vmatrix} a & b \\ c & d \end{vmatrix} = ad - bc$.

It is much harder to find the determinant of a 3×3 matrix. We use a technique called **expanding by minors**.

What is a **minor**?

In $\begin{vmatrix} a_1 & b_1 & c_1 \\ a_2 & b_2 & c_2 \\ a_3 & b_3 & c_3 \end{vmatrix}$, the minor of a_1 is $\begin{vmatrix} b_2 & c_2 \\ b_3 & c_3 \end{vmatrix}$.

Notice: $\begin{vmatrix} \not{a_1} & \not{b_1} & \not{c_1} \\ \not{a_2} & b_2 & c_2 \\ \not{a_3} & b_3 & c_3 \end{vmatrix}$

If you cross out the row and column of a_1, the minor of a_1 will be easier to see.

1. In $\begin{vmatrix} a_1 & b_1 & c_1 \\ a_2 & b_2 & c_2 \\ a_3 & b_3 & c_3 \end{vmatrix}$, what is the minor of b_1?

2. In $\begin{vmatrix} a_1 & b_1 & c_1 \\ a_2 & b_2 & c_2 \\ a_3 & b_3 & c_3 \end{vmatrix}$, what is the minor of c_1?

We use these minors to find the determinant of a 3×3 matrix:

$$\begin{vmatrix} a_1 & b_1 & c_1 \\ a_2 & b_2 & c_2 \\ a_3 & b_3 & c_3 \end{vmatrix} = a_1(\text{Minor of } a_1) - b_1(\text{Minor of } b_1) + c_1(\text{Minor of } c_1).$$

3. $$\begin{vmatrix} 1 & 0 & 2 \\ 0 & 3 & 2 \\ -2 & 1 & 1 \end{vmatrix} = \square \cdot \begin{vmatrix} \square & \square \\ \square & \square \end{vmatrix} - \square \cdot \begin{vmatrix} \square & \square \\ \square & \square \end{vmatrix} + \square \cdot \begin{vmatrix} \square & \square \\ \square & \square \end{vmatrix}.$$

$$= \square \cdot \square - \square \cdot \square + \square \cdot \square = \square - \square + \square = \square$$

4. $$\begin{vmatrix} 3 & -2 & 5 \\ -9 & 1 & 4 \\ 6 & 7 & 0 \end{vmatrix} = \square \cdot \begin{vmatrix} \square & \square \\ \square & \square \end{vmatrix} - \square \cdot \begin{vmatrix} \square & \square \\ \square & \square \end{vmatrix} + \square \cdot \begin{vmatrix} \square & \square \\ \square & \square \end{vmatrix}$$

$$= \square \cdot \square - \square \cdot \square + \square \cdot \square = \square - \square + \square = \square$$

By learning a pattern of signs (sometimes called the checkerboard pattern), we can use this method of expanding the minors for any row or column of the 3×3 determinant.

Here is the checkerboard pattern: $\begin{matrix} + & - & + \\ - & + & - \\ + & - & + \end{matrix}$

Look at the first expansion we learned:

$$\begin{vmatrix} \overset{+}{a_1} & \overset{-}{b_1} & \overset{+}{c_1} \\ a_2 & b_2 & c_2 \\ a_3 & b_3 & c_3 \end{vmatrix} = +a_1(\text{Minor of } a_1) - b_1(\text{Minor of } b_1) + c_1(\text{Minor of } c_1).$$

We can evaluate the same determinant using a minor expansion for the second row:

$$\begin{vmatrix} a_1 & b_1 & c_1 \\ \overset{-}{a_2} & \overset{+}{b_2} & \overset{-}{c_2} \\ a_3 & b_3 & c_3 \end{vmatrix} = -a_2(\text{Minor of } a_2) + b_2(\text{Minor of } b_2) - c_2(\text{Minor of } c_2).$$

5. Now try writing a minor expansion on the third column.

$$\begin{vmatrix} a_1 & b_1 & c_1 \\ a_2 & b_2 & c_2 \\ a_3 & b_3 & c_3 \end{vmatrix} = \_\_\_\_(\text{Minor of } \_\_\_)\_\_\_\_(\text{Minor of } \_\_\_)\_\_\_\_(\text{Minor of } \_\_\_)$$

Do all of these different minor expansions really result in the same determinant?

6. Evaluate $\begin{vmatrix} 2 & 8 & 4 \\ 5 & 3 & 0 \\ 6 & 1 & 7 \end{vmatrix}$ by expanding on the first row.

7. Evaluate $\begin{vmatrix} 2 & 8 & 4 \\ 5 & 3 & 0 \\ 6 & 1 & 7 \end{vmatrix}$ by expanding on the second row.

8. Evaluate $\begin{vmatrix} 2 & 8 & 4 \\ 5 & 3 & 0 \\ 6 & 1 & 7 \end{vmatrix}$ by expanding on the third column.

Student Activity B

Section 12.5: Discovering Cramer

Directions: Follow the directions for each step and fill in the boxes with the missing pieces as you go. Careful – as you move from line to line, the boxes might contain different expressions! At the end, you should get a formula that you might recognize!

For the system: $\begin{cases} ax+by=e \\ cx+dy=f \end{cases}$, $D=\begin{vmatrix} a & b \\ c & d \end{vmatrix}$, $D_x=\begin{vmatrix} e & b \\ f & d \end{vmatrix}$ and $D_y=\begin{vmatrix} a & e \\ c & f \end{vmatrix}$.

1. Simplify each determinant:

$D=ad-bc$ $\quad D_x=$ ______ $\quad D_y=$ ______

Now we work through the process of solving the system:

$\begin{cases} ax+by=e \\ cx+dy=f \end{cases}$ where $ad-bc\neq 0$.

2. First, we eliminate the variable *y* from the system:

Multiply both sides of the first equation by d: $\quad adx+\square y=\square$

Multiply both sides of the second equation by $(-b)$: $\quad \square x-\square=-bf$

3. Add the equations from step **1**: $\quad \square = \square$

4. Factor *x* from the left hand side of the equation: $\quad x(\quad)=ed-bf$

5. Solve for *x*:: $x=\dfrac{ed-bf}{\square}$

6. Thus $x=\dfrac{D_x}{\square}$

Still have time? Try to repeat steps 2-6 to solve for *y* (eliminate the variable *x*).

Student Activity C

Section 12.5: D-lightful Determinants

Directions: For each system of equations, find the desired determinants. You do **not** need to solve the system of equations. As you find the missing determinants, shade them in to paint by Determinants.

1. $\begin{cases} 2x+3y=-1 \\ y-x=8 \end{cases}$

Find D_x and D_y.

2. $\begin{cases} x+2y+4z=4 \\ 3x+2y+6z=1 \\ x+5y+4z=1 \end{cases}$

Find D_y and D_z.

3. $\begin{cases} 10x+12y+6z=2 \\ 5x+6y+3z=1 \\ x+3z=1 \end{cases}$

Find D, D_x, D_y and D_z.

4. $\begin{cases} x+z=4 \\ 3y+3z=12 \\ x+2y=6 \end{cases}$

Find D, D_x, D_y and D_z.

$\begin{vmatrix} 1 & 0 & 1 \\ 0 & 3 & 3 \\ 1 & 2 & 0 \end{vmatrix}$	$\begin{vmatrix} -1 & 3 \\ 8 & 1 \end{vmatrix}$	$\begin{vmatrix} 4 & 0 & 1 \\ 12 & 3 & 3 \\ 6 & 2 & 0 \end{vmatrix}$	$\begin{vmatrix} 2 & 3 \\ -1 & 1 \end{vmatrix}$	$\begin{vmatrix} 4 & 1 & 0 \\ 12 & 3 & 3 \\ 6 & 2 & 0 \end{vmatrix}$
$\begin{vmatrix} 2 & -1 \\ -1 & 8 \end{vmatrix}$	$\begin{vmatrix} 1 & 1 & 0 \\ 0 & 3 & 3 \\ 1 & 2 & 0 \end{vmatrix}$	$\begin{vmatrix} 1 & 2 & 4 \\ 3 & 2 & 6 \\ 1 & 5 & 4 \end{vmatrix}$	$\begin{vmatrix} 10 & 12 & 6 \\ 5 & 6 & 3 \\ 1 & 0 & 3 \end{vmatrix}$	$\begin{vmatrix} 10 & 12 & 6 \\ 5 & 6 & 3 \\ 1 & 3 & 0 \end{vmatrix}$
$\begin{vmatrix} 10 & 12 & 2 \\ 5 & 6 & 1 \\ 1 & 0 & 1 \end{vmatrix}$	$\begin{vmatrix} 10 & 12 & 2 \\ 5 & 6 & 1 \\ 1 & 3 & 1 \end{vmatrix}$	$\begin{vmatrix} 10 & 2 & 6 \\ 5 & 1 & 3 \\ 1 & 1 & 0 \end{vmatrix}$	$\begin{vmatrix} 10 & 2 & 6 \\ 5 & 1 & 3 \\ 1 & 1 & 1 \end{vmatrix}$	$\begin{vmatrix} 4 & 2 & 4 \\ 1 & 2 & 6 \\ 1 & 5 & 4 \end{vmatrix}$
$\begin{vmatrix} 1 & 0 & 4 \\ 0 & 3 & 12 \\ 1 & 2 & 6 \end{vmatrix}$	$\begin{vmatrix} 1 & 1 & 4 \\ 0 & 3 & 12 \\ 1 & 2 & 6 \end{vmatrix}$	$\begin{vmatrix} 1 & 3 \\ 8 & -1 \end{vmatrix}$	$\begin{vmatrix} 1 & 2 & 4 \\ 3 & 2 & 1 \\ 1 & 5 & 1 \end{vmatrix}$	$\begin{vmatrix} 2 & 3 \\ 1 & -1 \end{vmatrix}$
$\begin{vmatrix} 1 & 4 & 4 \\ 3 & 1 & 6 \\ 1 & 1 & 4 \end{vmatrix}$	$\begin{vmatrix} 2 & 12 & 6 \\ 1 & 6 & 3 \\ 1 & 0 & 3 \end{vmatrix}$	$\begin{vmatrix} 1 & 4 & 1 \\ 0 & 12 & 3 \\ 1 & 6 & 0 \end{vmatrix}$	$\begin{vmatrix} 2 & 12 & 6 \\ 1 & 6 & 3 \\ 1 & 3 & 0 \end{vmatrix}$	$\begin{vmatrix} 1 & 4 & 0 \\ 0 & 12 & 3 \\ 1 & 6 & 0 \end{vmatrix}$

Student Activity D

Section 12.5: Strange Determinants

Directions: Fill in the empty spaces in the table using Cramer's Rule. In the last column, write the solution (if one exists), or label the system as inconsistent or the equations as dependent. The first one has been done for you.

	$D =$	$D_x =$	$D_y =$	$D_z =$	**Solution?**
1.	2	3	-1	-4	$\left(\frac{3}{2}, -\frac{1}{2}, -2\right)$
2.	0	0	0	0	
3.	$\frac{1}{2}$	5	7	9	
4.	10	0	0	0	
5.	0	0	0	1	
6.	0	-1	-3	-5	
7.		15	25	30	Inconsistent
8.	0				Dependent
9.	3				$\left(6, -1, \frac{1}{2}\right)$
10.			12		$(-2, -3, 1)$

Assessment 12C

Chapter 12: End-of-chapter Assessment for Understanding

For each of the following, describe the type of problem and the strategies and key steps to remember while doing the problem. You do **not** have to complete the problems.

	Type of Problem	Strategies and Key Steps
1. Find D_x and D_y for the system: $\begin{cases} 3x-4y=7 \\ 5x-8y=8 \end{cases}$		
2. Solve the system using elimination: $\begin{cases} 3x-4y=7 \\ 5x-8y=8 \end{cases}$		
3. Solve the system using matrices: $\begin{cases} x+y+2z=1 \\ x+z=-1 \\ 3x-y-z=13 \end{cases}$		
4. If $D=0, D_x=2$, and $D_y=5$, describe the solution to the 2×2 system of equations.		
5. Solve the system below by graphing. $\begin{cases} y-3x=2 \\ 4x-y=-1 \end{cases}$		
6. Solve the system: $\begin{cases} x+y+2z=7 \\ 2x-y+z=5 \\ x+z=4 \end{cases}$		
7. Write the matrix in row-reduced form: $\left[\begin{array}{cc\|c} 1 & 3 & 3 \\ 1 & -4 & 10 \end{array}\right]$		
8. Is the pair of equations dependent or independent? $\begin{cases} 3x+y=8 \\ 6x+2y=8 \end{cases}$		

Assessment 12D

Chapter 12: Metacognitive Skills Assessment

Metacognitive skills refer to the ability to judge how well you have learned something and to effectively direct your own learning and studying. This is a self evaluation tool designed to help you focus your studying and to improve your metacognitive skills with regards to this math class.

Fill the 1st column out before you begin studying.
Fill the 2nd column out after you study and before you take the test.
Go back to this page after your test and circle any of the ratings that you would now change – this identifies the "disconnects" between what you think you know well and what you actually know well.

Use the scale below to assign a number to each topic.

5 I am confident I can do any problems in this category correctly.
4 I am confident I can do most of the problems in this category correctly.
3 I understand how to do the problems in this category, but I still make a lot of mistakes.
2 I feel unsure about how to do these problems.
1 I know I don't understand how to do these problems.

Topic or Skill	Before Studying	After Studying
Checking whether an ordered pair or ordered triple is a solution of a system of equations.		
Classifying a system of two or three equations as consistent or inconsistent.		
Classifying a system of two or three equations as dependent or independent.		
Graphing a system of two linear equations to find the solution.		
Understanding what the "solution" to a system of equations is graphically (both systems of two equations and systems of three equations).		
Finding or identifying a substitution equation for a system of two equations.		
Solving a system of two equations by substitution.		
Clearing the fractions or decimals from an equation.		
Identifying a variable to eliminate and performing the steps to set up that elimination in a system of two equations.		
Solving a system of two equations by elimination.		
Deciding whether substitution or elimination would be an easier method for solving a system of two equations.		
Solving a system of three equations with three variables.		
Declaring the variables for an application involving a system of equations.		
Writing the system of equations for a problem involving $d = rt$.		
Writing the system of equations for a value-mixture problem or a percent-solution problem.		
Writing the system of equations for an investment problem.		
Writing the system of equations for a break-point problem.		
Solving the system of equations from an application problem.		
Writing the conclusion to an application problem.		
Writing a system of equations as an augmented matrix.		
Applying row operations to a matrix.		
Using row operations to transform a matrix to row-reduced form.		

(*continued on next page*)

Topic or Skill	**Before Studying**	**After Studying**
Solving a system of equations by using matrices.		
Knowing what types of row-reduced matrices indicate inconsistent systems or dependent equations.		
Finding the determinant of a 2×2 matrix.		
Finding the determinant of a 3×3 matrix by expanding the minors.		
Knowing how to construct the determinants for using Cramer's Rule $(D, D_x, D_y, \ldots)$.		
Knowing Cramer's Rule.		
Solving a system of equations by using Cramer's Rule.		
Knowing what types of results from using Cramer's Rule indicate inconsistent systems or dependent equations.		

Student Workbook: Chapter 13

Table of Contents: *Conic Sections; More Graphing*

Assessment 13A

Pretest and Diagnostic Tool: Conic Sections; More Graphing

Directions: Complete this assessment without looking back at your notes or your book. **Do not use a calculator for this assessment.**

1. Simplify:
$x^2 - 12x + 36 + y^2 + 10y + 25$

2. Simplify: $(x-2)^2 + 3x - 4$

3. $x^2 - 10x + \_\_\_\_ = (x-5)^2$

4. $x^2 + 12x + 36 = (\_\_\_\_\_\_\_\_)^2$

5. Clear the fractions from $\dfrac{x^2}{4} + \dfrac{y^2}{16} = 1$

6. Clear the fractions from

$$\frac{(x-2)^2}{25} - \frac{(x+1)^2}{4} = 1$$

7. Solve by substitution: $\begin{cases} x + y = 3 \\ 3x + 2y = 5 \end{cases}$

8. Solve by elimination: $\begin{cases} 2x + y = 14 \\ 3x - y = 1 \end{cases}$

9. Solve: $2x^2 = 50$

10. Solve: $(x-2)^2 + 4 = (x+1)^2$

Student Activity A

Section 13.1: Constructing a Circle

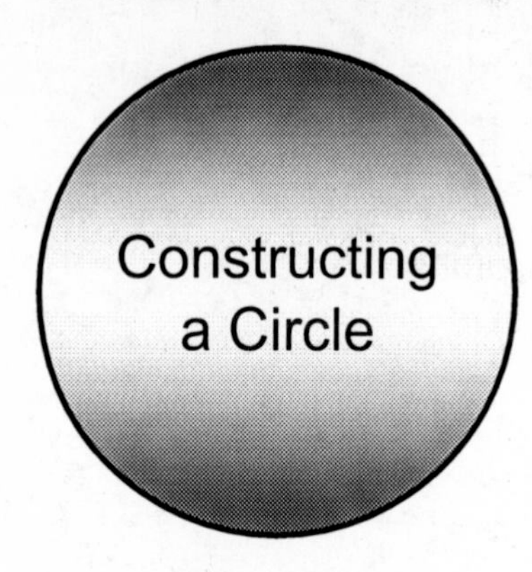

Supplies needed for one construction:

- Bulletin Board
- Tacks (3)
- Unlined paper (the bigger the better)
- String (length approx half the width of the paper)
- Crayon

Definition of a Circle: A **circle** is the *set of all points* in a *plane* that are a *equidistant* from a *fixed point* called its **center**. The fixed distance is called the **radius** of the circle.

Construct a Circle:

1. Use two tacks to affix your paper to the bulleting board.
2. Tie a loop on one end of the string around the crayon (near the tip).
3. Use the last tack to affix the other end of the string to the piece of paper (near the center of the paper).
4. Move the crayon to trace a curve, keeping the string taut.
5. Remove your drawing from the bulletin board.

Add labels:

- The crayon traced the *set of all points* equidistant from the center. In other words, the crayon traced a circle. Label your diagram **Circle**.
- The *fixed point* or *center* is represented by the hole where the center tack was. Label this point with the word **Center** and the point **(*h*, *k*)**.
- The *fixed distance* or *radius* is represented by the taut string. Draw a straight line from the center to the drawn circle. Label this line **radius** and the point where the line intersects the circle with **(*x*, *y*)**.

Student Activity B

Section 13.1: Triple Play on Circles

Instructions: In the table below, you have been given some of the information on a circle: the graph, the equation in standard form, or a description of a circle. Fill in the missing information to complete the table (and the triple play).

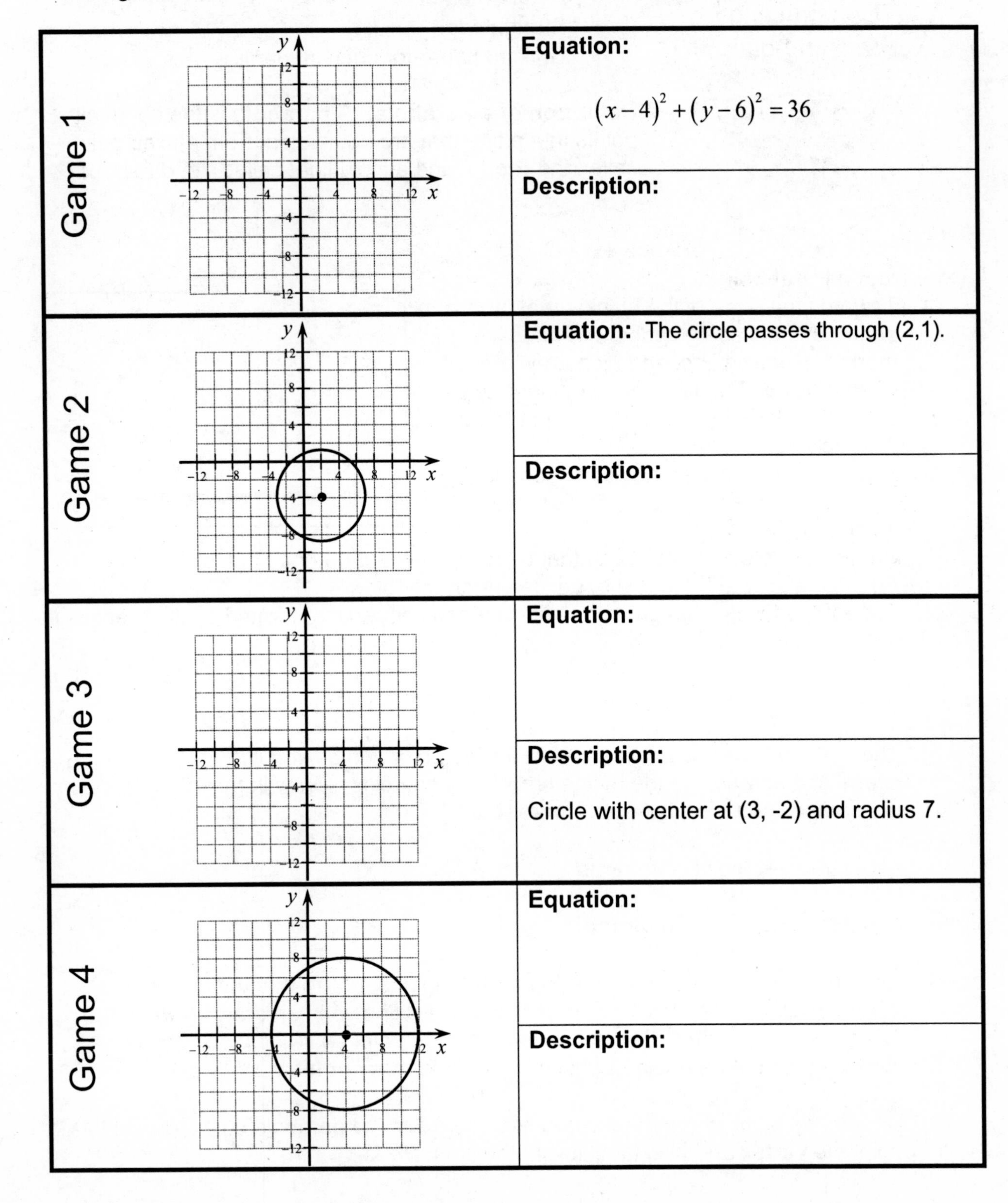

Game	Graph	Equation / Description
Game 1		**Equation:** $(x-4)^2+(y-6)^2=36$ **Description:**
Game 2		**Equation:** The circle passes through (2,1). **Description:**
Game 3		**Equation:** **Description:** Circle with center at (3, -2) and radius 7.
Game 4		**Equation:** **Description:**

Student Activity C

Section 13.1: Constructing a Parabola

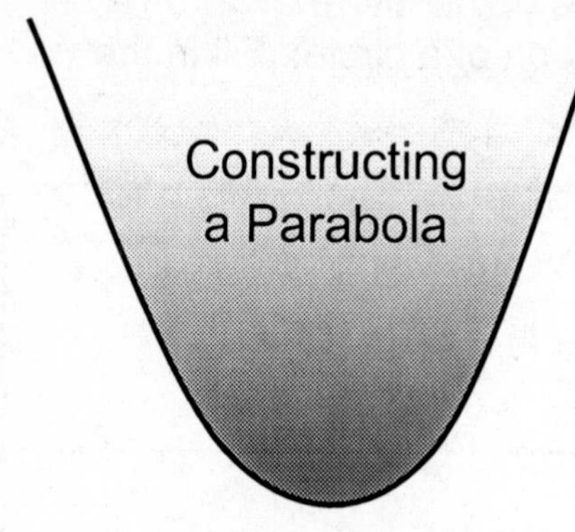

Supplies needed for one construction:

- A dark marker
- Straightedge
- Unlined paper for each student

Definition of a Parabola: A **parabola** is the set of all points in a plane that are equidistant from a fixed point, called the **focus**, and a fixed line, called the **directrix**.

Construct a Parabola:

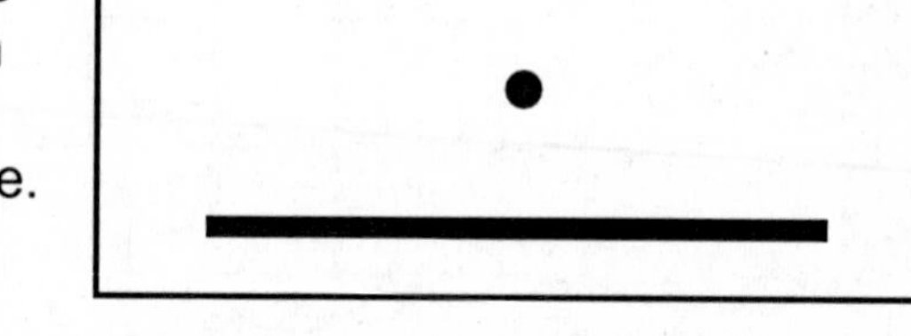

1. Towards the edge of the blank sheet of paper, use the straightedge and the dark marker to draw a line, and then draw a point off of the line. The line and point need to be dark enough that you can see them through the back side of the paper.
2. Fold the paper so that the point is on the line. Then make sure there is a good crease on the fold of the paper.
3. Unfold and refold the paper so that the point is in a different place on the line. Again, make sure there is a good crease on the fold.
4. Repeat the folding and creasing until you can see a curve formed by the creases in the paper.

Add labels:

- The *set of all points equidistant from a fixed point and a fixed line* are seen in the shape of a *parabola* made by the creases of the folds. Draw in the curve made by the creases and label this curve **parabola**.
- Label the fixed point with **focus**.
- Label the fixed line with **directrix**.
- Draw a dashed line between the focus and the directrix so that the line is perpendicular to the directrix (this is the shortest line you can draw between the focus and the directrix. Extend the line past the point so that it marks the line of symmetry on the parabola. Label this line **axis of symmetry.**
- Darken the point where the dashed line intersects the parabola curve. Label this point the **vertex** and also label it with the point **(*h*, *k*)**.

Student Activity D

Section 13.1: Triple Play on Parabolas

Instructions: In the table below, you have been given some of the information on a circle: the graph, the equation in standard form, or a description of a circle. Fill in the missing information to complete the table (and the triple play).

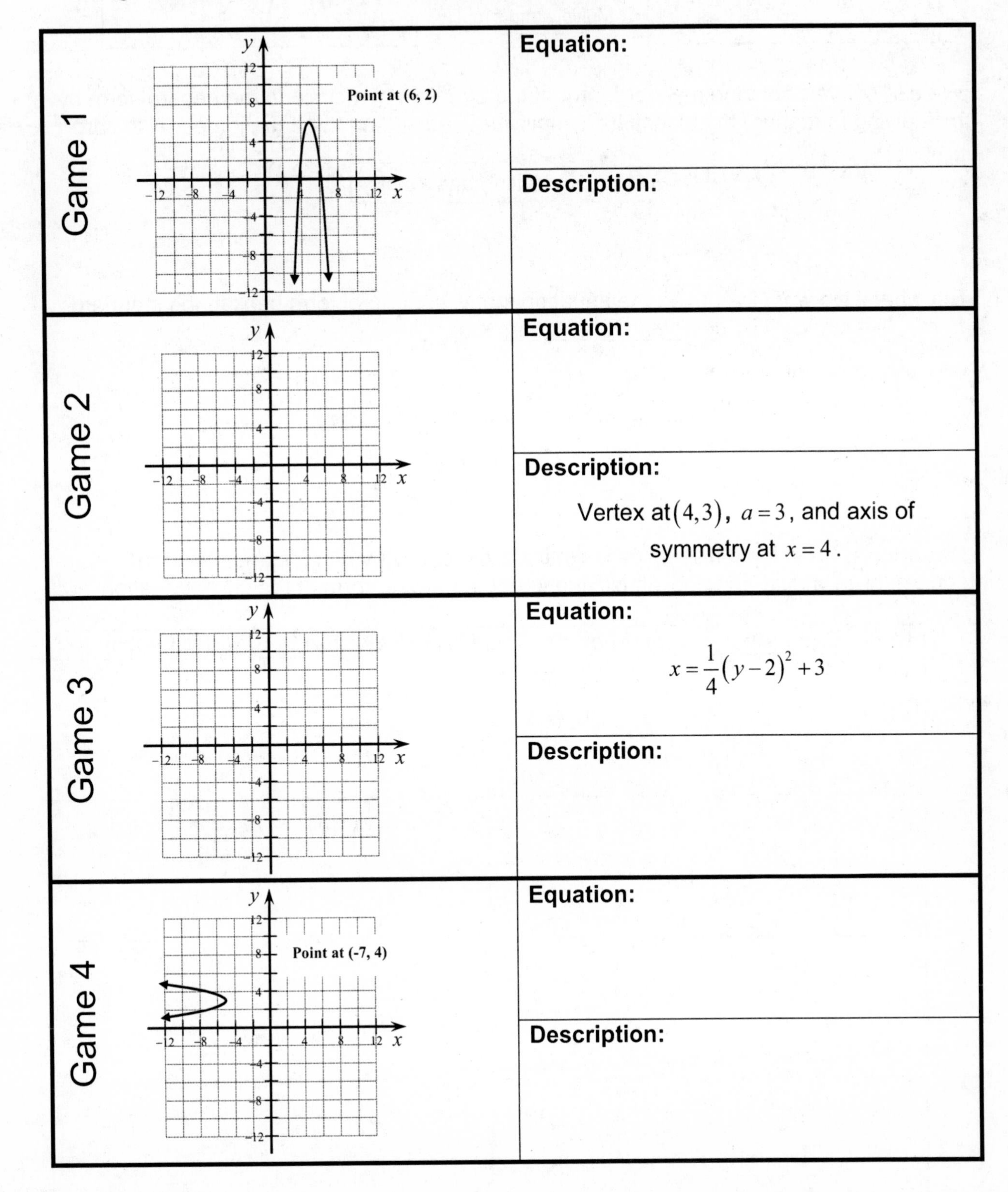

Guided Learning Activity

Section 13.1: Completing a Square to Get a Circle

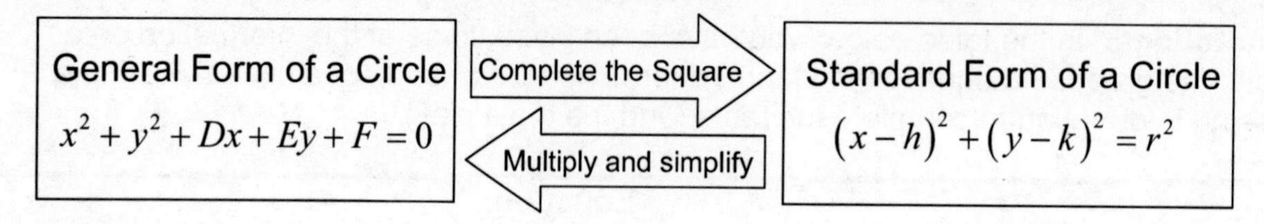

We can convert from the standard form of the equation of a circle to the general form by multiplying (squaring) the binomials, simplifying, and setting the equation equal to zero

$$(x-2)(x-2)+(y-3)(y-3)=25$$
$$x^2-4x+4+y^2-6y+9=25$$
$$x^2-4x+y^2-6y-12=0$$

Multiply and simplify ← $(x-2)^2+(y-3)^2=5^2$

But what if we want to convert the equation that is in general form back to the standard form? We can do it by <u>completing the square</u>.

$x^2+y^2+10x-12y+52=0$

Complete the Square →

$$x^2+10x+\_\_\_\_+y^2-12y+\_\_\_\_=-52$$
$$x^2+10x+\underline{25}+y^2-12y+\underline{36}=-52\underline{+25+36}$$
$$(x+5)^2+(y-6)^2=9$$

Directions: Fill in the missing spaces in the table below by either completing the square, or by multiplying and simplifying to get the proper form of the circle equation.

	Circle Equation in General Form	Circle Equation in Standard Form
1.	$x^2+y^2+2x-14y+34=0$	
2.		$(x+4)^2+(y-5)^2=4$
3.	$x^2+y^2-4x-10y-7=0$	
4.		$x^2+(y+3)^2=25$

Parabola General Form

$y = ax^2 + bx + c$

$x = ay^2 + by + c$

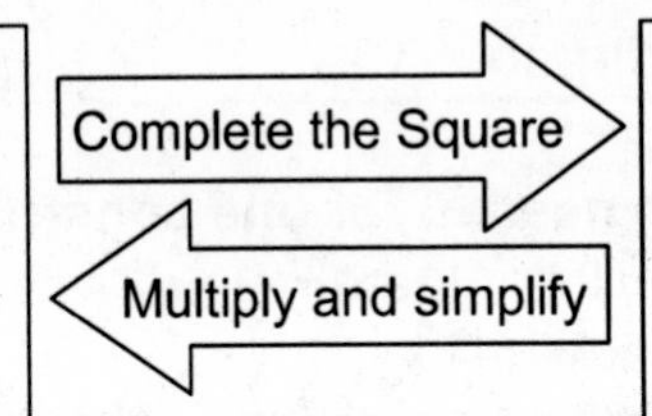

Parabola Standard Form

$y = a(x-h)^2 + k$

$x = a(y-k)^2 + h$

Again, we can convert from the standard form of a parabola to the general form by multiplying (squaring) the binomial and simplifying.

$y = 3(x^2 - 4x + 4) + 5$ ⟵ Multiply and simplify ⟵ $y = 3(x-2)^2 + 5$

$y = 3x^2 - 12x + 12 + 5$

$y = 3x^2 - 12x + 17$

To convert from the general form to the standard form, we complete the square.

$y = 3x^2 - 12x + 17$ ⟶ Complete the Square ⟶

$y = 3x^2 - 12x + 17$

$y = 3(x^2 - 4x + \_\_\_ - \_\_\_) + 17$

$y = 3(x^2 - 4x + \underline{4} - \underline{4}) + 17$

$y = 3(x^2 - 4x + 4) + 17 - 12$

$y = 3(x-2)^2 + 5$

	Parabola Equation in General Form	Parabola Equation in Standard Form
1.	$y = 2x^2 - 20x + 53$	
2.		$y = -2(x+4)^2 - 1$
3.	$x = y^2 + 12y + 41$	
4.		$x = -4(x-3)^2 + 2$

Student Activity A

Section 13.2: Constructing an Ellipse

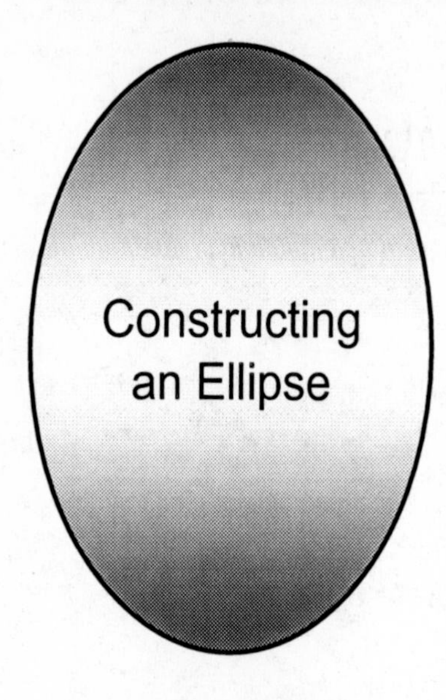

Supplies needed for one construction:

- Bulletin Board
- Tacks (4)
- Unlined paper (the bigger the better)
- String (length less than the length of the paper)
- Felt tip pen (ballpoint will puncture paper on bulletin board)
- Straightedge

Definition of an Ellipse: An **ellipse** is the *set of all points* in a plane for which the *sum of the distances from two fixed points* is a constant. The fixed points are called the **foci** of the ellipse. The **center** is at the midpoint of the two foci. The line segment that forms the longest distance across the ellipse is called the **major axis**. The line segment that forms the smallest distance across the ellipse is called the **minor axis**. The endpoints of the major axis are called **vertices**.

Construct an Ellipse:

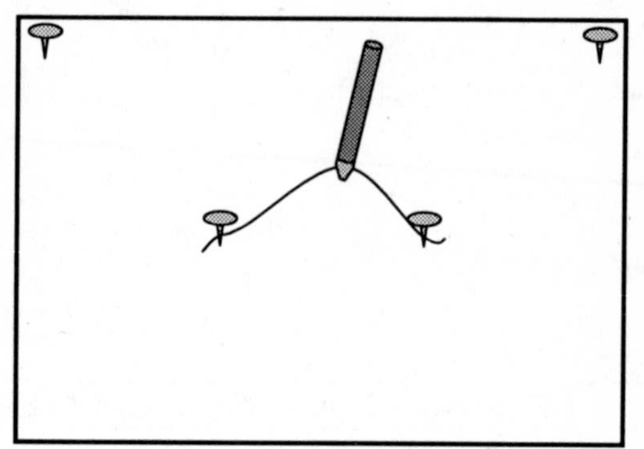

1. Use two tacks to affix your paper to the bulletin board.
2. Use the other two tacks to affix the ends of the string to the piece of paper. Place these to the left and right of the center of the paper (as shown). Leave some slack In the string (do not pull it taut).
3. Use the felt-tip pen to pull the string taut on the upper half of the paper (as shown). Move the pen to trace a curve, keeping the string taut.
4. After drawing half of the ellipse you will have to pick up the pencil an repeat the previous step on the lower half of the paper.
5. Remove your drawing from the bulletin board.

Add labels:

- The pen traced the *set of all points* in a plane for which the sum of the distances from two fixed points is a constant (the length of the string remained constant). Label your diagram with **Ellipse**.
- The points where the string was held down by a tack were the foci of the ellipse. Label each of these points with the word **focus**.
- Mark the point that is exactly halfway between the two foci and label it the **center**.
- Draw a line segment through the foci and center that extends to the ellipse curve. Label this line segment the **major axis**.
- Draw a line segment through the center that is perpendicular to the major axis that extends to the ellipse curve. Label this line segment the **minor axis**.
- Label each point at the end of the major axis with **vertex**.

Guided Learning Activity

Section 13.2: Graphing an Ellipse

To graph an ellipse from a standard equation: $\dfrac{(x-h)^2}{a^2}+\dfrac{(y-k)^2}{b^2}=1$

1. Find the center, (h,k), and plot it with a small $\times$.
2. Use a to find two points on the ellipse to the left and right of the center.
3. Use b to find two points on the ellipse above and below the center.
4. Draw a smooth curve to represent the ellipse.

Example: Graph $\dfrac{(x+2)^2}{36}+\dfrac{(y+4)^2}{64}=1$

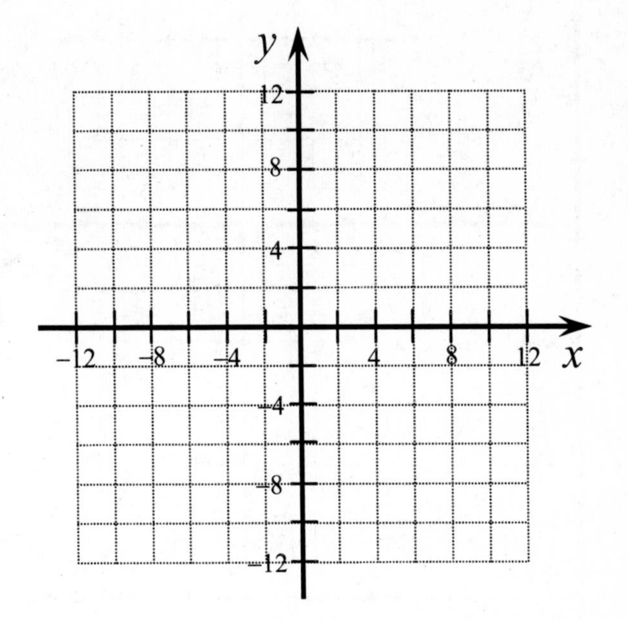

1. The center is at $(-2,-4)$. We plot that with a small $\times$.
2. Since $a^2=36$, we deduce $a=6$. Plot a point 6 units to the left of $(-2,-4)$ and another point 6 units to the right of $(-2,-4)$.
3. Since $b^2=64$, we deduce $b=8$. Plot a point 8 units above $(-2,-4)$ and another point 8 units below $(-2,-4)$.
4. Draw the smooth curve of the ellipse.

Now try these!

1. Graph: $\dfrac{x^2}{25}+\dfrac{(y+5)^2}{4}=1$

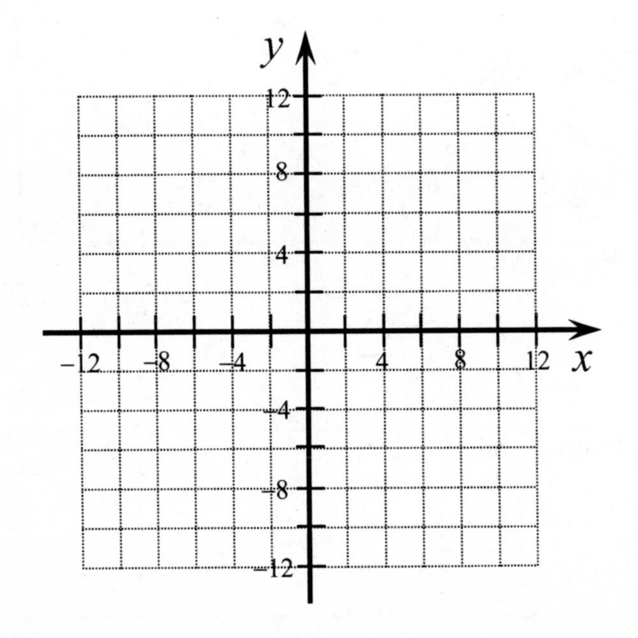

2. Graph: $\dfrac{(x-3)^2}{49}+\dfrac{(y+2)^2}{16}=1$

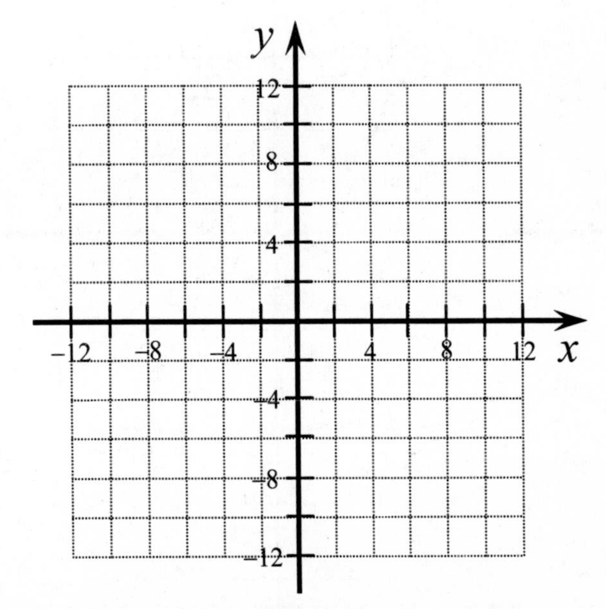

Student Activity B

Section 13.2: Triple Play on Ellipses

Instructions: In the table below, you have been given some of the information on an ellipse: the graph, the equation in standard form, or a description of a circle. Fill in the missing information to complete the table (and the triple play).

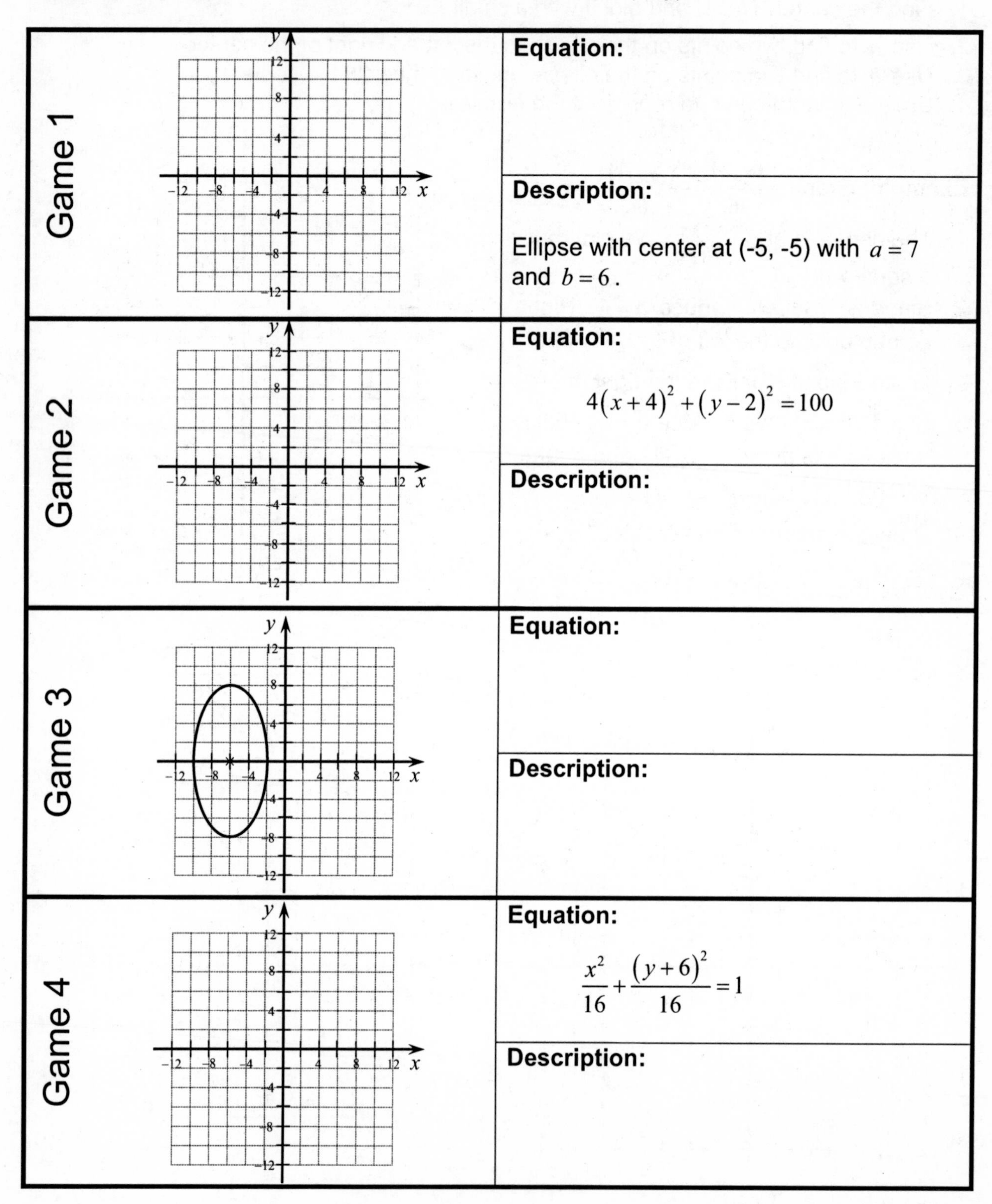

Game	Graph	Equation / Description
Game 1		**Equation:**
		Description: Ellipse with center at (-5, -5) with $a = 7$ and $b = 6$.
Game 2		**Equation:** $4(x+4)^2 + (y-2)^2 = 100$
		Description:
Game 3		**Equation:**
		Description:
Game 4		**Equation:** $\frac{x^2}{16} + \frac{(y+6)^2}{16} = 1$
		Description:

Assessment 13B

Chapter 13: Mid-chapter Assessment for Understanding

For each of the following, describe the type of problem and the strategies and key steps to remember while doing the problem. You do **not** have to complete the problems.

	Type of Problem	Strategies and Key Steps
1. Find the vertex of $y=2(x-4)^2+5$.		
2. Graph: $x=(y-1)^2+2$		
3. Graph: $x^2+(y-1)^2=16$		
4. Graph: $9(x-1)^2+25(y+2)^2=225$		
5. Write the circle $x^2+2x+y^2-10y=23$ in standard form.		
6. Write the parabola $y=-(x+3)^2-4$ in general form.		
7. An ellipse is centered at $(-10,4)$ with $a=9$ and $b=3$. Write the equation for the ellipse in standard form.		
8. A circle is centered at $(0,7)$ with radius $\sqrt{3}$. Write the equation for the circle in standard form.		

Student Activity A

Section 13.3: Constructing a Hyperbola

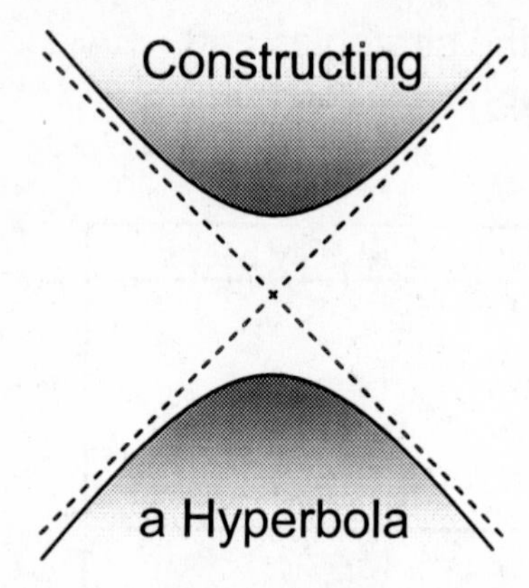

Supplies Needed:

- Unlined paper
- Dark marker or felt-tip pen (if tracing)
- Ruler (a ruler with millimeter markings will work best)
- Hyperbola template

Definition of a Hyperbola: A **Hyperbola** is the set of all points in a plane for which the difference of the distances from two fixed points is a constant.

Construct a Hyperbola:

1. Trace the template of the circle and center on your unlined paper (or use the one provided by your instructor.
2. Choose one of the points on the circle (the ones with the crosshairs) and fold the paper so that the crosshairs fall on point **P**. Crease well.
3. Choose the next point with crosshairs and fold the paper so that these crosshairs fall on point **P**. Crease well.
4. Repeat until you have folded every point with crosshairs to fall on **P**.
5. The creases in the paper will form two curved shapes. We call this a hyperbola. Sketch in the curves along the creases with a pencil.

Add labels:

- The *set of all points in a plane for which the difference of the distances from two fixed points is a constant* is the curve you have sketched (two curves actually). Label this drawing with **hyperbola**.

- The *two fixed points,* called the foci, are the center of the circle and the point drawn outside the circle. Label the two points (C and P) with the word **focus**.

- Pick a point on the hyperbola and label it $\mathbf{P_1}$. Use the ruler to draw a straight line from **C** to $\mathbf{P_1}$. Then measure the distance and label the line segment with its distance. Draw a straight line from the **P** to $\mathbf{P_1}$. Again, measure the distance and label the line segment with its distance.

- Repeat this procedure for a second point labeled $\mathbf{P_2}$.

- For $\mathbf{P_1}$, calculate the difference between the two distances. For $\mathbf{P_2}$, calculate the difference between the two distances. Are the differences constant?

Template: Constructing a Hyperbola

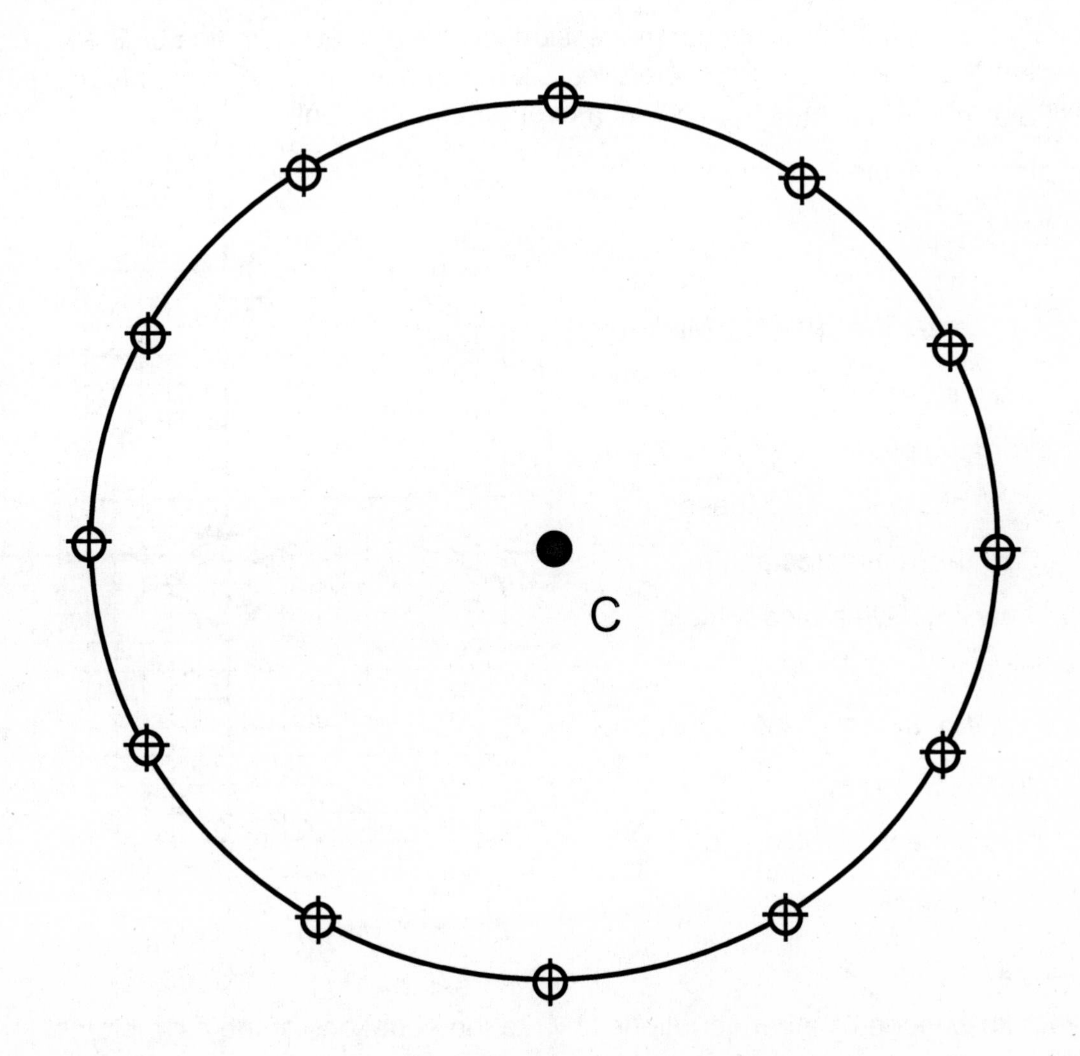

Guided Learning Activity

Section 13.3: Graphing a Hyperbola

Hyperbola centered at (h,k) opening left and right: $\dfrac{(x-h)^2}{a^2}-\dfrac{(y-k)^2}{b^2}=1$

To graph this hyperbola:

1. Plot the **center**, (h,k) with a small $\times$.
2. Draw the **central rectangle**.
   a. From the center, move a units left and plot a point. Then move a units right and plot a point. This marks the left and right sides of the rectangle. Also, these points mark the **vertices** of the hyperbola.
   b. From the center, move b units up and plot a point. Then move b units down and plot a point. This marks the top and bottom sides of the rectangle.
   c. Draw in the rectangle with a light dashed line.
3. Draw the **asymptotes** with dashed lines. The asymptotes are formed by the diagonals of the central rectangle.
4. Find the x- and y- **intercepts** (if they exist). Plot them.
5. Use the plotted points and the asymptotes to sketch a graph of the hyperbola.

Notice that a is under the variable expression involving x and it indicates the horizontal movement to create the central rectangle. Likewise, b is under the variable expression involving y and it indicates the vertical movement for the central rectangle.

Example: Graph the hyperbola:

$$\frac{x^2}{36}-\frac{(y-1)^2}{1}=1$$

1. The center is at $(0,-1)$. Mark this with a small $\times$.
2. From the equation, we see that $a=6$ and $b=1$. Use these to draw the central rectangle.
3. Draw the asymptotes with dashed lines.
4. Find the x- and y-intercepts and plot these points.
5. Sketch the hyperbola.

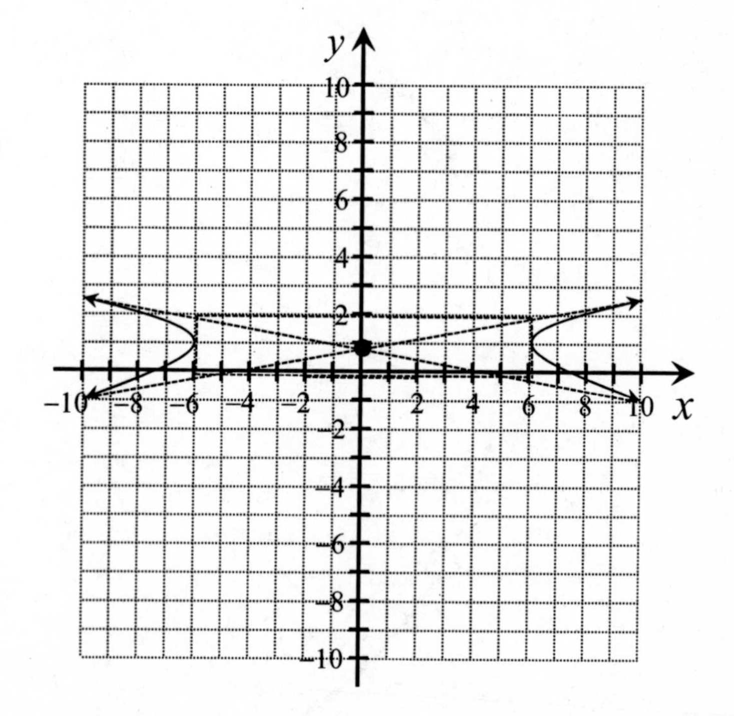

Challenge:

Use your knowledge of linear equations to write the equations of the asymptotes!

Now try these!

1. Graph: $\frac{(x-1)^2}{9}-\frac{(y+2)^2}{16}=1$

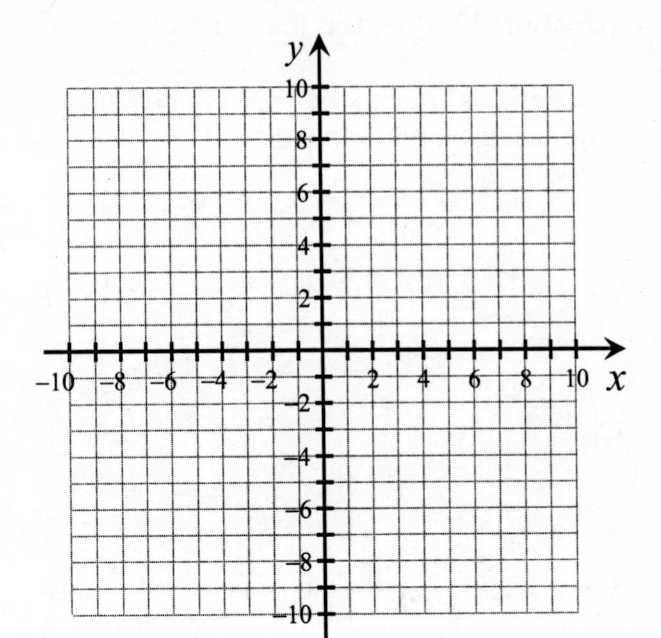

a. Center: ________

b. $a=$ ____ $b=$ ____

c. Vertices: _______ and _______

d. x-intercepts:

e. y-intercepts:

Hyperbola centered at (h,k) opening **up and down**: $\frac{(y-k)^2}{a^2}-\frac{(x-h)^2}{b^2}=1$

Everything is the same **except** that a now indicates the central rectangle by moving up and down from the center, and b is the number of units left and right of the center.

Notice that a is under the variable expression involving y and it indicates the vertical movement to create the central rectangle. Likewise, b is under the variable expression involving x and it indicates the horizontal movement for the central rectangle.

2. Graph: $\frac{y^2}{4}-\frac{(x-3)^2}{9}=1$

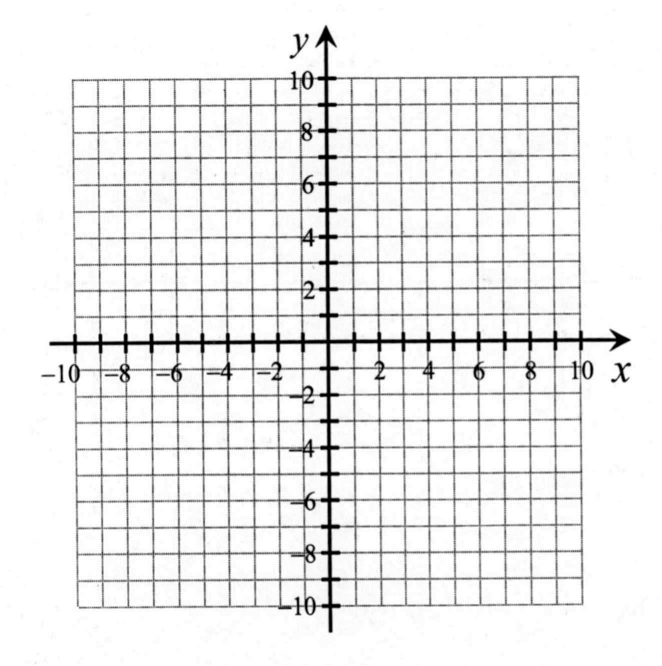

a. Center: ________

b. $a=$ ____ $b=$ ____

c. Vertices: _______ and _______

d. x-intercepts:

e. y-intercepts:

Student Activity B

Section 13.3: Graph that Conic!

Directions: Name the type of conic from the equation (circle, parabola, ellipse, or hyperbola), and then graph it!

1. $y = -2(x-3)^2 + 7$

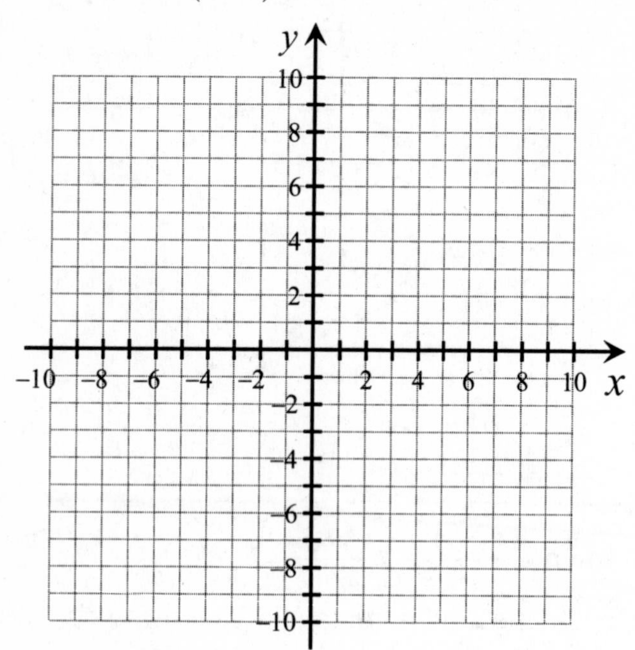

2. $4(x-4)^2 + 25(y+6)^2 = 100$

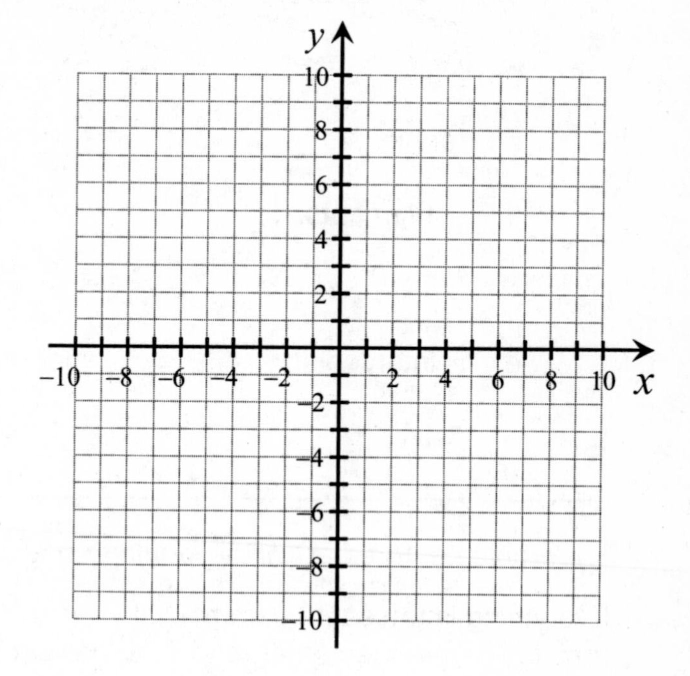

3. $\dfrac{x^2}{16} - \dfrac{y^2}{36} = 1$

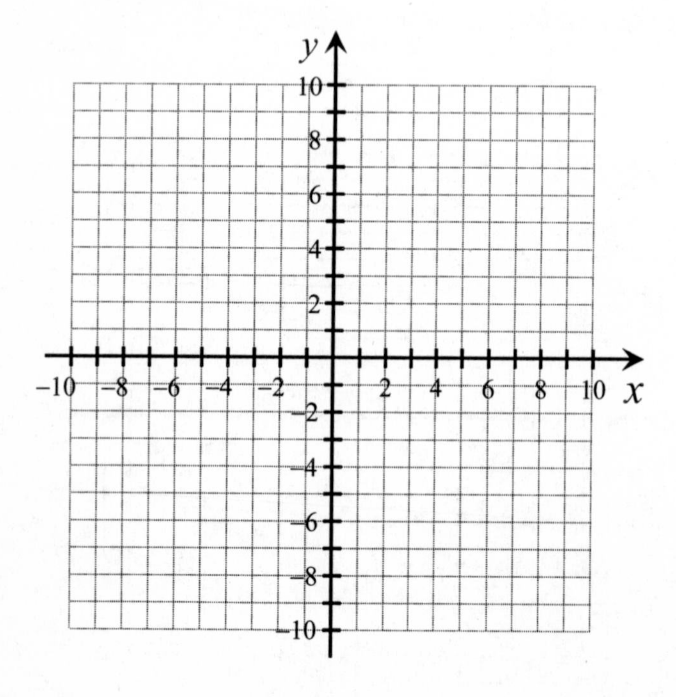

4. $(x+1)^2 + (y-5)^2 = 9$

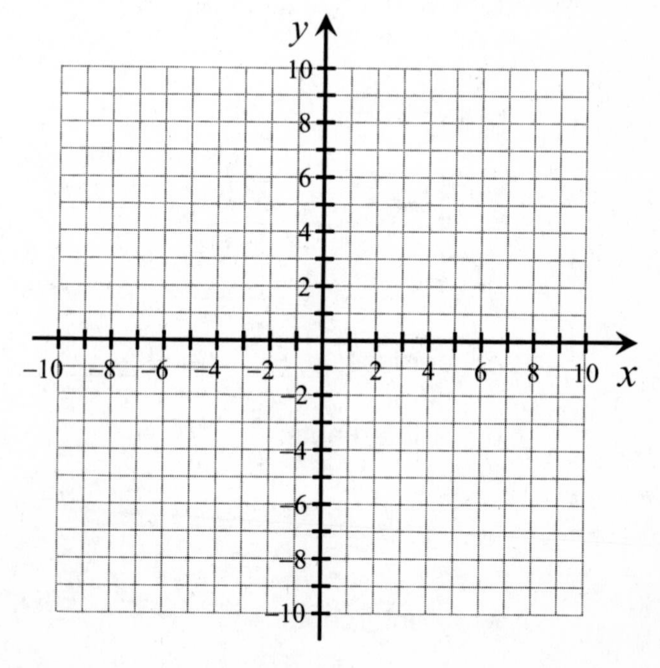

Student Activity C

Section 13.3: Name that Conic!

Match-up: Match each of the equations in the squares of the grid below with the conic classification. You may have to rearrange or simplify equations to determine their classification.

A Circle
B Ellipse
C Parabola opening up or down
D Parabola opening right or left
E Hyperbola opening right and left
F Hyperbola opening up and down
G Something else

$\frac{x^2}{25}-\frac{y^2}{9}=1$	$x=y^2+1$	$x^2+4y^2=4$
$x^2+3(y-3)^2=9$	$x^2+(y-3)^2=1$	$x^2+(y-3)^2=9$
$y-(x-4)^2=5$	$y=\frac{3}{x}$	$9(y-4)^2-4(x+1)^2=36$
$x+(y+9)^2=3$	$x^2+y^2=1$	$x^2+4x-y^2+2y=6$
$x^2+5x+3-(x^2+2x)=y$	$25x^2+100x-4y^2=100$	$\frac{x^2}{4}+\frac{y^2}{4}=1$

Guided Learning Activity

Section 13.4: Solving Nonlinear Systems Algebraically

Example 1: Solve: $\begin{cases} y = -x + 2 \\ y = -\frac{1}{2}(x-2)^2 + 4 \end{cases}$

First graph the equations to get an idea of where the points of intersection are.

$y = -x + 2$ is the equation of a __________________.

$y = -\frac{1}{2}(x-2)^2 + 4$ is the equation of a ____________________.

Estimate the solutions from the graph: (___,___) and (___,___)

Solve by substitution:

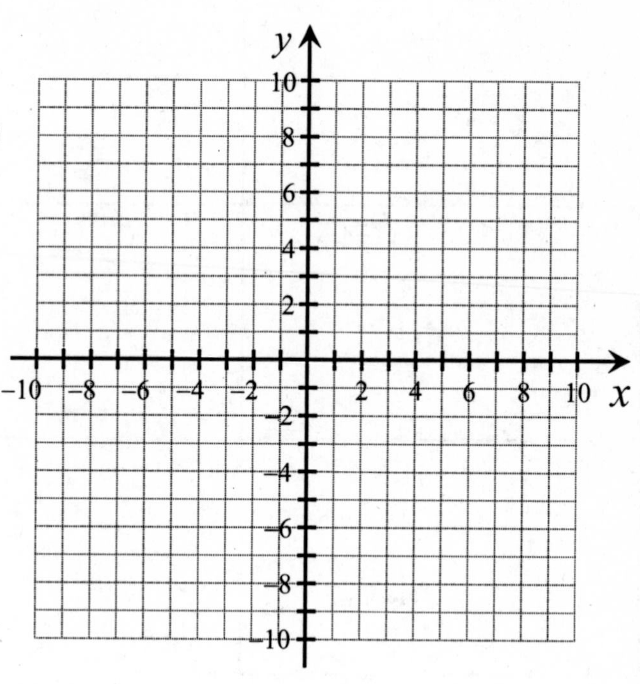

Substitute $y = -x + 2$ into the 2nd equation and solve for x:

$$-x + 2 = -\frac{1}{2}(x-2)^2 + 4$$

To find the solutions, substitute each value of x an original equation and evaluate for y:

When $x =$ ___, $y =$ ____.　　　　When $x =$ ___, $y =$ ____.

Algebraically, we find the solutions to the system are (___,___) and (___,___).

Example 2: Solve $\begin{cases} x^2 - y^2 = 1 \\ x^2 + y^2 = 49 \end{cases}$

First graph the equations to get an idea of where the points of intersection are.

$x^2 - y^2 = 1$ is the equation of a ________________.
$x^2 + y^2 = 49$ is the equation of a ________________.
Estimate the solutions from the graph: (___,___), (___,___), (___,___) and (___,___)

Solve by Elimination:

Notice that the y^2- terms can be eliminated by adding the equations:

$$\begin{array}{l} x^2 - y^2 = 1 \\ \underline{x^2 + y^2 = 49} \\ 2x^2 \qquad = 50 \end{array}$$

Solve this equation for x:
(HINT: There are two solutions.)

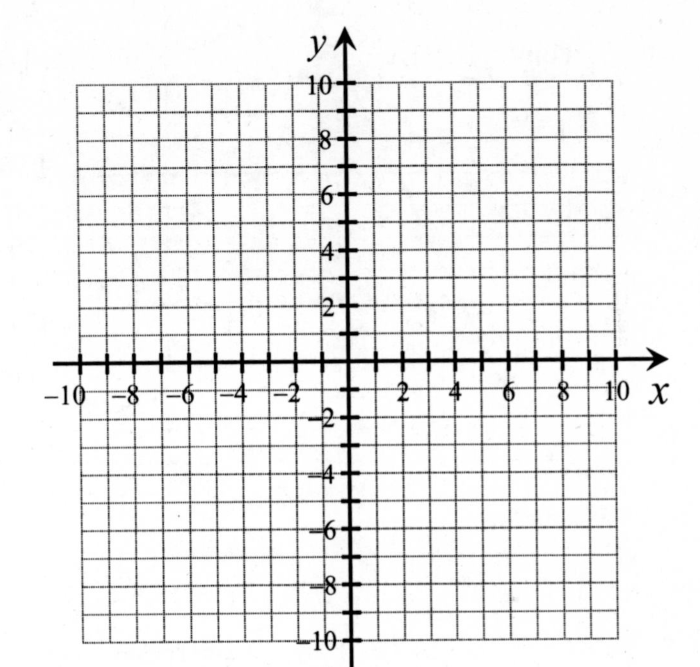

To find y, substitute each value of x into one of the original equations.
(HINT: Again, there are two solutions to each of these.)

When $x =$ ____, $y =$ ____. When $x =$ ____, $y =$ ____.

Algebraically, we find the solutions to the system are (___,___), (___,___), (___,___),

and (___,___).

Student Activity C

Section 13.4: Nonlinear System Detective

Instructions: Each graph shows a system of two equations where one of the equations is nonlinear. Write the equations then solve the system algebraically.

1.

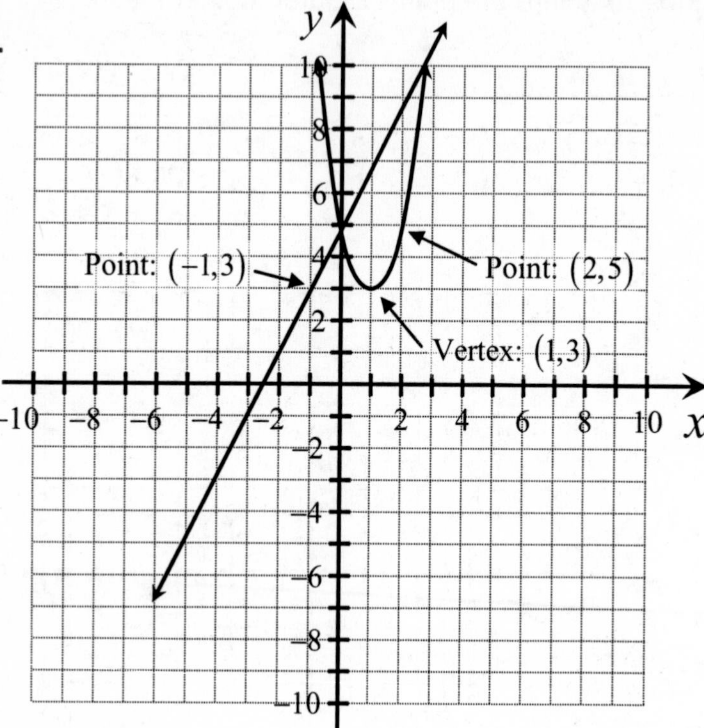

Equations:

Solve:

2.

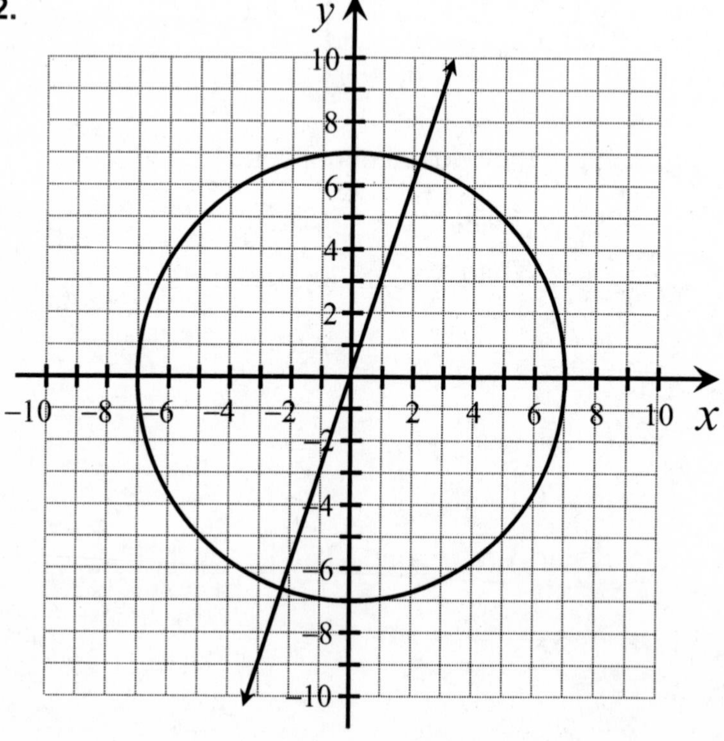

Equations:

Solve:

Assessment 13C

Chapter 13: End-of-chapter Assessment for Understanding

For each of the following, describe the type of problem and the strategies and key steps to remember while doing the problem. You do **not** have to complete the problems.

	Type of Problem	Strategies and Key Steps
1. Write the equation of the ellipse $9x^2 + y^2 - 6y = 27$ in standard form.		
2. Graph: $xy = 4$		
3. Solve: $\begin{cases} x^2 + y^2 = 12 \\ y = 2x - 2 \end{cases}$		
4. Graph: $\frac{(y-1)^2}{9} - \frac{x^2}{4} = 1$		
5. Write the equation of the hyperbola $9x^2 - 25y^2 = 225$ in standard form.		
6. Solve: $\begin{cases} 3x^2 + y^2 = 52 \\ x^2 - y^2 = 12 \end{cases}$		
7. Graph: $(x+2)^2 + (y-3)^2 = 1$		
8. Write the equation of the parabola $x = y^2 - 14y + 54$ in standard form.		

Assessment 13D

Chapter 13: Metacognitive Skills Assessment

Metacognitive skills refer to the ability to judge how well you have learned something and to effectively direct your own learning and studying. This is a self evaluation tool designed to help you focus your studying and to improve your metacognitive skills with regards to this math class.

Fill the 1st column out before you begin studying.
Fill the 2nd column out after you study and before you take the test.
Go back to this page after your test and circle any of the ratings that you would now change – this identifies the "disconnects" between what you think you know well and what you actually know well.

Use the scale below to assign a number to each topic.
5 *I am confident I can do any problems in this category correctly.*
4 *I am confident I can do most of the problems in this category correctly.*
3 *I understand how to do the problems in this category, but I still make a lot of mistakes.*
2 *I feel unsure about how to do these problems.*
1 *I know I don't understand how to do these problems.*

Topic or Skill	Before Studying	After Studying
Knowing the equations for the standard form of a circle and ellipse.		
Identifying the center of a circle from its equation in standard form.		
Graphing a circle or ellipse from the standard equations.		
Completing the square to build a squared binomial.		
Converting from the standard equation of a circle or ellipse to the general equation.		
Converting from the general equation of a circle or ellipse to the standard equation.		
Describing a circle or ellipse from its graph.		
Knowing the equations for the standard form of a parabola.		
Identifying whether a parabola will open up, down, left, or right by examining its equation.		
Graphing a parabola from the standard equation.		
Knowing the equations for the standard form of a hyperbola.		
Identifying whether a hyperbola will open left and right, or up and down.		
Graphing a hyperbola from its equation in standard form.		
Identifying the type of conic by inspecting the equation for the conic.		
Identifying $xy = k$ as an equation for a hyperbola.		
Estimating the solution(s) to a nonlinear system of equations by graphing the system.		
Solving a nonlinear system of equations by substitution.		
Solving a nonlinear system of equations by elimination.		

Student Workbook: Chapter 14

Table of Contents: *Miscellaneous Topics*

Assessment 14A

Pretest and Diagnostic Tool: Miscellaneous Topics

Directions: Complete this assessment without looking back at your notes or your book. **Do not use a calculator for this assessment.**

1. Multiply: $(x+3)(x+3)$

2. Multiply: $(a+b)^2$

3. Multiply: $(x+2)(x^2+4x+4)$

4. Simplify: $\dfrac{5\cdot4\cdot3\cdot2\cdot1}{3\cdot2\cdot1\cdot2\cdot1}$

5. Simplify: $10\cdot x\cdot(-2)^4$

6. Simplify: $56(2x)^3(-1)^5$

For 7-10: Identify the pattern and then fill in the blanks.

7. 2, 5, 8, 11, ___, ___

8. 250, 50, 10, ___, ___

9. $\frac{2}{3}, \frac{4}{9}, \frac{8}{27}$, ___, ___

10. Simplify $a+(n-1)d$ for $d=4$ and $a=-3$.

11. Solve for n: $40=4+6(n-1)$

12. Solve for d: $35=-5+(6-1)d$

13. Evaluate: $24\left(\frac{1}{2}\right)^5$

14. Solve for r: $16=54r^3$

15. If $r=\frac{2}{5}$, solve for a: $\frac{8}{25}=ar^3$

Guided Learning Activity:

Section 14.1: Pascal's Triangle

$$(a+b)^0 = \longrightarrow 1$$

$$(a+b)^1 = \longrightarrow 1a + 1b$$

$$(a+b)^2 = \longrightarrow 1a^2 + 2ab + 1b^2$$

$$(a+b)^3 = \longrightarrow 1a^3 + 3a^2b + 3ab^2 + 1b^3$$

$$(a+b)^4 = \longrightarrow 1a^4 + 4a^3b + 6a^2b^2 + 4ab^3 + 1b^4$$

$$(a+b)^5 = \rightarrow 1a^5 + 5a^4b + 10a^3b^2 + 10a^2b^3 + 5ab^4 + 1b^5$$

$$(a+b)^6 = 1a^6 + 6a^5b + 15a^4b^2 + 20a^3b^3 + 15a^2b^4 + 6ab^5 + 1b^6$$

Let's make some observations about the binomial expansions for $(a+b)^n$:

1. Relate the number of terms in each row to the power on the binomial.

2. In any row expansion, what is the degree of each term? How does that relate to the power on the binomial?

3. What is special about the first and last terms?

4. How can you predict the powers on a and b for each term in an expansion?

5. How are the coefficients in one row related to the coefficients in the next row?

Now let's examine the coefficients of each expansion. This array of numbers (forming a triangle) is called Pascal's triangle.

$$(a+b)^0 = 1$$

$$(a+b)^1 = 1a + 1b$$

$$(a+b)^2 = 1a^2 + 2ab + 1b^2$$

$$(a+b)^3 = 1a^3 + 3a^2b + 3ab^2 + 1b^3$$

$$(a+b)^4 = 1a^4 + 4a^3b + 6a^2b^2 + 4ab^3 + 1b^4$$

$$(a+b)^5 = 1a^5 + 5a^4b + 10a^3b^2 + 10a^2b^3 + 5ab^4 + 1b^5$$

$$(a+b)^6 = 1a^6 + 6a^5b + 15a^4b^2 + 20a^3b^3 + 15a^2b^4 + 6ab^5 + 1b^6$$

6. Fill in the missing coefficients for the expansion of $(a+b)^7$ and $(a+b)^8$.

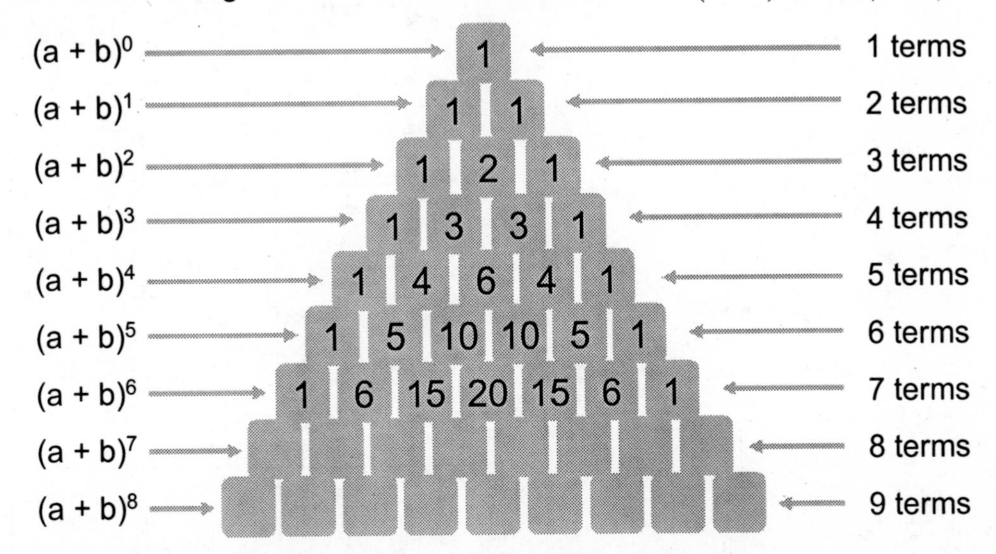

Use these patterns to fill in the terms of each expansion for problems 7-10:

7. $(a+b)^7 =$ ______ + ______ + ______ + ______ + ______ + ______ + ______ + ______

8. $(x+y)^5 =$ ______ + ______ + ______ + ______ + ______ + ______

9. $(a+2)^4 =$ ______ + ______ + ______ + ______ + ______

10. $(a+(-b))^6 =$ ______ + ______ + ______ + ______ + ______ + ______ + ______

Student Activity A

Section 14.1: Paint by Factorials

Directions: Evaluate each of the factorial expressions by simplifying first. Then shade in the box in the grid that corresponds to the simplified expression.

Example: $\frac{8!}{5!}=\frac{8\cdot7\cdot6\cdot5\cdot4\cdot3\cdot2\cdot1}{5\cdot4\cdot3\cdot2\cdot1}=\frac{8\cdot7\cdot6\cdot\cancel{5}^1\cdot\cancel{4}^1\cdot\cancel{3}^1\cdot\cancel{2}^1\cdot\cancel{1}^1}{\cancel{5}^1\cdot\cancel{4}^1\cdot\cancel{3}^1\cdot\cancel{2}^1\cdot\cancel{1}^1}==8\cdot7\cdot6=336$

1. Evaluate: $5!$

2. Evaluate: $\frac{100!}{99!}$

3. Evaluate: $\frac{99!}{100!}$

4. Evaluate: $\frac{4!}{0!}$

5. Evaluate: $\frac{10!}{8!(10-8)!}$

6. Evaluate: $4!\cdot3!$

7. Evaluate: $\frac{9!}{6!(9-6)!}$

8. Evaluate: $\frac{2!\cdot3!}{6!}$

9. Evaluate: $\frac{12!}{9!\cdot3!}$

10. Evaluate: $\frac{9!\cdot8!}{7!\cdot10!}$

1	100	24	3
5	$\frac{1}{100}$	$\frac{1}{60}$	$\frac{7}{2}$
0	220	120	2
$\frac{1}{3}$	$\frac{4}{5}$	144	12!
$\frac{1}{5}$	$\frac{1}{6}$	$\frac{1}{12}$	12
10!	84	45	undefined

Student Activity B

Section 14.1: Calculators and Factorials

For this activity you will need to find the factorial function on your calculator. First, check above the keys on your calculator for the exclamation point that symbolizes the factorial function (you may have to use a [2nd] or [SHIFT] to use it. Sometimes the factorial function is buried in some kind of [MATH] menu. If you are unable to locate the factorial function in any of these places, you can try looking in the catalog for your calculator. The catalog should give you a list of all of your calculator's available functions. Scroll through this list until you find **!** or **Factorial**, and then select the option.

Once you have found the factorial function on your calculator, use it to evaluate the following factorial expressions:

1. Evaluate: $10!$

2. Evaluate: $\frac{25!}{19!}$

3. Evaluate: $5!3!$

4. Evaluate: $8!0!$

Your calculator is able to evaluate some factorial expressions that result in **very** large numbers. However, even calculators run out of space eventually.

5. What does your calculator do if you try to evaluate $\frac{500!}{495!}$?

Most calculators are unable to calculate either $500!$ or $495!$. The numbers are simply too large! In this case, we need to first simplify the factorial by reducing common factors. **Then** we can use our calculator to evaluate the simplified expression.

$$\begin{aligned}\frac{500!}{495!} &= \frac{500\cdot 499\cdots 496\cdot 495\cdot 494\cdots 3\cdot 2\cdot 1}{495\cdot 494\cdots 3\cdot 2\cdot 1}\\ &= \frac{500\cdot 499\cdots 496\cdot \cancel{495}^{1}\cdot \cancel{494}^{1}\cdots \cancel{3}^{1}\cdot \cancel{2}^{1}\cdot \cancel{1}^{1}}{\cancel{495}^{1}\cdot \cancel{494}^{1}\cdots \cancel{3}^{1}\cdot \cancel{2}^{1}\cdot \cancel{1}^{1}}\\ &= 500\cdot 499\cdot 498\cdot 497\cdot 496\\ &= 30{,}629{,}362{,}512{,}400\end{aligned}$$

Now try evaluating these factorial expressions using the same technique:

6. Evaluate: $\frac{315!}{310!}$

7. Evaluate: $\frac{200!}{198!}$

Student Activity C

Section 14.1: Mathemagic Trick

Directions: Mathemagicians often use tricks learned from binomial expansions to speed up their "magic" calculations. In this activity, we're going to learn how to speedily square a 2-digit binomial. First, put away your calculators – you're not going to amaze your friends unless you can do this in your head!

We know the expansion for a binomial squared: $(a+b)^2 = a^2 + 2ab + b^2$

Let's use this to find 26^2:

Step 1: Break the two-digit number into its expanded form.

Example: $26 = 20 + 6 = a + b$

Step 2: Find the squared terms a^2 and b^2, add them and hold them in your memory.

Example: $20^2 = 400$ $\quad 6^2 = 36$ $\quad 400 + 36 = 436$

Step 3: Find the middle term, $2ab$ (multiply a and b, then double it).

Example: $20 \cdot 6 = 120$ $\quad 2 \cdot 120 = 240$

Step 4: Add the middle term to the number in your memory.

Example: $436 + 240 = 676$

Do the next three problems on paper to see if you've got the hang of it:

1. 41^2 **2.** 84^2 **3.** 73^2

Now try these in your head.

4. 28^2 **5.** 19^2 **6.** 65^2

Student Activity A:

Section 14.2: A Sequence of Tables

Directions: Fill in the missing information in the table below to demonstrate your knowledge of arithmetic sequences.

Level 1: Sequences in General

Sequence of numbers	Number of terms	State this term	General Term	Find this term
$-1, 4, 9, 14...$		$a_3 =$	$a_n = 5n - 6$	$a_{10} =$
$0, 5, 10, 15, 20, 25, 30$		$a_4 =$	$a_n =$	$a_6 =$
	5	$a_2 =$	$a_n = \frac{(-1)^n}{5^n}$	$a_5 =$
	Infinite	$a_5 =$	$a_n = 2^n - 1$	$a_8 =$

Level 2: Infinite Arithmetic Sequences

Sequence of numbers	$d =$	$a_1 =$	$a_n =$	$a_{20} =$
$6, 12, 18, 24,...$				
			$a_n = -2 + 5n$	
	5	8		
	3			$a_{20} = 66$

Level 3: Arithmetic Series

Series of numbers	$d =$	$a_1 =$	$a_n =$	Find this term	Find this sum
$4 + 12 + 20 + ...$				$a_{20} =$	$S_{20} =$
$50 + 55 + 60 + ...$				$a_{12} =$	$S_{12} =$
	3	-2		$a_{10} =$	$S_{10} =$
			$a_n = 4 + 3n$	$a_{100} =$	$S_{100} =$

Student Activity B:

Section 14.2: Discovering the Arithmetic Series Sum Formula

Directions: Fill in the missing steps to discover the formula for the sum of a finite arithmetic series.

Example 1:

$$S_8 = 1 + 7 + 13 + 19 + 25 + 31 + 37 + 43$$
$$S_8 = 43 + 37 + 31 + 25 + 19 + 13 + 7 + 1$$

Add: $2S_8 = 44 + 44 + 44 + 44 + 44 + 44 + 44 + 44$

Rewrite: $2S_8 = 8 \cdot 44$

Solve for S_8: $S_8 = \frac{8 \cdot 44}{2} = 176$

Example 2: (you try this one)

$$S_5 = -2 + 4 + 10 + 16 + 22$$
$$S_5 =$$

Add:

Rewrite:

Solve for S_5:

Now, in general:

$$S_n = a_1 + [a_1 + d] + \ldots + [a_1 + (n-2)d] + [a_1 + (n-1)d]$$
$$S_n = [a_1 + (n-1)d] + [a_1 + (n-2)d] + \ldots + [a_1 + d] + a_1$$

Add: $2S_n = [\quad] + [\quad] + \ldots + [\quad] + [\quad]$

Rewrite: $2S_n =$

Solve: $S_n =$

Now show how this is the same as $S_n = \frac{n(a_1 + a_n)}{2}$:

Student Activity A:

Section 14.3: Another Sequence of Tables

Directions: Fill in the missing information in the table below to demonstrate your knowledge of geometric sequences and series.

Level 1: Infinite Geometric Sequences

Sequence of numbers	$r =$	$a_1 =$	$a_n = a_1 \cdot r^{n-1}$	$a_8 =$
$4, 2, 1, \frac{1}{2}, ..$				
	5	$\frac{4}{25}$		
			$a_n = 12\left(\frac{1}{3}\right)^{n-1}$	
		$\frac{5}{64}$		1280

Level 2: Finite Geometric Series

Series of numbers	$r =$	$a_1 =$	$a_n = a_1 \cdot r^{n-1}$	Find this term	Find this sum
$24, 8, \frac{8}{3}, ...$				$a_6 =$	$S_6 =$
	10	3		$a_5 =$	$S_5 =$
			$a_n = 6 \cdot \left(\frac{1}{2}\right)^{n-1}$	$a_7 =$	$S_7 =$
		$\frac{1}{100}$		$a_8 = \frac{32}{25}$	$S_8 =$

Level 3: Infinite Geometric Series

Series of numbers	$r =$	$a_1 =$	$a_n = a_1 \cdot r^{n-1}$	Find this partial sum	Find the infinite sum (if possible)
$20, 10, 5, ...$				$S_5 =$	$S =$
	$\frac{3}{2}$	1		$S_5 =$	$S =$
			$a_n = 2\left(\frac{1}{3}\right)^{n-1}$	$S_5 =$	$S =$
		8		$S_5 =$	$S =$

Student Activity B:

Section 14.3: Discovering the Geometric Series Sum Formula

Directions: Fill in the missing steps to discover the formula for the sum of a finite geometric series.

Example 1:

$$S_6 = 25+5+1+\frac{1}{5}+\frac{1}{25}+\frac{1}{125}$$

Multiply by *r*: $S_6 \cdot \frac{1}{5} = \quad 5+1+\frac{1}{5}+\frac{1}{25}+\frac{1}{125}+\frac{1}{625}$

Subtract: $\frac{4}{5}S_6 = 25-\frac{1}{625}$

Solve: $S_6 = \frac{5}{4}\left(25-\frac{1}{625}\right)$

Example 2: (you try this one)

$$S_5 = \frac{1}{6}+\frac{1}{3}+\frac{2}{3}+\frac{4}{3}+\frac{8}{3}$$

Multiply by *r*: $\quad =$

Subtract:

Solve:

Now, in general: (assuming $r \neq 1$)

$$S_n = a_1 + a_1r + a_1r^2 + \ldots + a_1r^{n-2} + a_1r^{n-1}$$

$$S_n \cdot r = \quad a_1r + a_1r^2 + \ldots + a_1r^{n-2} + a_1r^{n-1} + a_1r^n$$

Subtract:

Some factoring:

Solve: $S_n =$

Directions: Fill in the missing steps to discover the formula for the sum of a infinite geometric series.

Example 3:

$$S = 25 + 5 + 1 + \frac{1}{5} + \frac{1}{25} + \frac{1}{125} + \ldots$$

Multiply by *r*: $S \cdot \frac{1}{5} = \quad 5 + 1 + \frac{1}{5} + \frac{1}{25} + \frac{1}{125} + \ldots$

Subtract: $\frac{4}{5} S_6 = 25$

Solve: $S_6 = \frac{5}{4}(25)$

Example 4: (you try this one)

$$S = \frac{8}{3} + \frac{4}{3} + \frac{2}{3} + \frac{1}{3} + \frac{1}{6} \ldots$$

Multiply by *r*: $=$

Subtract:

Solve:

Now, in general:

$$S_n = a_1 + a_1 r + a_1 r^2 + \ldots + a_1 r^{n-2} + a_1 r^{n-1} + a_1 r^n + \ldots$$
$$S_n \cdot r = \quad a_1 r + a_1 r^2 + \ldots + a_1 r^{n-2} + a_1 r^{n-1} + a_1 r^n + \ldots$$

Subtract:

Some factoring:

Solve: $S_n =$

Student Activity C:
Section 14.3: Infinitely Finite

It can be very difficult to grasp the idea that an infinite series can be bounded by a finite number. Here are two examples that might help you to clarify that, at the very least, it is possible.

Example 1: If each large square has an area of **1 square unit**, what is the size of each of the **smaller** squares or rectangles below?

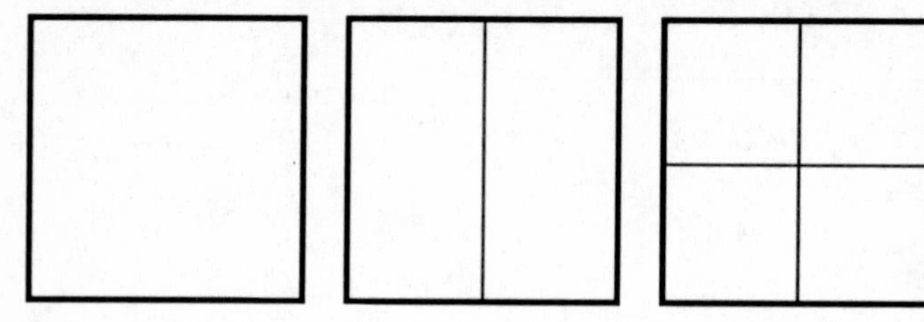

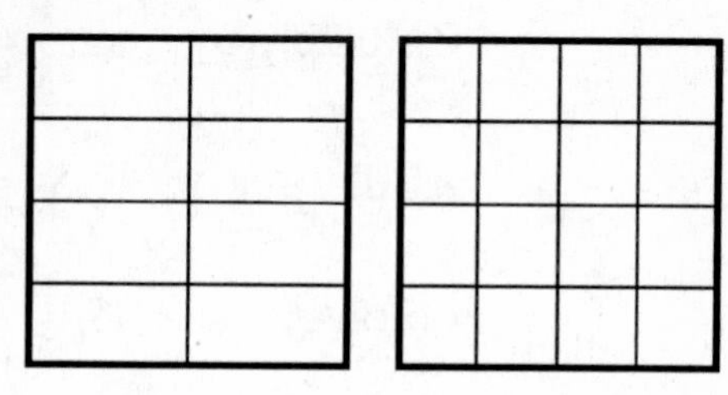

a. Area: _1_ Area: ___ Area: ___ Area: ___ Area: ___

Now we illustrate the series: $\frac{1}{2}+\frac{1}{4}+\frac{1}{8}+...$ by shading in the appropriate pieces of the square as we add each term of the series.

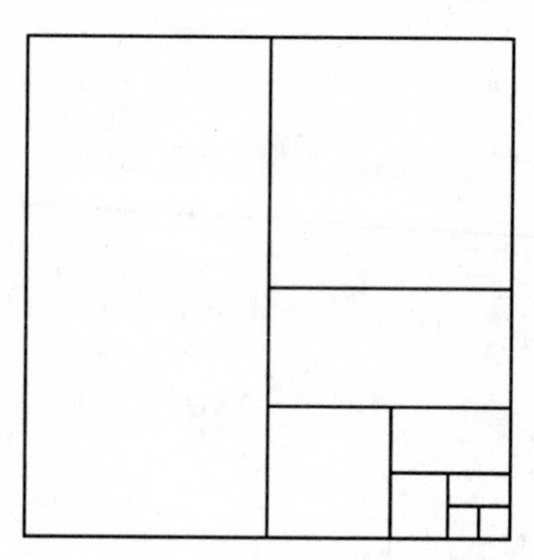

b. Based on this shading, what is the sum of the infinite series $\frac{1}{2}+\frac{1}{4}+\frac{1}{8}+...$? ____

c. Now show how to find this sum using $S=\frac{a}{1-r}$.

Example 2: Now we look at the series $\frac{3}{10}+\frac{3}{100}+\frac{3}{1,000}+\frac{3}{10,000}+...$

a. $a_1 =$ _____ and $r =$ _____.

b. Find the sum of the infinite series using $S=\frac{a_1}{1-r}$. _____

c. Now write the same series using decimals instead of fractions:

_____ + _____ + _____ + _____ + _____ + ...

d. You should be able to clearly see the sum of the series now without any sort of calculation. Write the sum of the series as a decimal: _________

Student Activity D:

Section 14.3: Value Remaining

Directions: Fill in the missing information in the table below for the application problems involving geometric sequences.

	What % is lost or gained every year.	**What is the value at the end of each year, as a percent of the previous year?**	$r =$	$a_1 =$
1. A bank account with a starting balance of $1,000 earns 5% interest every year.	5% gained	105%	1.05	1,000
2. An endowment of $1,000,000 plans to give away 10% of what's left of the original fund balance every year.				
3. The population of a town is 50,000 and is increasing by 6% a year.				
4. The population of a town is 20,000 and is decreasing by 2% a year.				
5. The value of a $120,000 house increases 3% every year.				
6. The value of an airplane that cost $1.2 million depreciates 12% every year.				
7. From studying a population of wolves, biologists determine that from year to year, only 80% of the remaining original population survives.				
8. Every year Mary spends 20% of the remaining balance of the $20,000 she inherited from her aunt.				

Assessment 14C

Chapter 14: End-of-chapter Assessment for Understanding

For each of the following, describe the type of problem and the strategies and key steps to remember while doing the problem. You do **not** have to complete the problems.

	Type of Problem	Strategies and Key Steps
1. Evaluate: $\frac{9!}{3!6!}$		
2. Find the sum of the first 10 terms of the arithmetic series with general term $a_n = -18 + 3n$.		
3. Write the expansion of $(x+y)^5$ using Pascal's triangle.		
4. Let a sequence be defined by $a_n = 6 + 5n$. Find a_{10}.		
5. Find the 7th term of the expansion of $(2x-1)^{12}$.		
6. Find the first four partial sums of the series: $1 - \frac{1}{3} + \frac{1}{9} - \frac{1}{27} + \ldots$		
7. Insert two arithmetic means between 6 and 30.		
8. An endowment containing \$200,000 plans to pay out 5% of the amount remaining in the original fund every year. What will the payment be in the 5th year?		
9. Write the general term for the sequence: $32, 8, 2, \frac{1}{2}, \ldots$		
10. Find S for the geometric series: $\frac{2}{5} + \frac{4}{25} + \frac{8}{125} + \ldots$		

Assessment 14D

Chapter 14: Metacognitive Skills Assessment

Metacognitive skills refer to the ability to judge how well you have learned something and to effectively direct your own learning and studying. This is a self evaluation tool designed to help you focus your studying and to improve your metacognitive skills with regards to this math class.

Fill the 1st column out before you begin studying.
Fill the 2nd column out after you study and before you take the test.
Go back to this page after your test and circle any of the ratings that you would now change – this identifies the "disconnects" between what you think you know well and what you actually know well.

Use the scale below to assign a number to each topic.

5 *I am confident I can do any problems in this category correctly.*
4 *I am confident I can do most of the problems in this category correctly.*
3 *I understand how to do the problems in this category, but I still make a lot of mistakes.*
2 *I feel unsure about how to do these problems.*
1 *I know I don't understand how to do these problems.*

Topic or Skill	Before Studying	After Studying
Finding a binomial expansion using multiplication.		
Replicating Pascal's Triangle.		
Using Pascal's Triangle to predict the coefficients on terms in a binomial expansion.		
Evaluating expressions involving factorial notation.		
Using the Binomial Theorem to write a binomial expansion.		
Predicting the next few terms in a sequence of numbers.		
Identifying the first term and common difference of an arithmetic sequence; writing the general term a_n.		
Identifying the first term and the common ratio of a geometric sequence; writing the general term a_n.		
Finding any term of an arithmetic or geometric sequence using the formula for the general term.		
Knowing the formulas to find the sum of a finite arithmetic series or a finite geometric series.		
Calculating the sum of a finite arithmetic or geometric series.		
Knowing the formula to find the sum of an infinite geometric series and when the formula cannot be used.		
Calculating the sum of an infinite geometric series (when it exists).		
Finding the arithmetic means or geometric means to insert between two terms.		
Using partial sums of a geometric series to make an educated guess about whether the series has a finite sum.		
Converting a repeating decimal into a common fraction.		
Solving application problems involving geometric series.